ESO ASTROPHYSICS SYMPOSIA
European Southern Observatory

Series Editor: Bruno Leibundgut

H.U. Käufl • C. Sterken (Eds.)

Deep Impact as a World Observatory Event: Synergies in Space, Time, and Wavelength

Proceedings of the ESO/VUB Conference held in Brussels, Belgium, 7-10 August 2006

Springer

Volume Editors

Hans Ulrich Käufl
European Southern Observatory
Karl-Schwarzschild-Str. 2
85748 Garching, Germany
hukaufl@eso.org

Christiaan Sterken
Vrije Universiteit Brussel
Astronomy Group
Pleinlaan 2
1050 Brussels, Belgium
csterken@vub.ac.be

Series Editors

Bruno Leibundgut
European Southern Observatory
Karl-Schwarzschild-Str. 2
85748 Garching, Germany

ISBN: 978-3-642-09564-1 e-ISBN: 978-3-540-76959-0

ESO Astrophysics Symposia ISSN: 1431-2433

Cover design: WMXDesign GmbH, Heidelberg

Printed on acid-free paper

9 8 7 6 5 4 3 2 1

springer.com

Preface

In the context of the NASA Deep Impact space mission, comet 9P/Tempel1 has been at the focus of an unprecedented worldwide long-term multi-wavelength observation campaign. The comet was also studied through its perihelion passage by various spacecraft including the Deep Impact mission itself, the Hubble Space Telescope, Spitzer, Rosetta, XMM and all major ground-based observatories in a wavelength band from the cm-wave radio astronomy to x-rays.

The purpose of this meeting was to bring together an audience of theoreticians and observers – across the electromagnetic spectrum and from different sites and projects – to make full use of the massive ground-based observing data set. Synergy between the many data sets can only be achieved, if observers share their data and arrive at a consistent interpretation. A coherent presentation of all data sets in an open forum should allow the theoreticians to fully appreciate all observational constraints on modelling the cometary nucleus, the cometary gas, the cometary plasma, the cometary dust, and the comet's surface and its activity.

This Proceedings book contains papers presented at the meeting in oral or poster format. The organisation of the book, though not a faithful record of the talks, closely reflects the structure of the Workshop, with topics grouped by discipline. Unfortunately, several speakers have not submitted their manuscript for publication.

The Scientific Organising Committee consisted of Chris Sterken (Chair), Ulli Käufl (co-chair), Mike A'Hearn, Hermann Boehnhardt, Michael Combi, Yan Fernandez, Marco Fulle, Luisa Lara, Casey Lisse, Jean Manfroid, Karen Meech, Javier Licandro, Heike Rauer, Rita Schulz, Gerhardt Schwehm, and Diane Wooden.

The Local Organising Committee consisted of Chris Sterken, Yacine Bouzid, Alessandro Ederoclite, Taavi Tuvikene and Michael Weiler.

The Chairs of the Sessions were Ulli Käufl, Javier Licandro, Yan Fernandez, Diane Wooden, Carey Lisse, Luisa Lara, and Hermann Boehnhardt.

The meeting took place in the Palace of the Royal Academies for Science and the Arts of Belgium, in Brussels.

We thank the active SOC members for their help in putting together a well-balanced, attractive and solid scientific programme.

We thank LOC members for their continuous help during the meeting. We also appreciated the dedicated technical help of Mr. Kris Brossé during the meeting. Special thanks go to Lieve Truwant (Mrs. C. Sterken), for her help with all the work going from the making of badges to preparing the conference bags, for being with us throughout the meeting, and for invaluable moral and practical support to participants in need of help. Last but not least our thanks go also to Mrs. Asya Mikova for support in the technical editing of these proceedings.

We thank the various bodies and agencies who have made this workshop possible due to their generous support, viz.

- The European Southern Observatory
- Vrije Universiteit Brussel
- The Royal Academies for Science and the Arts of Belgium
- Research Foundation Flanders (FWO)
- Brussels Hoofdstedelijk Gewest – Région de Bruxelles-Capitale
- The Flemish Ministry for Foreign Policy, European Affairs, Science and Technology

Ulli Käufl and Chris Sterken

Editors

Contents

Part V Cometary Plasma, Cometary Space Missions, and the Future

List of Contributors

Afanasiev, V. L.
Special Astrophysical Observatory of RAS

Ageorges, N.
European Southern Observatory

A'Hearn, Mike
University of Maryland
ma@astro.umd.edu

Aldering, G.
Lawrence Berkeley Lab

Bagnulo, S.
European Southern Observatory

Barber, R. J.
University College London
bob@theory.phys.ucl.ac.uk

Barrera, L.
Universidad Metropolitana de Ciencias y de la Educacion

Baugh, Nicole
Lunar and Planetary Lab, University of Arizona
nbaugh@lpl.arizona.edu

Beer, Harrari Elealeal
NASA Ames, Moffet Field
ealeal44@yahoo.com

Belton, M. J. S.
Belton Space Exploration Initiatives

Bensch, Frank
Universitat Bonn
fbensch@astro.uni-bonn.de

Biele, Jens
Deutsches Zentrum für Luftund, Raumfahrt e.V. (DLR)
jens.biele@dlr.de

Biver, Nicolas
Observatoire de Paris, Section de Meudon
Nicolas.Biver@obspm.fr

Bockelée-Morvan, D.
Observatoire de Paris

Bodewits, Dennis
University of Groningen
bodewits@kvi.nl

Boehnhardt, Hermann
Max-Planck Institut für Sonnensystemforschung
boehnhardt@mps.mpg.de

Böhnhardt, H.
Max-Planck-Institut für sonnensystemforschung

Boissier, J.
Observatoire de Paris

Bonev, Tanyu
Institute of Astronomy,
Bulgarian Academy of Sciences
tbonev@astro.bas.bg

Bouzid, Yacine
Vrije Universiteit Brussel
ybouzid@vub.ac.be

Burton, M. G.
University of New South Wales

Butner, H. M.
Joint Astronomy Centre

Carillo, E.
Universidad Católica del Norte

Charnley, S. B.
NASA Ames Research Center

Chubko, Larissa
Shevchenko National University, Kyiv
chubko@observ.univ.kiev.ua

Churyumov, Klim
Shevchenko National University, Kyiv
klim.churyumov@observ.
univ.kiev.ua

Claeys, Philippe
Vrije Universiteit Brussel
phclaeys@vub.ac.be

Cochran, Anita
McDonald Observatory University of Texas at Austin
anita@barolo.as.utexas.edu

Codella, C.
INAF-Istituto di Radioastronomia

Colom, P.
Observatoire de Paris

Combi, Mike
University of Michigan, Ann Arbor
mcombi@umich.edu

Córdoba Barahona, C.
Astronomical Observatory University of Narino

Coulson, I. M.
Joint Astronomy Centre

Crockett, C. J.
University of Maryland

Crovisier, Jacques
Observatoire de Paris Section de Meudon
Jacques.Crovisier@obspm.fr

De Keyser
Belgian Institute for Space Aeronomy

Delanoye, Sofie
Belgian Institute for Space Aeronomy
sofie.delanoye@aeronomie.be

Delsanti, A.
University of Hawaii

DiSanti, Michael
NASA Goddard Space Flight Center
disanti@ssedmail.gsfc.nasa.gov

Ederoclite, Ale
Vrije Universiteit Brussel
aederocl@vub.ac.be

Erd, C.
ESA Science Payloads and Advanced Concepts Office

Etzel, P. B.
San Diego State University

Farnham, Tony
University of Maryland
farnham@astro.umd.edu

Feldman, Paul
Johns Hopkins University
pdf@pha.jhu.edu

Fernandez Yan
University of Central Florida, Orlando
yfernandez@physics.ucf.edu

Filipović, M. D.
University of Western Sydney

Filonenko Vitaly
Kharkov National University, Kharkov
filonenko@astron.kharkov.ua

Fink Uwe
Lunar and Planetary Lab, University of Arizona
uwefink@lpl.arizona.edu

Florén, H.
Stockholm Observatory

Fornasier, S.
University of Paris

Frisk, U.
Swedish Space Corporation

Fukushi, H.
The University of Tokyo

Furusho, Reiko
Waseda University, Shinjuku, Tokyo
furshork@cc.nao.ac.jp

Groussin Olivier
University of Maryland
groussin@astro.umd.edu

Gutiérrez, P. J.
Instituto de Astrofisica de Andalucía

Hadamcik, Edith
Université Pierre et Marie Curie, Paris 6
Edith.Hadamcik@aerov.jussieu.fr

Hainaut, O.
European Southern Observatory

Hampton, D. L.
Ball Aerospace and Technology

Harker, David E.
University of California, San Diego
dharker@ucsd.edu

Harrington, David
University of Hawaii, Honolulu
dmh@ifa.hawaii.edu

Harwit, M.
Cornell University

Hjalmarson, Å.
Onsala Space Observatory

Ho Tra-Mi
University of Bern
tra-mi.ho@phim.unibe.ch

Hodapp, K. W.
University of Kawaii

Honda, M.
Japan Aerospace Exploration Agency

Hviid, S. F.
Max-Planck-Institut für Sonnensystemforschung

Ivanov, V.
European Southern Observatory

Jackson, William
University of California Davis
wmjackson@ucdavis.edu

Jehin Emmanuel
Université de Liège
ejehin@ulg.ac.be

Jones, Paul
University of New South Wales, Sidney
Paul.Jones@atnf.csiro.au

Jorda, L.
Laboratoire d'Astrophysique de Marseille

Käufl, Hans Ulrich
European Southern Observatory, Garching b. München
hukaufl@eso.org

Kadono, Toshihiko
Japan Agency for Marine-Earth Science and Technology Yokohama
kadono@jamstec.go.jp

Kaneyasu, N.
Meisei University

Kandori, R.
National Astronomical Observatory of Japan

Kawadu, T.
Kyoto University

Kawakita Hideyo
Kyoto Sangyo University, Motoyama
kawakthd@cc.nao.ac.jp

Keller, H. U.
Max-Planck Institut für Sonnensysternforschung

Kelley Michael
University of Minnesota, Minneapolis
msk@astro.umn.edu

Kerber, F.
European Southern Observatory

Kleshchonok, V. V.
Shevchenko National University

Knollenberg, J.
Institute of Planetary Research

Kolokolova, L.
University of Maryland

Kruchynenko, V. G.
Shevchenko National University

Kueppers, Michael
Max-Planck Institut für Sonnensystemforschung
kueppers@linmpi.mpg.de

Kuhn, J. R.
University of Kawaii

Kührt, E.
DLR, Institute for Planetary Research

Kwok, S.
University of Calgary

Lara, Luisa
Instituto de Astrofísica de Andalucía Granada
lara@iaa.es

Lecacheux, A.
Observatoire de Paris Lederer, Susan California State University, San Bernardino
slederer@csusb.edu

Levasseur-Regourd, Anny-Chantal
Université Pierre et Marie Curie, Paris 6
Chantal.Levasseur@aerov.jussieu.fr

Li, Aigen
University of Missouri, Columbia
LiA@missouri.edu

Licandro Javier
Isaac Newton Group of Telescopes, Santa Cruz de La Palma
licandro@ing.iac.es

Lis, D. C.
Caltech, Pasadena

Lisse Carey
Johns Hopkins University Applied Physics Laboratory, Laurel
carey.lisse@jhuapl.edu

LoCurto, G.
European Southern Observatory

Lowry, S. C.
Queen's University Belfast

Lukyanyk, I. V.
Shevchenko National University

Lyke, J. E.
W.M. Keck Observatory

Magee-Sauer, K,
Rowan University

Manfroid, J.
University of Liege

Marco, O.
European Southern Observatory

Martin, D.
ESA Science Payloads and Advamced Concepts office

Matsunaga, N.
The University of Tokyo

McFadden, L. A.
University of Maryland

Meech, Karen
University of Hawaii, Honolulu
meech@ifa.hawaii.edu

Melnick, G. J.
Harvard-Smithsonian Center for Astrophysics

Miller, S.
University College London

Mitchell, Tyler
San Diego State University
tmitchel@sciences.sdsu.edu

Miyata, T.
University of Tokyo

Möhlmann, D.
DLR, Institute for Planetary Research

Moreno, R.
Observatoire de Paris

Mori Yuki
Tokyo University
moriyuki@impact.k.u-tokyo.ac.jp

Moriarty-Schieven, G.
Joint Astronomy Centre

Mumma Michael
NASA Goddard Space Flight Center
mmumma@ssedmail.gsfc.nasa.gov

Nakajima, Y.
National Astronomical Observatory of Japan

Neufeld, D. A.
Johns Hopkins University

Okouma, P.
University of Cape Town

Olberg, M.
Onsala Space Observatory

Olson, Roberta
Museum of The New-York Historical Society
rolson@nyhistory.org

Ootsubo, Takafumi
Nagoya University
ootsubo@u.phys.nagoya-u.ac.jp

Palagi, F.
INAF-Istituto di Radioastronomia

Pantin, E.
CEA Saclay

Pasachoff, Jay
Williams College, Williamstown
Jay.M.Pasachoff@williams.edu

Patten, B. M.
Harvard-Smithsonian Center for Astrophysics

Paubert, G.
IRAM, Avd. Divina Pastora

Perenguez López, J.
University of Nario

Pittichova Jana
University of Hawaii, Honolulu
jana@ifa.hawaii.edu

Pompei, E.
European Southern Observatory

Poppi, S.
INAF-Osservatorio Astronomico di Cagliari

Potter, S.
South African Astronomical Observatory

Quijano Vodniza, Alberto
University of Nario
aqv-ct@computronix.com.co

Quijano Vodniza, A. J.
University of Nario

Rauer Heike
Deutsches Zentrum für Luft- und Raumfahrt e.V. (DLR)
Heike.Rauer@dlr.de

Reach, W. T.
California Institute of Technology

Rengel, Miriam
Max-Planck Institut für Sonnensystemforschung
Rengel@mps.mpg.de

Richter, L.
DLR, Institute for Space Simulation

Rodgers, Steve
NASA Ames Research Center,
rodgers@dusty.arc.nasa.gov

Rojas Pereira, M.
University of Nario

Sako, S.
University of Tokyo

Sandqvist, A.
Stockholm Observatory

Sarkissian, J. M.
Australia Telescope National Facility

Saviane, I.
European Southern Observatory

Schulz Rita
ESA Research and Scientific Support Department, Noordwijk
rschulz@rssd.esa.int

Sekiguchi, T.
National Astronomical Observatory of Japan

Selman, F.
European Southern Observatory

Shafi, N.
University of Cape Town

Smit, H.
ESA Science Payloads and Advamced Concepts office

Snell, R. L.
University of Massachusetts

Stallard, T. S.
University College London

Sterken Chris
Vrije Universiteit Brussel
csterken@vub.ac.be

Stüwe, J.
Leiden University

Sugita, Seiji
University of Tokyo
sugita@k.u-tokyo.ac.jp

Sumner, M.
Caltech, Pasadena

Sunshine, Jessica
University of Maryland
sunshinej@alumni.brown.edu

Tamura, M.
National Astronomical Observatory of Japan

Tennyson, J.
University College London

Tolls, V.
Harvard-Smithsonian Center for Astrophysics

Tozzi, Gian Paolo
INAF-Osservatorio Astrofisico di Arcetri, Firenze
tozzi@arcetri.astro.it

Tuvikene, Taavi
Vrije Universiteit Brussel
ttuviken@vub.ac.be

Ulamec, Stephan
Deutsches Zentrum für Luft- und Raumfahrt e.V. (DLR)
stephan.ulamec@dlr.de

Villanueva Geronimo
NASA Goddard Space Flight Center
villanueva@ssedmail.gsfc.nasa.gov

Vogt, N.
Universidad de Valparaiso

Vora, P.
College of William & Mary

Voronkov, M. A.
Australia Telescope National Facility

Watanabe, J.
National Astronomical Observatory of Japan

Weiler Michael
Institut für Planetenforschung, German Aerospace Center (DLR), Berlin
Michael.Weiler@dlr.de

Weissman Paul
Jet Propulsion Laboratory, Pasadena
Paul.R.Weissman@jpl.nasa.gov

Wellnitz, Dennis
University of Maryland
wellnitz@astro.umd.edu

Welsh, W. F.
San Diego State University

Whitman, K.
University of Hawaii

Winnberg, A.
Onsala Space Observatory

Wirth, Kristin
ESA Research and Scientific Support Department, Noordwijk
Kristin.Wirth@esa.int

Wooden, Diane
NASA Ames Research Center, Moffet Field
dwooden@mac.com

Woodney, L. M.
California State University

Woodward, Chick
Spitzer Science Center, Pasadena
chelsea@physics.umn.edu

Participants in the Palace of the Royal Academies for Science and the Arts of Belgium (photo courtesy T. Tuvikene)

Part I

The Deep Impact Event

Previously Unobserved Water Lines Detected in the Post-Impact Spectrum

R. J. Barber, S. Miller, T. S. Stallard, and J. Tennyson

University College London, Gower Street, London WC1E 6BT

1 Water Lines in Tempel 1

The team from UCL, monitored the Deep Impact event using the CGS4 spectrometer on the United Kingdom Infrared Telescope, UKIRT. Our principal objective was to determine the temporal development of solar pumped fluorescent (SPF) transitions of H_2O following impact, and to interpret the results using the recently published ab inito water line list, BT2 [1]. BT2 was produced at UCL using the DVR3D suite of programs [5], and is the most complete and most accurate water line list in existence.

Normal observing techniques were employed: these and other aspects of our work are fully reported in Barber et al. [2]. We obtained spectra in the wavelength range centred on 2.894 μm, with a spectral range of ±0.040 μm. Apart from containing a number of SPF transitions [3], it is largely devoid of other molecular lines (such as CO), which makes it possible to model the region using the BT2 line list without having to include other species. However, in the subsequent examination of the data, in order to maximise the S/N ratio, we restricted our analysis to a narrower wavelength range: 2.8945 - 2.8985 μm.

It had been our intention to obtain spectra of Tempel 1 on the night prior to impact, on impact night, and on the night after impact. However, prior to impact the comet was not sufficiently bright for us to be able to obtain useful data, and on the night after impact, reduced intensity and a deterioration in the observing conditions prevented us obtaining high quality data. Our results therefore relate only to the period of 143 minutes immediately following impact.

Our signal was effectively confined to one pixel row (we estimate that this contained ~65% of the signal), and it attests to the accuracy of the UKIRT tracking system that there was no detectable drift from this position over the whole observing session. No attempt was made to recover the small amount of signal from adjacent rows, as an analysis of the data revealed that this would have resulted in a reduction in the overall S/N ratio. Moreover, because of the weakness of the signal, it was necessary to combine the data for the whole observing run in order to obtain a useful S/N ratio, which meant that, taking the spectral region as a whole, no temporal resolution could be obtained (see below for comments on temporal resolution for specific groups of lines).

The spectrum for the period July 4 05:54-08:17 U.T. is shown in Fig. 1. The fact that some of the intensities are negative may in part be due to our having over-corrected for the continuum, or more likely, is due to noise, which we estimate to be in the region of 0.4×10^{-16} $\mathrm{Wm^{-2}\ \mu m^{-1}}$.

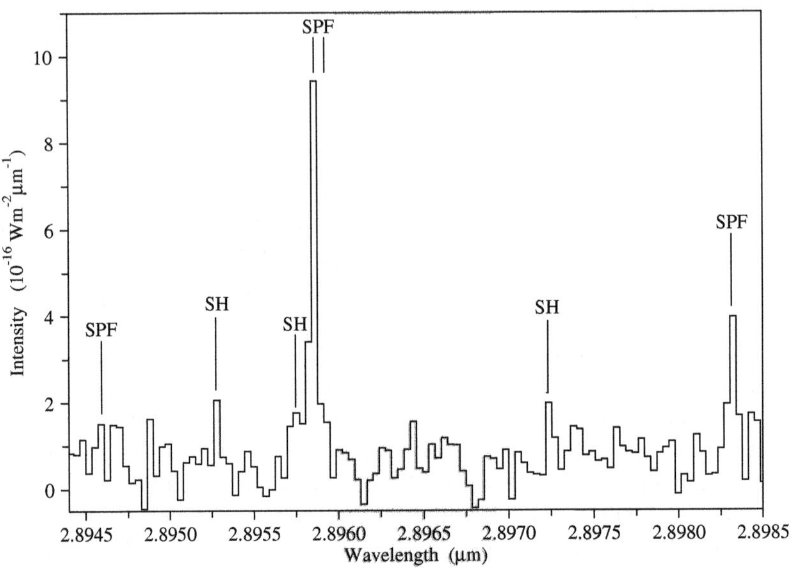

Fig. 1. Observed post-impact spectrum Tempel 1. Wavelengthis in the rest frame. Reproduced from Barber et al. [2].

Two bright fluorescent transitions at 2.89580 and 2.89831 $\mathrm{cm^{-1}}$, stand out against a background of weaker features. Many of these weaker features are close to, or below, the noise threshold and these we disregard. However, there are a several where we estimate the S/N ratio to be greater than 4 and which, unlike noise, appear in the same place (albeit with differing intensities) in many of the individual frames. We interpret these as also being genuine signals and have marked them either SPF (where their wavelengths correspond to known fluorescent transitions), or SH, (in those cases where transitions are thought to be by some other route). In the first column of Table 1 we give the observed wavelength of each of the features (adjusted for red shift).

The position of many of the spectral features in Fig. 1 can be replicated using the BT2 synthetic water line list. Some of these are identified as being due to SPF transitions (sometimes blended). However, some of the features did not correspond to known SPF transitions, and moreover, as far as we are aware, have not previously been recorded in cometary spectra. We have labelled these 'SH' (the acronym is convenient as it can stand for 'solar heating', which is definitely true, or 'stochastic heating', which may also be an apt description). It should be noted that although the line positions derived from BT2 agree well with those of the observed features, the intensities are only approximate guides, as the LTE assumption used by BT2 is not generally valid for cometary spectra. Moreover, since the mechanism behind the formation of the SH transitions is not known, a precise match between the intensity of the observed features and the BT2 synthetic spectra is not to be expected.

Table 1. Assignments of SPF and previously unobserved SH lines in the post-impact spectrum of Tempel 1. The first column gives the observed wavelength of each of the features (adjusted for red shift), the next column identifies the transition: the vibrational quantum numbers are given in round brackets and the rotational quantum numbers in square brackets. The last three columns give: the experimentally-determined wavelength of the listed transition, the Einstein A coefficient computed using BT2, and our designation of type: SPF or SH. Reproduced from Barber et al. [2].

$\lambda_{observed}$ µm ±0.00005	Identification (see text for notation)	$\lambda_{laboratory}$ µm	A_{if} s^{-1}	Type
2.89458	(101)[211] - (001)[220]	2.89462	1.9	SPF
2.89527	(103)[110] - (102)[110]	2.89526	53.5	SH
2.89527	(211)[322] - (210)[211]	2.89528	8.5	SH
2.89573	(210)[101] - (011)[000]	2.89570	5.1	SH
2.89580	(200)[110] - (100)[221]	2.89578	4.7	SPF
2.89591	(101)[202] - (100)[321]	2.89590	1.7	SPF
2.89723	(220)[212] - (021)[111]	2.89728	4.4	SH
2.89831	(200)[110] - (001)[111]	2.89830	6.6	SPF

In attempting to identify the non-SPF features in our observed spectrum, we generated a series of synthetic spectra using BT2, some of which are shown in Figs. 3 and 4. In order to improve the signal to noise ratio and also to lessen the significance of errors in the wavelength calibration of our detector, estimated to be in the region of 0.00004 µm (slightly greater than a single pixel width, which is 0.000033 µm), we artificially reduced the resolution of our observed data from an instrument-limited ~37 000 to a pixel-averaged limited value of ~17 500 by taking a moving average of five pixels in the wavelength dimension in order to assist in matching the observed and synthetic data. Figure 2 shows our observed spectrum with this artificially de-graded spectral resolution. In this figure the vertical scale is terminated at a level well below the peak intensities of the two strongest SPF transitions. This was done to aid identification of the weak transitions.

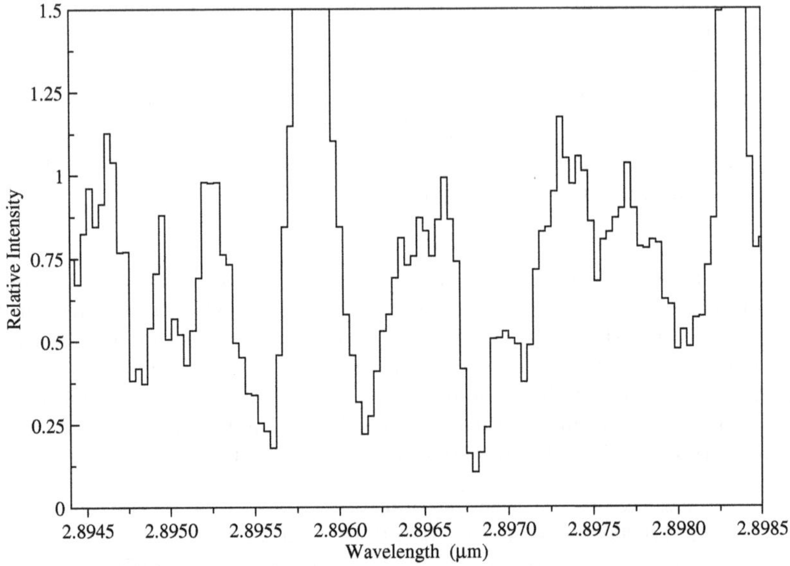

Fig. 2. Observed post-impact spectrum Tempel 1, moving average of 5 pixels. Wavelength is in the rest frame. The vertical scale has been limited to assist identification of the weaker features. Reproduced from Barber et al. [2].

Figure 3 shows three synthetic spectra generated at 3 000, 4 000 and 5 000 K (assuming LTE), using the BT2 line list (with some corrections to the calculated wavelengths based on the available experimental data). In producing these spectra, we applied the restriction that only states with $J \leq 3$ are included. The resolution in Fig. 3 was set to be the same as Fig. 2. Fig. 4 shows another set of synthetic BT2 spectra generated at the same three temperatures, but this time including all J levels (up to 50). Some of the same features are observed as in Fig. 2, but we note that Fig. 3 matches the observed spectrum better than Fig. 4 does.

Figure 3 is therefore a high-ν, low-J spectrum and it reproduces well the position (less-so, the intensities) of many of the non-SPF features in the low-resolution observed spectrum, Fig. 2. It should be noted that similar features are observed in synthetic spectra generated for temperatures greater than 3 000 K and the appearances of the spectra vary little once $T_{vib} > 4\,500$ K, except for

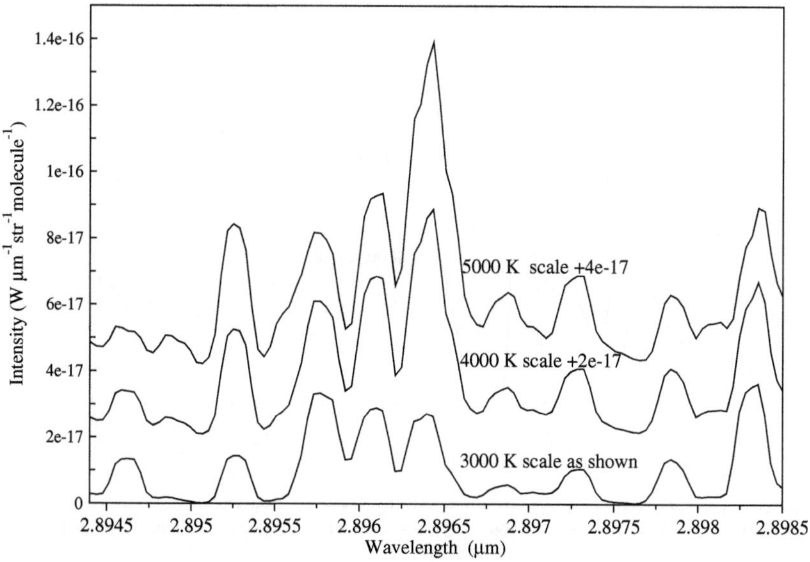

Fig. 3. BT2 synthetic spectra at 3 000, 4 000 and 5 000 K, J_{max}=3. Reproduced from Barber et al. [2].

differences in the general levels of intensity. It is also noted that there are other features in our observed spectrum Fig. 2 that are replicated in the synthetic spectrum Fig. 3. These are also due to water emission and are produced by the blending of many overlapping transitions from different vibrational manifolds. They have not been included in Table 1 as it is not possible to assign them to one or two individual transitions.

2 Assigning the Features

Among the assigned features are fluorescent emission lines from levels that have two quanta of H_2O stretching, such as (2 0 0)→(1 0 0) and (1 0 1)→(1 0 0): in the former transition, the emission involves one quantum of ν_1, in the latter it involves one quantum of ν_3. A quantum of ν_1 and of ν_3, have similar energies ~3 450 cm^{-1}. In both cases, the final state is the 1ν_1 state. Because a quantum of ν_2 carries less than half the energy of a ν_1 or ν_3 quanta, (~1 500 cm^{-1}), transitions involving a change of one quantum of ν_2 are not observed in our selected wavelength range.

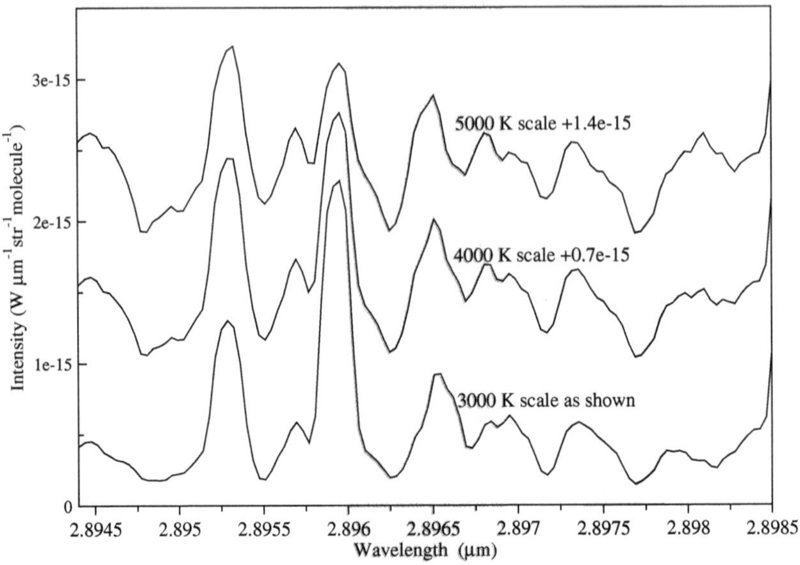

Fig. 4. BT2 synthetic spectra at 3 000, 4 000 and 5 000 K, J_{max}=50. Reproduced from Barber et al. [2].

Our post-impact spectrum of Tempel 1 also includes several transitions from states that include one or more quanta of ν_2, or that involve a total of 4 vibrational quanta. These are not SPF spectral features. It will be seen from Table 1 that these include the blend of (1 0 3)→(1 0 2) and (2 1 1)→(2 1 0) at 2.8953 µm, and (2 2 0)→(0 2 1) at 2.8972 µm. It seems likely that upper levels with more than two vibrational quanta have not been populated by solar pumping from ground vibrational states, but by another mechanism. Whereas SPF transitions originate from upper states having energies in the region of 7 300 cm^{-1}, the transitions that we have labelled 'SH' all originate from higher energy states (those upper states having four vibrational quanta are in the energy range 10 300–14 400 cm^{-1}. It is possible that the production route involves H_2O molecules that have sublimed from the freshly exposed icy grains ejected by the impact. However, an understanding of the precise nature of these SH lines will require further research. We have recently learned that Villanueva et al. [6] observed some previously unrecorded lines in the 2.8313 µm region in the post-impact spectrum of Tempel 1, which may be due to transitions from higher energy ro-vibrational states of H_2O.

2.1 Comparison with Other Spectra

We have compared our results with the spectra obtained by Mumma et al. [4].

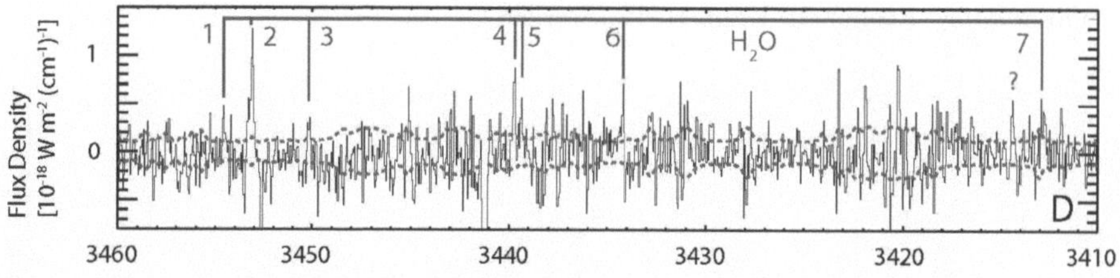

Fig. 5. Spectrum of Tempel 1 post-impact (abscissa is frequency in units of cm^{-1}). Reproduced from Mumma et al. [4] - spectrum D.

As far as the positions of the observed SPF transitions are concerned, our results agree well with those of Mumma et al.. However, none of the SH features that we identify appear in Mumma et al.'s spectrum D. One possible reason for the difference between our results and those of Mumma et al. is the difference in times when the spectra were obtained. Our results were obtained between 05:54 and 08:17 U.T. on impact night, whilst Mumma et al.'s spectrum in Fig. 5 were obtained between 6:43 and 7:25 U.T. By summing the intensities of all the SH features, we were able to achieve a degree of temporal resolution that was not possible for the individual lines. We observed that the total intensity of the SH features was particularly strong during the period 20–40 minutes after impact., but by 50 minutes after impact had declined to a level that was only slightly above that of the background noise. This could be the reason why the features are not observed in Mumma et al.'s spectra..

Recently we have obtained UKIRT spectra of Comet 73P/Schwassmann-Wachmann, fragment-C in order to investigate whether this recently fragmented comet exhibits similar SH features to those that we observed in Temple 1. Our preliminary investigation of the 73P spectra does not show any of the SH features. Work on this is continuing.

References

1. Barber, R.J., Tennyson J., Harris G.J., Tolchenov R. 2006a, MNRAS 368, 1087
2. Barber, R.J., Miller S., Stallard T., Tennyson J. et al. 2006b 'UKIRT Deep Impact observations: light curve, ejecta expansion rates and water spectral features', Icarus (submitted).
3. Dello Russo N., DiSanti M.A., Magee-Sauer K., Gibb E.L., Mumma M.J., Barber R.J., Tennyson J. 2004, Icarus 168, 186
4. Mumma M.J., Disanti M.A., Magee-Saurer K., Bonev B.P., Villanueva G.L., et al. 2005, Science 310, 270

5. Tennyson J., Kostin M., Barletta P., Harris G.J., Polyansky O.L., Ramanlal J., Zobov N.F. 2004, Comp. Phys. Comm. 163, 85
6. Villanueva G.L., Bonev B.P., Mumma M.J., DiSanti M.A., Magee-Sauer K., 2007, 'Infrared spectral survey of the ejecta of comet Tempel 1 using NIRSPEC/Keck-2' Springer-Verlag ESO Astrophysical Symposia Series (Deep Impact as a World Observatory Event) ed. Käufl H. U., Sterken C.

Keck 1 HIRES Observations of Tempel 1 at the Time of Impact

W. M. Jackson[1] and A. Cochran[2]

[1] Chemistry Department, University of California, Davis wmjackson@ucdavis.edu
[2] McDonald Observatory, University of Texas anita@barolo.as.utexas.edu

Abstract The Keck 1 HIRES observations of the Deep Impact event on July 4 have been qualitatively re-examined in an effort to isolate the changes in the HIRES spectra that can be attributed to the event itself. To accomplish this goal, background spectra due to ambient emission that was already present in the comet were extracted from the pixels at the end of the slit. This material can not have any emission from material that was produced at the time of the impact until at least several thousand seconds after the impact because it takes this long for the material to reach this distance from the nucleus. With the data extracted in this manner it can be shown that the spectra from OH, $O(^1S)$, CN, NH, NH_2, C_2, and C_3 changed as a function of time after impact. In the spectra of all of these species there is evidence for a prompt change in the emission intensity that probably is due to the hot plume formed initially during the impact. After the first 600 s the time dependence of the intensities of the emission spectra species appear to be consistent with the photochemical formation and destruction of parents and daughters. The spectral changes with time associated with the CH emission appear to be unique in that the only change that is observed is right after the impact; at times after this the emission intensity is constant. All of these changes are discussed in terms of the photochemical mechanisms that have been thought to form the observed free radicals in comets.

1 Observations and Reductions

On 4 July 2005 UT, we used the HIRES instrument on the Keck 1 telescope[1] to observe the impact of comet 9P/Tempel 1 with the Deep Impact spacecraft. The impact speed of 10.3 km sec^{-1} resulted in an impact kinetic energy of 19 GJ [2]. The impact occurred at 05:44:36 UT (Earth-received time of 05:52:02 UT).

The HIRES instrument is a cross-dispersed, high spectral resolution spectrograph. We used the blue cross-disperser for our observations, resulting in a

[1] The data presented herein were obtained at the W. M. Keck Observatory, which is operated as a scientific partnership among the California Institute of Technology, the University of California and the National Aeronautics and Space Administration. The Observatory was made possible by the generous financial support of the W. M. Keck Foundation.

nominal resolving power $\lambda/\Delta\lambda = 47,000$, with the slit projecting to 4.8 pixels on the detector. The bandpass of the three CCD detectors covered 3047–3951Å, 3971–4926Å, and 4977–5894Å, in multiple orders per detector. The slit was set at the parallactic angle, which means that it rotated on the sky during the night. As a result, any atmospheric dispersion was along the slit. We used a slit which was 7×0.86 arcsec on the sky or 4570×562 km at the comet's distance. This slit was much smaller than the coma of the comet so we were imaging only the inner coma. A complete description of the observations, including a log of the time of observations and the angle of the slit on the sky, can be found in Cochran et al. (2006; hereafter Paper 1) [3]. In addition, observations were obtained with the HIRES instrument on 30 May 2005 and 5 and 6 July 2005 UT (the observations of 30 May and 4 July were obtained under Director's time and are available to the public from the Keck archive).

Had their been no impact, we would have observed the ambient gas outflow of the comet. In that case, we would normally extract the spectrum of the comet along the whole 7 arcsec slit. The continuum would be removed by the use of solar analogue stars scaled to the comet's continuum. This would then result in the cometary gas spectrum of the inner coma. However, the impact of the comet and the spacecraft caused additional material to be released from the nucleus and it is this additional material that we wished to monitor.

As a first step, we performed the extraction of the spectrum along the complete length of the slit and attributed any changes in the spectrum with time to the effects of the impact. A complete description of this process, along with sample spectra are included in Paper 1. While this technique is instructive, it is difficult to understand the magnitude of the changes from the impact because the ambient spectrum is so dominant in the earliest times after impact. Therefore, we needed to find a way to remove the ambient spectrum from the spectra recorded on 4 July.

The removal of the ambient spectrum proved difficult because we did not have a representative ambient spectrum prior to impact. We obtained one spectrum of the comet just before the impact on 4 July. However, that spectrum was obtained prior to astronomical twilight; the 720 sec observation was started at 05:36:15 (eight degree twilight was at 05:29UT; 12 degree twilight (nautical) was at 05:58UT; 18 degree twilight (astronomical) was at 06:28). Thus, the sky was still very bright. The sky spectrum is also a solar spectrum of a different color and with a different Doppler shift than the comet's continuum. When we removed the solar continuum from this pre-impact spectrum, we were left with a very noisy spectrum.

We obtained three spectra of the comet on 30 May 2005. These too proved unsuitable to be representative of the ambient comet in July because they were obtained at different heliocentric (1.55 AU vs. 1.51 AU) and geocentric (0.75 AU vs. 0.91 AU) distances than the 4 July data. In addition, the comet was observed to undergo many outbursts between 30 May and 4 July.

To try to minimize the contribution of the ambient coma, we attempted to "isolate" the nucleus within the slit (of course, we did not resolve the nucleus with ground-based observations). To do this, we extracted all of the spectra over the size of the seeing disk, 0.7 arcsec or 457 km. These spectra were generally of much lower S/N than the full slit extractions. These extractions were reported in Paper 1 and example spectra are shown in that work.

While the approach of using a small aperture minimizes the contribution of the outer (ambient) coma to the observations, the impact was an impulse of short duration and the material flowed quickly out of the small seeing disk-based aperture over which we extracted spectra. In this short distance, photodissociation will be unimportant and the spectrum would not show very much change. Thus, our approach of extracting spectra over only a relatively small aperture would not gain the most information from the impact.

Instead of arbitrarily limiting the aperture to the size of the seeing disk, we decided to follow the impact gas as it flowed outwards. After the impact, we expected to see the amount of gas released from the nucleus enhanced over the ambient output. The gas then flowed outwards. When our aperture was filled, the excess emission would again decrease. We used the filling of the aperture to derive a minimum outflow velocity for the CN of 0.51 km sec^{-1} in Paper 1. This value is in agreement with the OH outflow velocity of 0.5 km sec^{-1} determined by Schulz (2006), [14] with XMM Optical observations.

By assuming that material was released at the instant of impact, adopting an outflow velocity of 0.55 km sec^{-1} and assuming that the material flows outwards symmetrically, we can predict the distance the gas would have traveled from the nucleus by the mid-time of each observation. This is shown in Table 1.

Table 1. Post-impact gas outflow

File number	Start time	Mid time	Exposure time (sec)	Elapsed time (sec)	Km† traveled	Pixels† traveled
66	05:55:18	06:08:18	600	496	273	1.7
67	06:06:12	06:11:12	600	1150	633	4.1
68	06:17:05	06:24:35	900	1953	1074	6.9
69	06:32:59	06:40:29	900	2907	1599	10.2
70	06:48:53	06:56:23	900	3861	2124	13.6
71	07:04:47	07:12:17	900	4815	2648	17.0
72	07:20:42	07:28:12	900	5770	3174	20.3
73	07:36:36	07:44:06	900	6724	3698	23.7
74	07:52:29	07:59:59	900	7677	4222	27.1
75	08:08:25	08:15:55	900	8633	4748	30.4
76	08:24:19	08:31:49	900	9587	5273	33.8
77	08:40:13	08:55:13	1800	10991	6045	38.7
78	09:11:10	09:26:10	1800	12848	7066	45.3

† Value at the mid-time

To "follow the gas", we extracted the spectra centered on the optocenter and with an extraction length equal to plus-or-minus the pixel length indicated in the last column of Table 1. The material would reach the full length of the slit (±14.5 pixels) after the fifth observation (file number 70). After this point, denoted with a small gap in Table 1, we were no longer able to continue to follow the impact-derived gas completely. For these subsequent spectra, we extracted data over the complete length of the slit.

For the first four observations, the ends of the slit contained only the ambient cometary spectrum. We extracted spectra covering the first and last 3.5 pixels along the slit for each of the first four spectral files obtained (66–69). Then we averaged these spectra to increase the S/N. We assumed that these averages were representative of the ambient cometary spectrum.

We were concerned that the photodissociation of the species would cause the amount of gas measured at the ends of the slit to be lower than would be true near the nucleus and thus would not be representative of the ambient spectrum on the optocenter. However, not much photodissociation should take place in 2000 km. To check that this was a small effect, we extracted the pre-impact spectrum from 4 July with varying sized apertures corresponding to the values in the last column of Table 1. We verified that, within the noise, the spectra were the same. Thus, we felt justified in using the spectra from the ends of the slit as representative of the ambient spectrum of the comet. These spectra had higher S/N than the pre-impact spectrum, with its bright blue sky.

Once the spectra were extracted, we used observations of a ThAr lamp to assign a dispersion curve to each order. Typical rms errors of the wavelength solution were 0.004 Å. The cometary spectra were then Doppler shifted to the spectrograph rest frame using the orbital information. We next used Mauna Kea Observatory mean extinction coefficients to correct the spectra for signal outside the atmosphere. For the highest airmass observations at the bluest wavelengths (typically below about 3300 Å) at the end of the night, these corrections are not accurate since no flux was received through the atmosphere to correct.

In order to be able to intercompare observations, we needed to normalize the observations to a unit time and pixel. In practice, we normalized to 15 minute observations, since this was the most common exposure time. In addition, we normalized to a unit of 2 pixels (i.e. we divided by the pixel value in the last column of Table 1).

Next, we removed the contribution of the continuum which results from the solar light being reflected off the dust. Finally, we removed the ambient comet spectrum by subtracting the spectral average (normalized similarly to the other spectra) which we derived from the ends of the slit. These ambient spectra were generally devoid of any solar continuum so the solar spectrum did not need to be removed (the dust is the only component which showed a significant change along the length of the slit). At this point, we had a series of spectra which were the gas component that resulted from the impact, per unit time and spatial interval, outside the atmosphere.

2 Behavior of the Gaseous Species

The molecules which are typically observed within our bandpass in spectra for comets at ~1.5 AU would include OH, NH, CN, C_3, CH, C_2 and NH_2. The emission spectra are generally formed by resonance fluorescence. All of these species are daughters or granddaughters, resulting from photodissociation of their parents. We detected emissions from all of these species. In addition, we observed the O (^1S) atomic line. This line is a prompt emission (it is formed in the excited state) and is generally hidden by the emission of a C_2 band. The C_2 emission from comet Tempel 1 was weak, allowing us to detect the atomic oxygen.

Observations with the instruments on board the Deep Impact spacecraft indicated that, of the 19 GJ impact energy, less than 1 GJ went into a change of the orbit of the comet. In the first 1–2 sec, a hot plume was emitted, carrying about 1 ton of hot material. After that time, an additional 10^4 ton of cold material was emitted. Most of the excess impact energy went into kinetic energy of the hot plume. There was a large increase in H_2O detected by the spacecraft. An even larger amount of CO_2 was detected [1].

It is reasonable to expect that there will be thermal decomposition of some of the parent molecules in this hot plume. Typically, when molecules that are thought to be the parents of the observed cometary emission are subjected to large temperatures they form the same kind of radicals that are observed when these same molecules undergo photodissociation. The effect of this hot plume should be limited to our first time step because the Deep Impact Satellite observations indicated that the plume only lasted for a very short time. We would expect that the effect of the hot plume would manifest itself in the production of species essentially instantaneously on our time scale and in the production of new cometary emissions.

The major molecular component of the volatile ice in a cometary nucleus is H_2O. The photodissociation of H_2O produces OH as a dominant daughter product. The (0,0) band of OH has transitions in the UV, at 3080 Å. With the blue cross disperser, we were reasonably sensitive to OH as long as the airmass was not too high. Using our spectra which have the ambient coma removed, we were able to measure the effect of the impact on the production of OH. Figure 1 shows the OH behavior with time. We see that there is a large rise in the intensity of the OH emission until about 2000 sec after impact and after that the signal decays.

Some of the H_2O will dissociate to atomic oxygen, as will the OH. Oxygen has three observed transitions in the visible part of the spectrum: the "green" line at 5577.339 Å is an O (^1S) transition; the "red doublet" at 6300.304 and 6363.776 Å are O (^1D) transitions. Our bandpass included the O (^1S) but not the red doublet. Unlike the molecular bands which we observe, the oxygen lines are not formed by resonance fluorescence or dissociative recombination [4]. Instead, they are prompt emissions and arise from atoms produced directly into the excited ^1S or ^1D states. These transitions are "forbidden" transitions since the ground state is a ^3P state. The red doublet can be present without the green

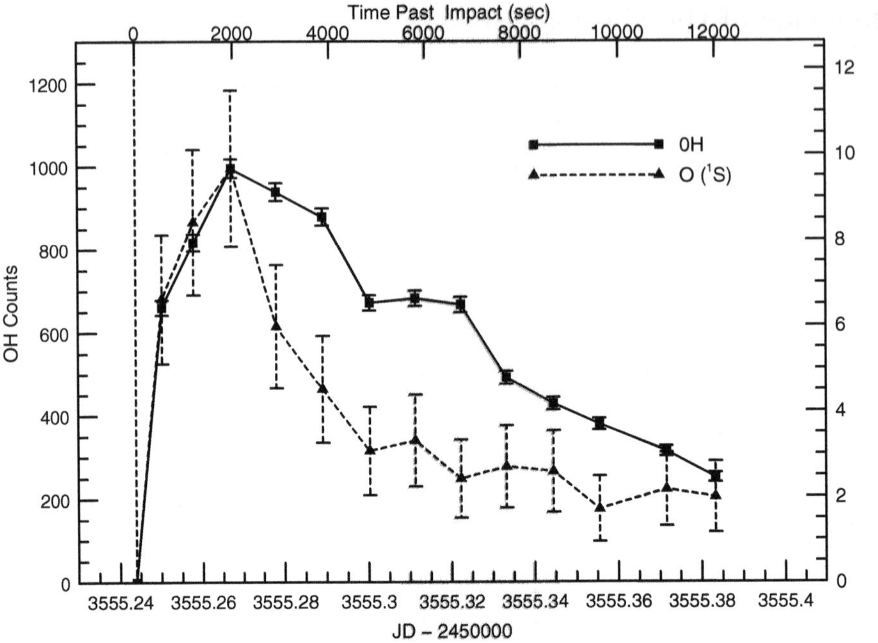

Fig. 1. The integrated flux of 8 lines of the OH (0,0) band are shown along with the deconvolved flux from O (^{1}S). The two species rise together but then the O (^{1}S) decreases more rapidly than does the OH.

line. However, if the green line is present, the red doublet will also be present because 95% of the green line atoms decay via the red doublet (5% decay in the UV). The lifetime of these oxygen lines is believed short, with the ^{1}D lines lasting 150 sec and the green line lasting a mere 1 sec. Thus, if we detect the green ^{1}S line, we are directly tracing the parent. The branching ratios to produce oxygen from H_2O or OH are extremely uncertain. Indeed, it is not clear if the O (^{1}S) line derives from water as that decay product has not been measured in the laboratory. Other potential parents for the O (^{1}S) are CO and CO_2.

Measuring the O(^{1}S) is more complicated than measuring other species. Normally, it is well hidden within the C_2 $\Delta v = -1$ band. Since this C_2 band was weak in comet Tempel 1, we were able to detect the cometary line with little C_2 contamination. However, at the Doppler shift of the comet, the cometary line was partially blended with the telluric O(^{1}S) line. We separated the cometary and telluric oxygen lines by using a 2-Gaussian deconvolution. We show the intensity of the cometary O(^{1}S) in Fig. 1.

By comparing the behavior of the O (^{1}S) with the OH in Fig. 1, we can see that the O (^{1}S) peaked at the same time as the OH but decreased in intensity much more rapidly than did the OH. Since the O (^{1}S) has such a short lifetime and its detection traces its parent, we can conclude that OH is not a major parent for the O (^{1}S). If we plot the O (^{1}S) intensity as a function of $1/\rho^2$, where ρ is the cometocentric distance, we find a linear trend with a slope of 2. This confirms that we are tracing a parent species with the O (^{1}S).

The trend of the OH with cometocentric distance is different than that for O (^{1}S). Once the peak has been reached, the OH decreases as would be expected for the decay of a daughter.

Since the decay of the O (^{1}S) emission follows the parent decay we can conclude that this is a measure of the time dependence of the influx of gas resulting from the impact. It represents the impulse function for the parents that should be present in all of the emissions if all of their parents are produced at the same time and no additional loss mechanism is present. The time dependences of many of the other emissions that we have observed differ from the impulse response of the O atoms. This is as expected since the different species will have different rate constants for formation and destruction.

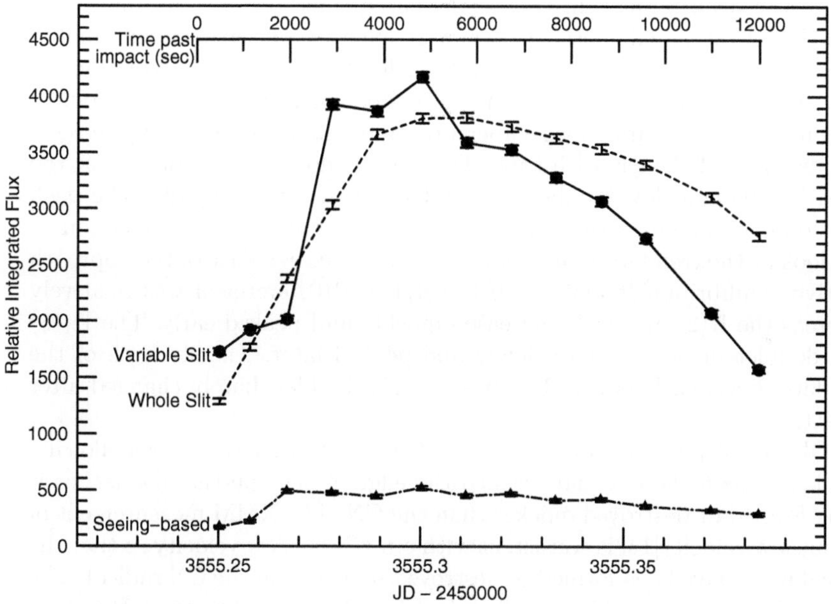

Fig. 2. The integrated line intensities of 19 R and P branch CN lines. The results of this integration for the three different methods of extraction of the spectra (see text) are shown. In the variable slit mode there is an immediate increase in the intensity of the emission followed by a slower rise to a peak intensity at 2907 s. Thus, the intensity in the CN band in the variable slit peaks earlier than the intensity of these bands in the complete slit extraction.

The CN (0,0) band is one of the best studied cometary emission features because it is generally quite strong. The OH (0,0) band is inherently stronger but it occurs at 3080 Å, in the region of strong atmospheric absorption, making it a challenge to detect. The CN band is a $\Sigma-\Sigma$ transition so it has no Q branch. Further, each of the CN rotational lines are unresolved doublets at the comet's outflow velocity and the spectral resolution of HIRES. Each of the unresolved doublet pair rotational lines are well separated from its neighbor, making them easy to measure. The lowest J-level lines of the P branch are to the blue of the bandhead and the P branch wavelengths increase (frequencies decrease) with higher J-levels until about $J = 27$, then the wavelengths decrease, superposing the higher J-level lines on the lower J-level lines. However, lines with $J > 20$ are rarely seen in cometary spectra. The R branch lines have decreasing wavelength with increasing J-level. The (1,1) band overlaps with parts of the R branch but its lines are much weaker than the (0,0) band lines.

We measured the combined intensity of 19 lines of the P and R branch of the (0,0) band of CN by integrating the flux above the local continuum. Figure 2 shows the results of this integration for our three methods of extracting the data. The whole-slit extraction shows the effect of the gas flowing out of the aperture. The rise to the peak is smooth. Our seeing-based (method 2) trend line is the line with the lowest signal/noise. As would be expected, because of the smaller "aperture", the peak emission is earlier than for the whole slit extraction. In addition, the rise to the peak is not as steep from point 1 to 2 as for point 2 to 3. When we use our variable-size aperture to follow the gas outflow, we see a different response of the gas with time. The CN gas does not rise rapidly at the beginning but remains low for the first 2100 s and then rises rapidly. The peak CN emission comes 900 s earlier than it does for the full aperture extraction.

If we look at the trend for individual CN lines we see evidence of the approach to radiative equilibrium. It looks as if the R(1)+ R(0) emission was relatively flat, whereas the R(2) and R(1) increased quickly and peaked early. The $R_1(8)$ + $R_2(7)$ doublet increased more slowly and peaked later. The strength of the higher J-level lines, such as the $R_1(12)$ + $R_2(11)$ doublet, barely changed after the impact.

The OH signal peaked sooner than the CN signal, so either it was flowing outward faster than the CN and reached the edge of our aperture sooner, or it was being formed or destroyed quicker than the CN. The XMM measurement of the OH outflow velocity[14] is consistent with our CN outflow velocity so that the OH radical must have been formed or destroyed sooner than the CN radical. The decay time of the CN radical is actually a factor of 2 shorter than the OH radical [7, 8]. In fact, as Jackson has pointed out, even though the rate constant for the loss of OH radicals can vary with the radial velocity of the comet with respect to the Sun [10], these values are all of the same order or slower than the loss rate for CN. This decay could be due to the decay of the parent rather than the decay of the radical itself because the time response curves can not distinguish between loss of the daughter and loss of the parent. This implies that the parent of CN has to have a photochemical dissociation rate constant that is slower than the corresponding rate constant for water. Fray et al. have recently reviewed the photodissociation rate constants for molecules that are possible parents of CN [5]. When these rate constants are compared to the photodissociation rate constant for the loss of H_2O, all of them but HCN are faster than water. Even HCN decays at about the same rate as water. None of these proposed parent molecules explain the observed differences in the time dependence of OH and CN. It is very unlikely that the parent of OH is anything other than water so the only way that this can be explained is that the emission that comes from OH after the Deep Impact encounter has to have come from the direct production of the radical during the impact event. If ongoing laboratory experiments show that the O (^{1}S) atom is not produced from photodissociation of water then another mechanism will be needed to explain both emissions.

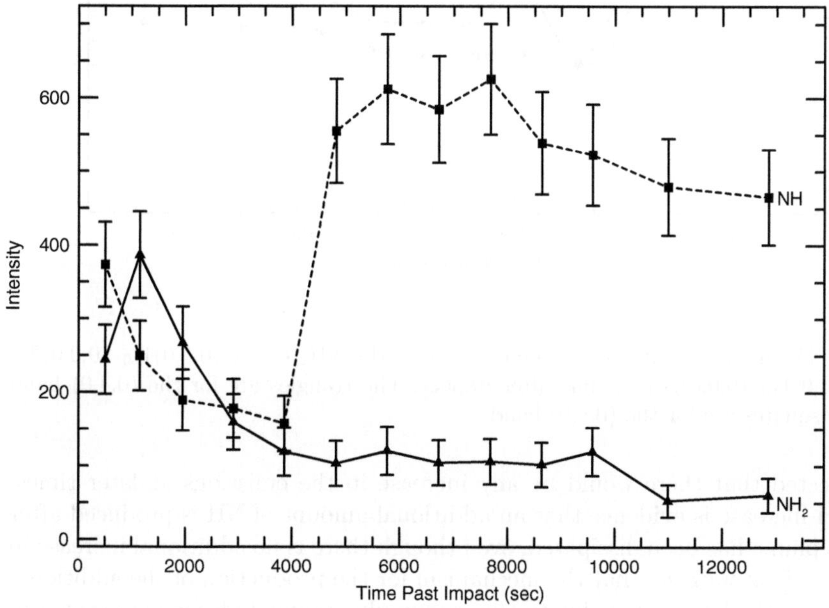

Fig. 3. The integrated flux of the emission from the NH_2 and NH rotational lines. The closed diamonds are for NH_2 and the open squares are for NH emissions. The error bars are the three sigma error bars for a 95 confidence limit. The NH_2 and NH emissions peak more quickly than the OH and O^1S. The NH peaks even quicker than the NH_2.

Figure 3 shows the time dependence on the NH and NH_2 emissions derived from the inner pixels corrected for steady state emission of these emissions in the coma. The NH_2 emission peaks at 1200 s, which is 900 s earlier than both the O (^{1}S) and the OH emission. It is also considerably faster than the CN emission. This implies that the rise of these emissions are not just due to the increase in the concentration of the parents caused by the Deep Impact collision. NH shows a sharp increas around 4000 sec after impact but NH_2 does not increase at later times. It could be that both NH and NH_2 are initially produced in the hot plume but if that is the only way that they were produced then it would not

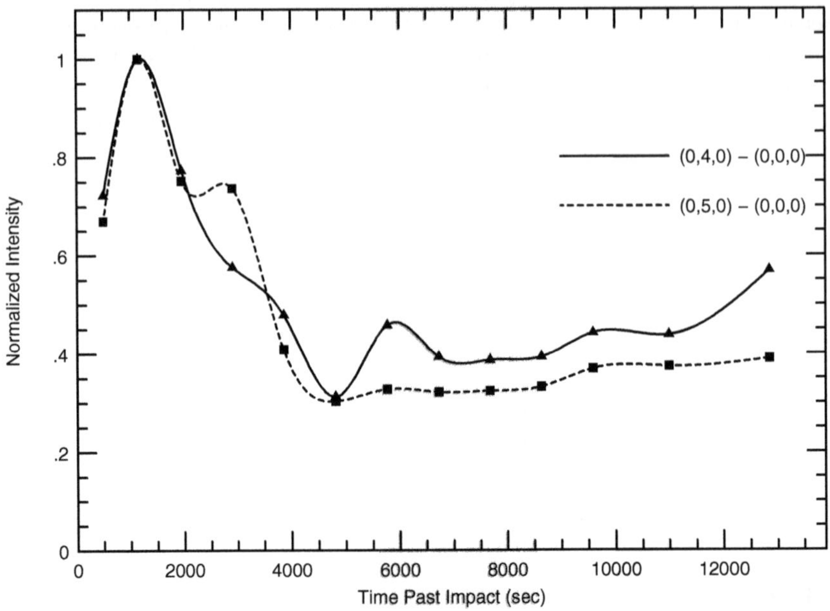

Fig. 4. The integrated flux of the emission from the NH_2 of the $(0,5,0)1_{10}$–$(0,0,0)2_{20}$ and $(0,4,0)1_{10}$–$(0,0,0)2_{20}$ vs time after impact. The trangles are for the (0,4,0) band and the squares are for the (0,5,0) band.

be expected that there would be any increase in the emissions at later times. The NH increase is evidence that an additional amount of NH is produced after the hot plume has been dissipated, even though there is no additional increase in the NH_2. This suggests that the mechanism for the production of the additional NH that peaks later is not due to the commonly assumed mechanism shown in reactions (1) and (2) and implies that NH_3 cannot be the sole parent of both NH and NH_2 since no rise is seen in the NH_2 emission that corresponds to the later rise of NH emission.

$$NH_3 + h\nu_{UV} \rightarrow H(^1S) + NH_2 \quad (1)$$
$$NH_2 + h\nu \rightarrow H(^1S) + NH \quad (2)$$

The random errors for the time dependence of the nitrogen emission are small enough so that time dependence of the curves can be trusted if there are no unknown systematic errors. There is a way of testing for systematic errors by comparing the time behavior of two emission lines that originate from the same lower state. A comparison of such a pair of lines for NH_2 is shown in Fig. 4. The first three points are clearly varying in the same manner, while there are some small differences at later times. Nevertheless the overall trends are almost identical. It also confirms that the overall trends observed in the temporal curves of a species should be a good indication of the true temporal behavior of that species.

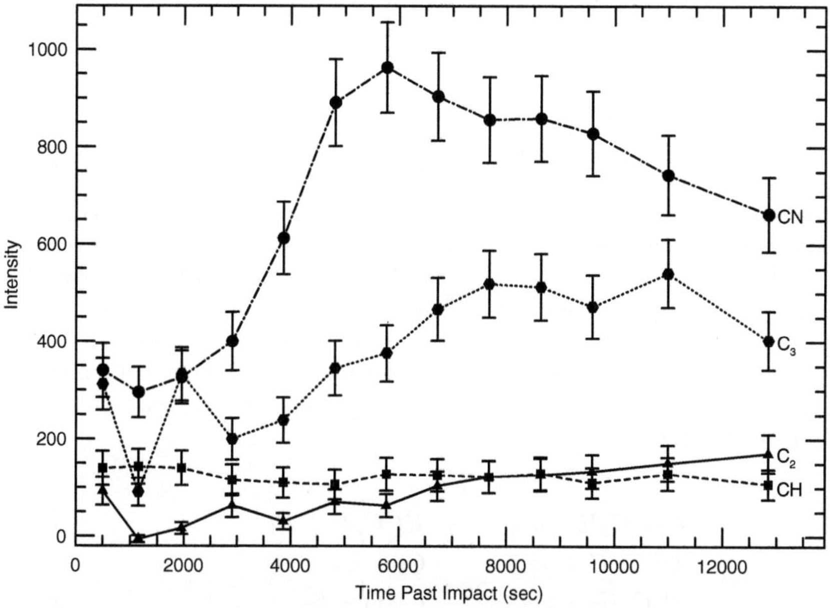

Fig. 5. The integrated flux of the emissions from all of the observed carbon species. C_2 and C_3 show an initial increase and decrease then a steady increase with time, as would be expected for a granddaughter species. CN increases until the aperture is filled and then decreases. CH shows no change of intensity after the impact.

The time dependence of all of the carbon emissions are summarized in Fig. 5. It is clear that there are distinct differences among the different radicals. The CH emission does not change with time, while the others do. However, the CH emission is above the background emission before the event. It may be that this is the signature of a distributed event where the radicals are somehow produced from dust that is emitted as a result of the Deep Impact event.

One of the proposed parents for CH emission is CH_4 via reaction (3).

$$CH_4 + h\nu_{121.6nm} \rightarrow CH + H + H_2 \tag{3}$$

The Keck NIRSPEC team for the Deep Impact collision reported that there was an increase in the CH_4 [13] so the fact that no increase is seen in the CH emission after impact implies that this can not be a significant source of CH in comets. Laboratory studies that involve the direct detection of CH could confirm this observation. The rate constant for the photochemical CH loss is 9.20 x 10^{-3}, which is much faster than the loss rates for OH, CN, and NH_2, so that the rise and fall of this radical should have been even faster than those. It is clear that the CH emission does not change as would be predicted.

C_2 emission is present right after impact and then it decreases and rises at later times. This is consistent with the direct production of C_2 by the hot plume followed by the two step production of the radical from a parent such as acetylene that was excavated as a result of the impact (reactions 4 and 5) [9, 16].

$$C_2H_2 + h\nu \rightarrow C_2H + H \tag{4}$$
$$C_2H + h\nu \rightarrow C_2 + H \tag{5}$$

The temporal behavior of the C_3 emission is similar to the C_2 emission in that it is present right after impact then it decreases, rises again, only to decrease again before rising to a plateau at later times. This is consistent with the proposal that both of these radicals arise as granddaughters rather than daughters. Just as with C_2, the C_3 has an initial rise suggesting that some of it is produced by the hot plume. The material produced at later times is consistent with a two step mechanism involving allene, $H_2C_3H_2$, or propyne, CH_3C_2H via reactions (6) through (9) [11, 12].

$$CH_3C_2H + h\nu_{VUV} \rightarrow HCC_2H + H_2 \tag{6}$$
$$HCC_2H + h\nu \rightarrow C_3 + H_2 \tag{7}$$
$$H_2C_3H_2 + h\nu_{VUV} \rightarrow C_3H_2 + H_2 \tag{8}$$
$$C_3H_2 + h\nu \rightarrow C_3 + H_2 \tag{9}$$

Laboratory studies show that allene produces C_3 more efficiently than propyne which suggests that reactions 8 and 9 may be the most important mechanism for the production of C_3 [15, 6].

To date we have no evidence for the presence of new emissions in the analyzed cometary spectra.

3 Conclusions

The Keck HIRES spectra have been processed to extract the temporal behavior of the CN, NH, NH_2, O (^1S), OH, C_2, CH, and C_3 that results from the Deep Impact encounter with Comet 9P/Tempel. The OH and O (^1S) have the same initial rise but the later decays faster than the former. The faster decay of the O (^1S) confirms that a pulse of gas passes through the coma. The slower decrease of the OH has to be connected with the formation and decay processes that form this radical. The overall temporal behavior of the observed species certainly show that these curves have embedded in them information about how the radicals are being formed. It appears that the hot plume produced in the initial impact initially forms some of the radicals. After the hot plume has passed through the coma the radicals are formed by different mechanisms. In some cases the photochemical processes normally present in comets form some of these radicals during this period. In others there are new mechanisms, yet to be discovered, that are responsible for the observed time dependences. To firmly establish the chemistry requires a detailed evaluation of the range of rate constants that are needed to fit each curve. This is the next stage in the evaluation of the Keck HIRES data.

Acknowledgements We thank Dr. Fred Chaffee for making Director's Discretionary time available pre-impact. Our thanks to Dr. Hien Tran for obtaining the observations of 30 May and 4 July and for help on 5 and 6 July. This work was funded by NASA Grant NNG04G162G (ALC), NASA Grant NNG06A67G (WMJ) and NSF Grant CHE-0503765 (WMJ).

References

1. M. F. A'Hearn. Comets after Deep Impact. Presentation at *Deep Impact as a World Observatory Event – Synergies in Space, Time*, H. U. Käufl and C. Sterken, editors, Springer-Verlag, 2006.
2. M. F. A'Hearn, M. J. S. Belton, W. A. Delamere, J. Kissel, K. P. Klaasen, L. A. McFadden, K. J. Meech, H. J. Melosh, P. H. Schultz, J. M. Sunshine, P. C. Thomas, J. Veverka, D. K. Yeomans, M. W. Baca, I. Busko, C. J. Crockett, S. M. Collins, M. Desnoyer, C. A. Eberhardy, C. M. Ernst, T. L. Farnham, L. Feaga, O. Groussin, D. Hampton, S. I. Ipatov, J.-Y. Li, D. Lindler, C. M. Lisse, N. Mastrodemos, W. M. Owen, J. E. Richardson, D. D. Wellnitz, and R. L. White. Deep Impact: Excavating Comet Tempel 1. *Science*, 310:258, 2005.
3. A. L. Cochran, W. M. Jackson, K. J. Meech, and M. Glaz. Observations of comet 9P/Tempel 1 with the Keck 1 HIRES instrument during Deep Impact. *Icarus*, page In press, 2006.
4. M. C. Festou and P. D Feldman. The forbidden oxygen lines in comets. *Astr. and Ap.*, 103:154, 1981.
5. N. Fray, Y. Bénilan, H. Cottin, M.-C. Gazeau, and J. Crovisier. The origin of the CN radical in comets: A review from observations and models. *Plan. Space Sci*, 53:1243, 2005.
6. J. D. Huang, Y. T. Chen, A. H. Kung, W. M. Jackson, and C.-K. Ni. Photodissociation of propyne and allene at 193 nm with vacuum ultraviolet detection of the products. *J. Chem. Phys.*, 110:3320, 1999.
7. W. F. Huebner and C. W. Carpenter. Solar photo rate coefficients. Informal Report LA-8085-MS, Los Alamos Scientific Lab., Los Alamos, NM, October 1979.
8. W. F. Huebner, J. J. Keady, and S. P. Lyon. Solar photo rates for planetary atmospheres and atmospheric pollutants. *Astrophys. Space Sci.*, 195:1, 1992.
9. W. M. Jackson. The photochemical formation of cometary radicals. *J. Photochem.*, 5:107, 1976.
10. W. M. Jackson. The lifetime of the OH radical in comets at 1 AU. *Icarus*, 41:147, 1980.
11. W. M. Jackson, D. S. Anex, R. E. Continetti, B. A. Balko, and Y. T. Lee. Molecular beam studies of the photolysis of allene and the secondary photodissociation of the C_3H_x fragments. *J. Chem. Phys.*, 95:7327, 1991.

12. W. M. Jackson, Y. Bao, R. S. Urdahl, X. Song, J. Gosin, and C. Lu. Cometary implications of recent laboratory experiments on the photochemistry of the C_2H and C_3H_2 radicals. In A. W. Harris and E. Bowell, editors, *Proceedings of the International Conference held at Northern Arizona University, Flagstaff, USA, Asteroids, Comets, Meteors 1991*, page 253, Houston, 1992. Lunar and Planetary Institute.
13. M. J. Mumma, M. A. DiSanti, K. Magee-Sauer, B. P. Bonev, G. L. Villanueva, H. Kawakita, N. Dello Russo, E. L. Gibb, G. A. Blake, J. E. Lyke, R. D. Campbell, J. Aycock, A. Conrad, and G. M. Hill. Parent volatiles in comet 9P/Tempel 1: Before and after impact. *Science*, 310:270, 2005.
14. R. Schulz. Detection of icy grains with XMM optical monitor. Presentation at *Deep Impact as a World Observatory Event – Synergies in Space, Time*, H. U. Käufl and C. Sterken, editors, Springer-Verlag, 2006.
15. X. Song, Y. Bao, R. S. Urdahl, J. N. Gosine, and W. M. Jackson. Laser-induced fluorescence studies of C_3a formation and isomerization in the 193 nm photolysis of allene and propyne. *Chem. Phys. Lett.*, 217:216, 1994.
16. O. Sorkhabi, V. M. Blunt, H. Lin, M. F. A'Hearn, H. A. Waver, C. Arpigny, and W. M. Jackson. Using photochemistry to explain the formation and observations of C_2 in comets. *Plan. Space Sci.*, 45:721, 1997.

Deep Impact Groundsupport Observations of Comet 9P/Tempel 1: a Student Contribution

C. Sterken[1], Y. Bouzid[1], T. Tuvikene[1], N. Shafi[2], P. Okouma[2], E. Carillo[3], N. Vogt[3,4], and S. Potter[5]

[1] Vrije Universiteit Brussel, Pleinlaan 2, 1050 Brussels, Belgium
[2] National Astrophysics and Space Science Programme, Department of Mathematics and Applied Mathematics, University of Cape Town, Rondebosch 770, South Africa
[3] Instituto de Astronomia, Universidad Católica del Norte, Av. Angamos 610, Antofagasta, Chile
[4] Department of Physics and Meteorology, Universidad de Valparaiso, Avda. Gran Bretaña 1111, Valparaiso, Chile
[5] South African Astronomical Observatory, P.O. Box 9, Observatory, 7935 South Africa

Abstract The encounter of Deep Impact with comet 9P/Tempel 1 led to an unprecedented coordinated observational ground-based observing campaign which attracted many non-specialists. This paper discusses aspects of observational accuracy achieved by untrained observers using standard CCD cameras at 0.5 and 1-m class telescopes in South Africa and Chile. Our experience illustrates that calibrated photometry of comets cannot be accomplished without dedicated observing.

1 The Observations

The encounter of Deep Impact with comet 9P/Tempel 1 led to an unprecedented coordinated observational ground-based observing campaign which attracted many non-specialists. This paper discusses aspects of observational accuracy achieved by untrained observers using standard CCD cameras at 0.5 and 1-m class telescopes in South Africa and Chile.

Observations were carried out by MSc and PhD students without any experience in cometary imaging and without any assistance by their supervisors. Only routinely-available instrumental configurations were used.

1.1 Observations at SAAO

The observations were made with the 1.0-m telescope at South African Astronomical Observatory in Sutherland. The camera has a back-illuminated 1024 × 1024 SITe CCD detector and covers a field of view of 5 × 5 arcminutes with a pixel scale of 0.31 arcsec pix^{-1}. Integration times were kept short as no guiding was possible. No standard stars were observed. As these observations preceded the monitoring of a field for which the CCD camera had to be rotated by 45 degrees, all SAAO cometary images are rotated by 135 degrees counterclockwise (Fig. 1).

Fig. 1. V-band image of the comet obtained at SAAO on June 28, 2005 and at OCA on July 07, 2005 (*left* and *middle*, respectively). *Right:* V-band image of the the open cluster Trumpler 24 with identifications for the variable V (GSC 07872-01052) and constant stars A, B and C (N is up and E is left). GSC 07872-02362 (D) is a relatively bright eclipsing binary used as comparison star.

1.2 Observations at Observatorio Cerro Armazones

Data were obtained with the 84-cm telescope at Observatorio Cerro Armazones (OCA), Chile during 6 nights in July 2005. We used an SBIG ST-9E CCD-camera with a 512 × 512 KAF 0261E chip with a field of view of 11.6 arcminutes (pixel scale 1.36 arcsec pix^{-1}), see Fig. 1 (north is up-right and east is down-right).

Additional observations were made with the OCA 41-cm telescope using an SBIG ST-7 CCD-camera with 765 × 510 pixels and *RGB* filters. No bias and flats were obtained, but the images were calibrated with calibration frames obtained during a subsequent observing run.

2 Observational Accuracy

We evaluate the quality of our data by comparing variable star results from Trumpler 24, an open cluster almost simultaneously observed with the same instruments. In order to obtain the assessment independently of data reduction procedures, we derived stellar magnitudes in two different ways: aperture photometry and combined PSF/aperture photometry.

Aperture photometry procedures based on DAOPHOT algorithms included in the IDL Astronomy User's Library were used. PSF/aperture photometry was carried out with the reduction package MOMF (Multiple Object Multiple Frame) which applies a combined point-spread function and aperture photometry. We refer to [1] for a complete description of the procedures.

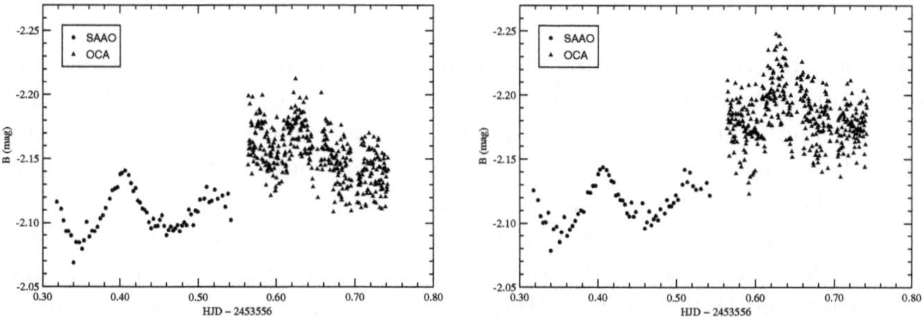

Fig. 2. Differential photometry of GSC 07872-01052 observed at SAAO and OCA. *L*eft: pure aperture photometry, *R*ight: combined PSF/aperture photometry.

Figure 2 shows the differential light curve of a variable in Trumpler 24 observed at SAAO and OCA. In the left panel the magnitudes were derived by pure aperture photometry, whereas the right panel shows the result of PSF/aperture photometry for both datasets. Both diagrams show much larger scatter and zero-point shifts for the OCA results, indicating that the effect is mainly to be ascribed to differences of the local instrumental systems. Note, however, that the scatter is not independent of the data reduction approach: pure aperture photometry leads to about 0.02 mag less dispersion than when involving PSF fitting. The light-curve morphology is also different.

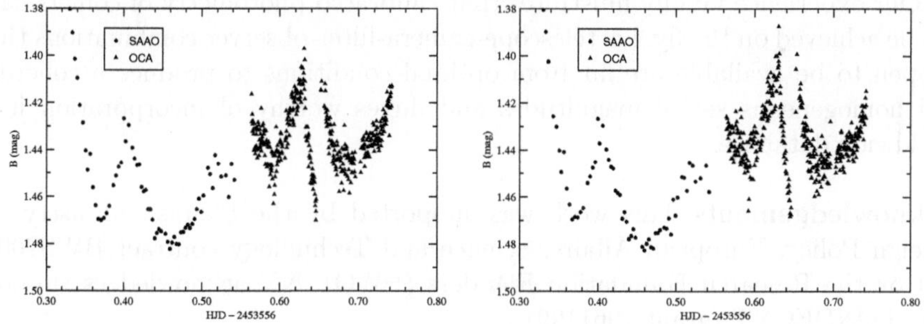

Fig. 3. Same as Fig. 2, but using a brighter comparison star (star D).

The large difference in scatter is mainly due to the relative faintness of the comparison stars, combined with mediocre image quality: as the exposure levels decrease, poor image quality enhances the dispersion.

Figure 3 shows similar differential light curves when using a brighter (eclipsing binary) comparison star. The result is less dependent on the data reduction method. The sharp decline during HJD 245 3556.63–3556.64 is caused by dome obstruction; the effect almost escapes notice when using faint comparison stars. Note the strong decline during the time interval HJD 245 3556.30–3556.35 which is due to an eclipse phenomenon of the comparison star.

3 Conclusions

Photometry of comets is mainly the photometry of (moving) extended objects with a composite (i.e. emission-line and continuum) spectrum. Broad-band filters are widely applied to isolate a portion of a cometary spectrum, though their use is limited because broad pass bands smear out contributions from the continuum and from the strong emission lines. A Working Group established by IAU Commission 15 at the 1979 General Assembly recommended specific narrow-band filters which isolate particular spectral regions, which led to the conception of the International Halley Watch (IHW) filter system. Many of these filters have now physically deteriorated to the point that they are no longer usable, and some of the original IHW continuum filters suffer from substantial contamination by emission lines, so the issue of optimal filters is not quite settled yet.

Cometary photometry is different from stellar photometry. The primary limitation of accuracy is not the detector, but the handling of sky-background, the choice of standard stars (which have to be changed along the path of the moving comet), extinction determination, the magnitude-extraction algorithms, etc. The measurements are frequently collected at high air masses and often occur in twilight, when sky brightness changes rapidly, and when few standard stars can be observed. In addition, two totally different classes of standard stars are required: G-type standard stars similar to the Sun (for accurately determining the continuum due to the reflectance of the cometary dust), and standard stars (for absolute flux calibration of the emission features), usually chosen as B stars because of their few spectral lines. For more details, see [2].

Our experience vividly illustrates that calibrated photometry of comets cannot be achieved on the fly, i.e. telescope-camera-filter-observer combinations that happen to be available are far from optimal conditions to produce a coherent and homogeneous set of magnitudes and fluxes worthy of incorporation in a worldwide database.

Acknowledgements This work was supported by the Flemish Ministry for Foreign Policy, European Affairs, Science and Technology contract BWS2005, and by the Research Foundation Flanders (FWO). NV acknowledges support from FONDECYT (grant 1061199).

References

1. T. Tuvikene, M. Y. Bouzid, A. Ederoclite, C. Sterken: A Comparative Test of CCD Reduction Procedures. In: *The Future of Photometric, Spectrophotometric and Polarimetric Standardization,* (ASP Conf Series 2007), 364, 579
2. A.T. A'Hearn: Photometry of Comets. In *Solar System Photometry Handbook*, ed by R.M. Genet (Willmann-Bell, Richmond 1983)

Observations of Comet 9P/Tempel 1 and Deep Impact by the OSIRIS Cameras onboard Rosetta

M. Küppers[1], H. U. Keller[1], S. Fornasier[2,3], P. J. Gutiérrez[4], S. F. Hviid[1], L. Jorda[5], J. Knollenberg[6], S. C. Lowry[7], M. Rengel[1], and The OSIRIS Team

[1] Max-Planck-Institut für Sonnensystemforschung, Germany kueppers@mps.mpg.de
[2] Dipartimento di Astronomia and CISAS, Università di Padova, Italy
[3] University of Paris 7, France
[4] Instituto de Astrofísica de Andalucía - CSIC, Spain
[5] Laboratoire d'Astrophysique de Marseille, France
[6] Institute of Planetary Research, DLR, Germany
[7] Astrophysics Research Centre, Queen's University Belfast, United Kingdom

Abstract The OSIRIS cameras onboard ESA's Rosetta spacecraft observed comet 9P/Tempel 1 near-continuously from 28 June until 14 July 2005. The Narrow Angle Camera (NAC) monitored the cometary dust, while the wide angle camera (WAC) observed through filters sensitive to emissions of OH, CN, OI, Na, and the associated continuum.

The brightness of the dust increased by a factor of 4.5 within 30 minutes, measured in a circular aperture of 3000 km radius centered on the nucleus. The brightening is likely caused by ejection of material from the crater, decreasing optical depth of the ejecta, and the evaporation of icy grains in the ejecta cloud. The dust moved with a typical velocity of 160 ms^{-1}, suggesting acceleration of the dust by gas in the inner cometary coma. Images of the dust cloud taken several days after the impact show the effect of solar radiation pressure. The size distribution derived from an analysis of the radiation pressure is comparable to that commonly observed in cometary comae.

The increase of the intensity of the OH emission due to the impact corresponds to a water production of (1.5–3)$\times 10^{32}$ molecules. The dust production depends on the uncertain amount of large dust grains ejected by the impact, but the dust/gas ratio is likely to be larger than one.

1 Introduction

Rosetta, a spacecraft of the European Space Agency launched on 2 March 2004, is currently on its way to a rendezvous with comet 67P/Churyumov-Gerasimenko (hereafter C-G) in 2014. In June and July 2005, it was located well to observe comet 9P/Tempel 1 (hereafter Tempel 1) around the Deep Impact event: its distance from the comet was 0.53 AU (compared to a geocentric distance of Tempel 1 of 0.89 AU), and the solar elongation angle of just above 90 degrees minimized straylight problems for remote sensing instruments.

OSIRIS is the scientific camera system on Rosetta [17]. It comprises a narrow angle camera (NAC), designed to study the nucleus and the near-nucleus dust at comet C-G, and a wide angle camera (WAC), built to investigate the gas and dust environment of comet C-G. Several goals were identified for the OSIRIS observations of comet Tempel 1:

- Monitoring the dust production and the morphology of the dust coma with the NAC
- Photometry of the comet after the impact (light curve determination) with high (about 40 seconds) time resolution with the NAC
- Monitoring the water (with emissions of OH and OI) and CN production with the WAC
- Search for sodium emission with the WAC

First results of the OSIRIS observations were published in [16, 20]. A more comprehensive presentation of the data and additional results are given in [18]. Here we summarize our results and focus on a comparison with findings by other researchers.

2 Observations and Data Reduction

2.1 Instruments and Observations

The OSIRIS cameras are unobstructed mirror systems. The focal length of the NAC is $f = 72$ cm and the focal ratio $f/D = 8$. For the WAC, $f = 14$ cm and $f/D = 5.6$. Both cameras are equipped with 2048×2048 pixels CCD detectors with a pixel size of 13.5 μm and with two filter wheels with 8 positions each. The NAC contains 11 medium band-width filters (23–87 nm FWHM), 4 re-focussing plates, and a neutral density filter. A combination of two re-focussing plates allows to observe effectively with a clear filter. For the WAC, one filter wheel is always in empty position, allowing to choose between 12 narrow band filters (4–14 nm FWHM) and 2 broad band filters (green and red).

The observations of comet Tempel 1 took place between 28 June and 14 July 2005. On 4 July, the spatial scale at the comet was 1500 kmpixel^{-1} with the NAC and 7800 kmpixel^{-1} with the WAC. The phase angle of comet Tempel 1 was 69 degrees, at a distance of 0.53 AU from Rosetta and 1.51 AU from the Sun. The filters used for the observations of Tempel 1 are listed in Table 1. More details are given in [18].

2.2 Data Reduction

Image Calibration

OSIRIS images are processed by a standard calibration pipeline which performs the following steps:

Table 1. Filters used for the OSIRIS observations of comet Tempel 1.

Camera	Filter	Central wavelength	bandwidth [nm]	Typical exposure time [s]
NAC	orange	647.5	86.9	600[a]
	red	743.5	64.2	900
	near-IR	882.0	66.0	1200
	IR	987.3	37.8	1800
	clear	620	470	300[b]
WAC	OH	308.6	3.9	600
	CN	386.9	5.1	600
	UV 375[c]	375.2	9.6	600
	OI	630.6	4.2	1200
	Na	589.4	4.8	1200
	Vis. 610[d]	611.2	9.9	1200

[a] Reduced to 45 s close to the impact time
[b] Reduced to 35 s close to the impact time
[c] Continuum for OH and CN
[d] Continuum for OI and Na

- The CCDs are read out using a dual 14 bit Analogue to Digital Converter (ADC). DN values created by the "upper" ADC (DN values $\geq 10^{14}$) get an additional offset added by the read-out electronics. This offset is subtracted
- A noise filter removes coherent noise produced by the power converter from the data
- The bias is subtracted
- A bad column in the NAC CCD is filtered out
- The images are divided by flat created pre-launch in the ground calibration
- The data values are converted into radiometrically calibrated data in units of W m^{-2} nm^{-1} $sterad^{-1}$ using observations of spectrophotometric standard stars and solar analogs.

Extraction of Light Curves for Gaseous Emissions

The radiometrically calibrated data were used as input for the computation of the WAC light curves in the gas filters. Since the narrow band images are frequently of low signal/noise, the position of the nucleus was determined from images taken in the red broad-band filter. Cosmic ray events and stars were removed by comparing each image with a median stack of 7 images (the image together with the last 3 images taken before it and the next 3 images taken after it). Pixels deviating by more than 2 standard deviations were replaced by the median value. For OH and CN, two images were taken closely spaced in time. To remove possible remnants of cosmic rays or stars, the minimum of both images was taken as representative of the time step. Then the continuum image was multiplied by a correction factor accounting for the different passband between the two filters and the different radiance of the sun. The corrected continuum image was then subtracted from the gas filter image. Finally, a standard photometry routine was used to sum up the signal in circular apertures centered on the cometary nucleus.

3 Results

3.1 Light Curve of the Dust

The light curve is shown in Fig. 1 on time scales of hours and days, while Fig. 2 shows the first few minutes after the impact. The light curve shows various features:

- By far the most prominent change of radiance with time is a steep increase starting in the first minute after the impact and lasting about 40 minutes, followed by a shallower increase until about 90 minutes after the impact. Later on the brightness decreased because the impact created material left the photometric aperture. This basic shape was discussed in [20] and seen by various observers, e.g. [24, 25, 28, 2, 3, 7, 9, 11, 19, 30]. The increase took substantially longer than the time scale of crater formation. The likely reasons are decreasing optical depth and fragmentation of (icy) grains [20, 6].
- The first few minutes showed several changes in brightness slope, possible connected to the end of crater formation after about 4 minutes, followed by the ejecta becoming optically thin [16]. However, while slope changes are seen by other observers (e.g. [9]), the hiatus at about 4 minutes is seen in the OSIRIS data only. The reason is currently unknown. The different viewing geometry from Rosetta as compared to the Earth may play a role. The steep increase in the first few minutes implies that the dust ejected from the crater was quickly accelerated in the cometary coma.
- Apart from the impact, little time variation is seen. The overall brightness decreased by $\approx 10\%$ during the two weeks of observations. Apart from that, the variation of the data is only 4%, correlated with the rotation of the comet [18]. The result of little long-term variations (and no long-term effect of the impact) is consistent with results from many other observations (e.g. [28, 14, 26, 8]). Large, not impact related, variations are seen in sub-mm emission of water [4] on a timescale of days. See [22] a strong increase in HI Lyman α emission that they interpret as a factor of two increase of water production a week after the impact. The discrepancy between the large variability in cometary production seen by [4] and [22] and little variability in most other dats sets is currently unexplained.

3.2 Size-Velocity Distribution of Impact-Ejected Dust

Since the temporal variation in the coma of Tempel 1 as observed by OSIRIS is dominated by the impact, the background coma can be subtracted from the cloud created by the impact with high accuracy (Fig. 3). The resulting images of the impact created dust cloud clearly show that material is pushed in anti-solar direction by solar radiation pressure. Since the ratio between radiation pressure

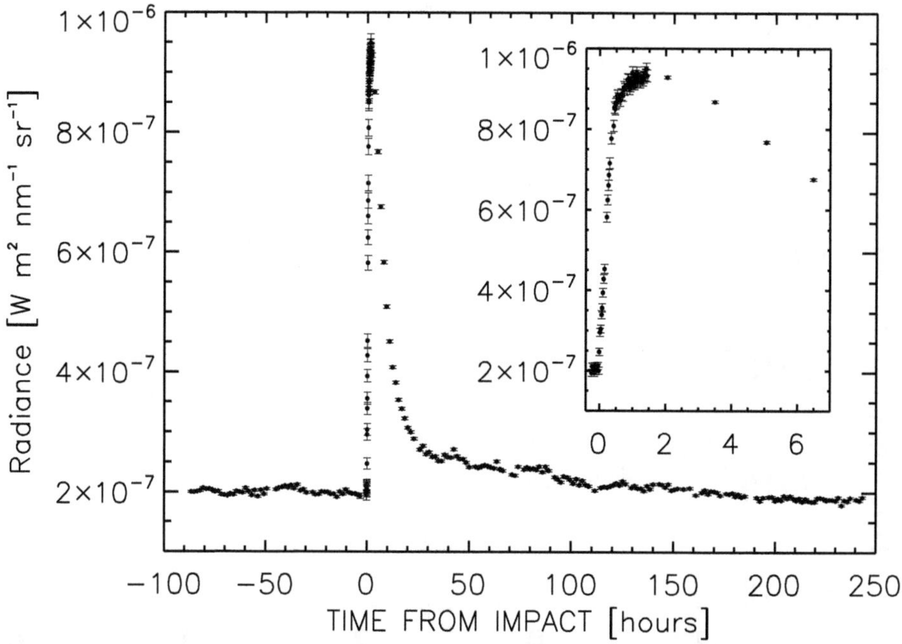

Fig. 1. Light curve of comet Tempel 1 around Deep Impact with the orange filter of the OSIRIS NAC. The radius of the aperture is 2 pixels or 3000 km. Taken from [20].

and solar gravity increases for decreasing particle size, the evolution of the cloud is sensitive not only to particle velocities, but also to dust grain size. Reference [15] used a Monte-Carlo model to derive the size and velocity distribution of the dust from the evolution of the morphology of the impact cloud. The main results are:

- The velocity distribution shows a broad maximum at about 190 m s^{-1} with a FWHM of 150 m s^{-1}. This agrees with most other determinations of dust velocities in the impact created dust cloud (e.g. [29, 7, 30]). The dust is much faster than would be expected for impact ejecta ([13]), suggesting acceleration by gas in the cometary coma. A different approach to the determination of the velocity distribution from the OSIRIS data is presented by [27].
- For particles smaller than 20 μm, the dust size distribution exponent is -3.2 ± 0.2. Small, μm-sized particles dominate the brightness and the cross section of the dust, while larger particles dominate the mass. The size distribution is not different from that observed in cometary dust tails ([15]).
- The total mass of particles $<$ 1.4 μm is $1.5\pm0.2 \times 10^5$ kg, that of particles $<$ 100 μm is (1–14) $\times$ 10^6 kg. The total mass cannot be determined from the dust coma morphology because both the contribution to the total brightness and the radiation pressure force are small for large particles.

Observations in the infrared spectral range ([29, 10, 21]) show large differences between the Deep Impact dust cloud and the dust emitted by Tempel 1 before the

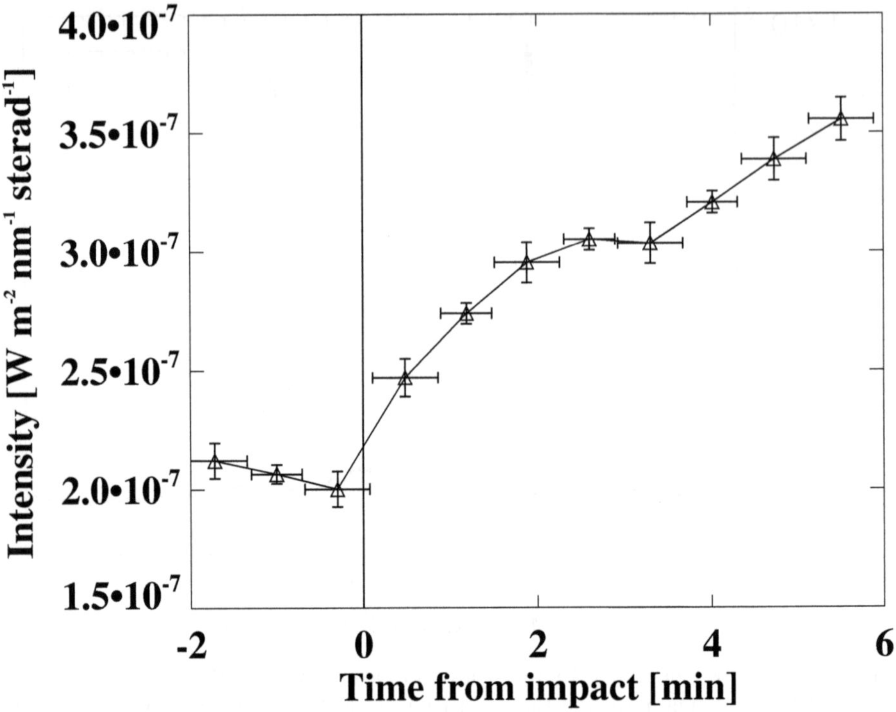

Fig. 2. The first minutes after the impact. Images taken through the clear filter are scaled and shown together with images obtained with the orange filter. The aperture radius is 3000 km. Taken from [18].

impact. In particular, the appearance of the 10 µm emission feature suggests that a large fraction of the impact created dust mass consists of small particles. Either the dust grains emitted by Tempel 1 before the impact were unusually large, or the dust ejected by the impact appeared different in different wavelength regions. If cometary dust is made of fluffy aggregates, it may be possible that their optical properties are dominated by the aggregate in the infrared, while the monomers forming the aggreagate determine the scattering properties in the visible range. Then, if such aggregates are fragmented by the impact, the impact created dust may look similar to the dust from normal activity in the visible range, but not in the IR. However, the radiation pressure on the fragmented dust should be different from that on the aggregates, so the size distribution derived from the dust morphology should reflect the changes due to fragmentation. More detailed modelling is necessary to determine the physical properties of the dust ejected by the impact.

3.3 Gas Production

Figure 4 shows the light curves measured in the emissions of OH and CN. They show a steady brightness increase following the impact. The shape of the increase

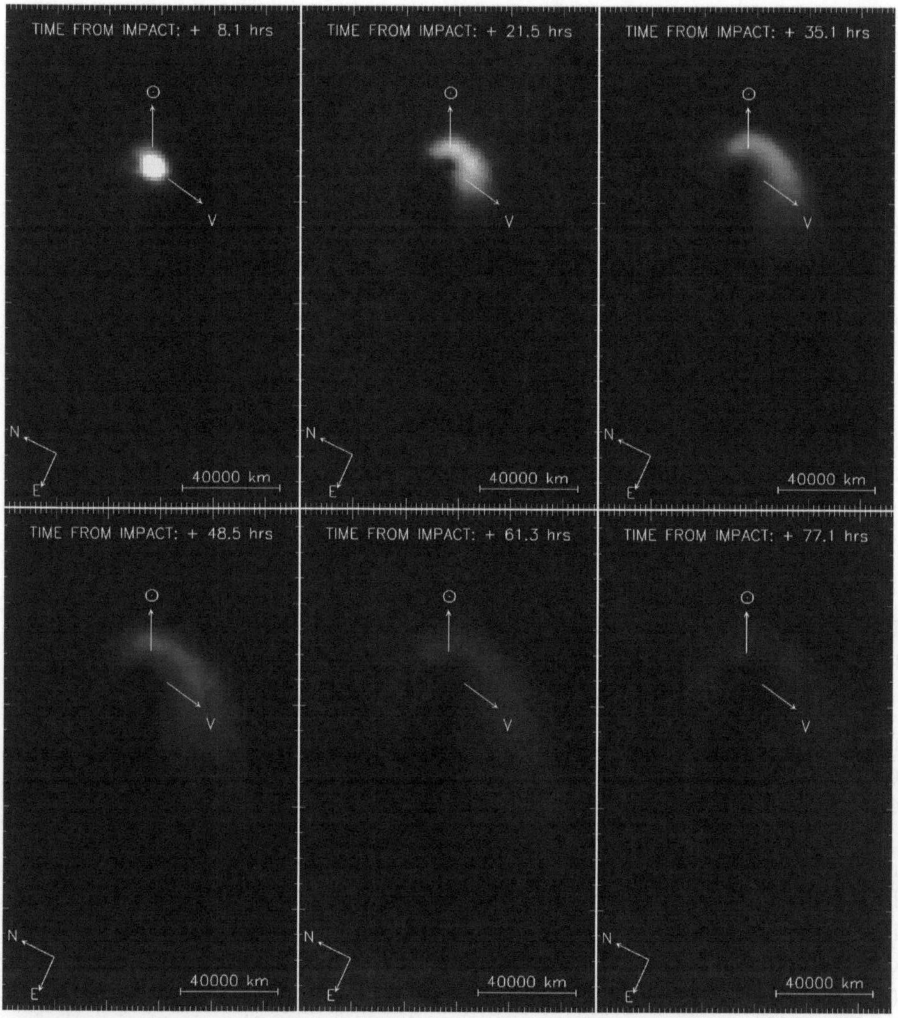

Fig. 3. Six images of the impact cloud taken in the orange filter. The average pre-impact coma is subtracted. Each image is an average of approximately 12 hours of data. The direction of the sun and of the motion of the comet are indicated.

is consistent with creation by photodissociation of parent molecules (water for OH, unknown for CN) that were created instantaneously by the impact. The later brightness decrease is caused by molecules leaving the photometric apertures and by dissociation of OH and CN into its atomic components. Apart from the additional radiance created by the impact, the light curves show very little time variation, in agreement with the dust observations by OSIRIS.

The production rates of water and of the CN parent were calculated with the method described in [20, 18]. The results are shown in Table 2, together with production rates taken from the literature. Both gas production rates before

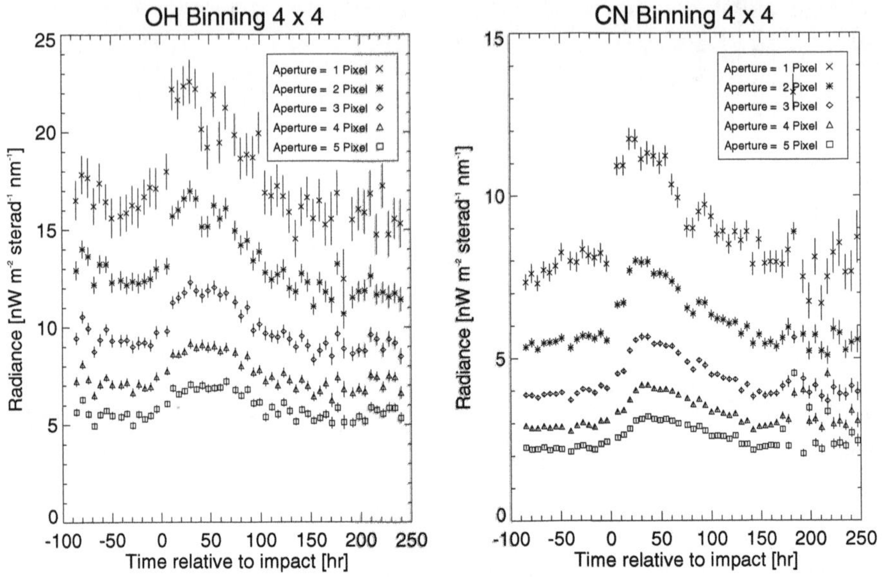

Fig. 4. Light curve of OH and CN emission, measured with the WAC. One binned pixel corresponds to 31200 km at the comet.

impact and the number of impact related molecules generally agree well between different measurements, given the uncertainties associated with the calculation of production rates from photometric measurements.

The impact produced water of (1.5–3) $\times$ 10^{32} molecules corresponds to a mass of (4.5–9) $\times$ 10^6 kg. Since the energy needed to sublime that much water by far exceeds the kinetic energy of the impactor, most of the water left the comet in icy form, probably as ice grains, and was sublimated by sunlight. The light curve of the dust in Fig. 1 as well as infrared observations of parent molecules ([25, 6]) suggest that sublimation of most grains took less than an hour.

The CN/water mass ratio appears to be higher in the ejecta than before impact, although an unchanged ratio is marginally consistent with the data. An increased CN/water ratio would be consistent with observations of an enhanced abundance of organics in the impact ejecta ([1, 25, 6]).

Sodium emission was searched for both in the cometary coma and in the impact ejecta. As expected at a heliocentric distance of 1.5 AU, no sodium was found in the background coma. Some positive residuals in the first few hours after the impact are under investigation. A previous report about a clear detection of sodium emission is incorrect due to an error in the cosmic ray removal.

Observations of the OI line at 630 nm provide additional information about the water production. OI emission is clearly visible in the data, however the quantitative reduction turns out to be problematic and is still ongoing.

Table 2. Water and CN production rates of comet Tempel 1 measured before impact and number of impact created water and CN molecules. Units are moleculessec^{-1} for pre-impact production rates and number of molecules for gas creation by the impact.

Reference	H_2O (pre-impact)	H_2O (impact)	CN (pre-impact)	CN (impact)
This work[a]	4.4×10^{27}	$1.5–3 \times 10^{32}$	1.0×10^{25}	$5–9 \times 10^{29}$
[28][a]	6×10^{27}	$< 4 \times 10^{32}$	$\approx 10^{25}$	-
[8][b]	9×10^{27}	-	-	-
P. Feldman[c]	-	2.3×10^{32}	-	-
[5][d]	8×10^{27}	$(1.7 \pm 0.4) \times 10^{32}$	-	-
[4][e]	6.6×10^{27}	$< 6 \times 10^{32}$	-	-
[6][f]	9.4×10^{27}	-	-	-
[23][g]	-	$(1.4 \pm 0.2) \times 10^{32}$	-	-
[26][h]	-	-	1.1×10^{25}	$(3.7 \pm 1.9) \times 10^{29}$

[a] from OH (308 nm) and CN (388 nm)
[b] from OI (130 nm)
[c] priv. comm., from OH (308 nm)
[d] from H_2O ground state transition
[e] from H_2O ground state transition
[f] from H_2O infrared transitions
[g] from OH (308 nm)
[h] from CN (388 nm)

3.4 Dust to Gas Mass Ratio

The measured dust/gas mass ratio in cometary comae and dust tails is typically around unity. However, it is unclear how representative that value is, because it is conceivable that there may be more ice relative to dust in active parts of the surface than in inactive regions. The Deep impact ejecta cloud provides a unique opportunity to measure the dust/ice ratio in an inactive region of a comet.

While the gas production of the impact, essentially equal to the water production, can be estimated relatively accurate from our measurements of the OH emission, the dust mass is more difficult to determine. For a dust size distribution with a size exponent > -4 as in the case of Deep Impact (Sect. 3.2), the most critical parameter for conversion of dust brightness into mass is the upper size cutoff of the distribution, i.e. the size of the largest particle. We use a 1d gasdynamical calculation ([18]) to determine 3 mm as the size of the largest particles which will leave the nucleus of Tempel 1 due to acceleration by collisions with gas molecules (note that most impact ejecta are expected to be launched from the crater with a velocity below the escape velocity of the nucleus).

Assuming that the size distribution of small dust particles determined by [15] can be extrapolated to larger particles, that only particles smaller than 3 mm leave the nucleus, and that the dust particles are spheres with an albedo of 0.1 and a density of 1000 kgm^{-3}, the dust/ice mass ratio in the impact ejecta is between 2 and 20. This would indicate that most of the mass of the comet is in the dust component (Fig. 5).

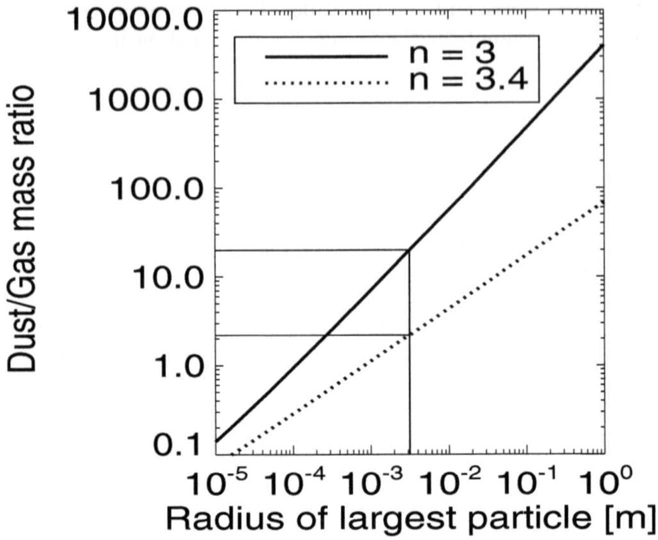

Fig. 5. Dust/ice mass ratio as a function of the size of the largest particle. The dust size distribution power law exponent of −3.4 and −3 is the minimum and maximum from [15]. The vertical line is drawn at a dust size of 3 mm.

4 Discussion

The OSIRIS results provide the following impression of the sequence of events after the impact of the projectile into comet Tempel 1: The impact excavated 5000–10000 tons of gas, and a presumably larger mass of dust. Assuming that all excavated material left the comet, the mass of ejected gas and dust corresponds to a lower limit for the crater size of about 30 m ([18]) and an upper limit for the strength of the cometary material of a few kPa.

The measured velocity of the dust cloud and model calculations both suggest that the impact ejecta were quickly accelerated by the gas in the cometary coma. Therefore information about the density or strength of the target material cannot be derived from observations of the ejecta if they are taken more than a few minutes after the impact ([18, 12]).

The brightness of the ejecta cloud increased for $\approx$ 40 minutes after the impact due to decreasing optical thickness and fragmentation of grains. Most of the water ice was ejected in solid form and sublimated by sunlight. No long-term effect of the impact was visible. However, the crater formed by the impact is small compared to the surface area of the comet. So if a new active area was created, it would not be detectable by OSIRIS.

References

1. A'Hearn, M.F., Belton, M.J.S., Delamere, W.A., et al. (2005): Science, **310**, 258
2. Barber, R.J., Miller, S., Stallard, T. et al. (2007): Icarus, **191**, 371
3. Bauer, J.R., Weissman, P.R., Choi, Y.-J., et al. (2007): Icarus, **191**, 296
4. Bensch, F., Melnick, G.J., Neufeld, D.A., et al. (2006): Icarus, **184**, 602
5. Biver, N., Bockelée-Morvan, D., Boissier, et al. (2007): Icarus, **187**, 253
6. DiSanti, M.A., Villanueva, G.L., Bonev, B.P., et al. (2007): Icarus, **191**, 481
7. Feldman, P.D., McCandliss, S.R., Route M., et al. (2007): Icarus, **191**, 276
8. Feldman, P.D., Stern, S.A., Steffl, A.J., et al. (2007): Icarus, **191**, 258
9. Fernández, Y.R., Lisse, C.M., Kelley, M.S., et al. (2007): Icarus, **191**, 424
10. Harker, D.E., Woodward C.E., Wooden D.H. (2005): Science, **310**, 278
11. Hodapp K.W., Aldering G., Meech K.J. et al. (2007): Icarus, **191**, 389
12. Holsapple K.A. and Housen K.R. (2007): Icarus, **187**, 345
13. Housen K.R., Schmidt R.M., Holsapple K.A. (1983): J. Geophys. Res. **88**, 2485
14. Jehin E., Manfroid J., Hutsmékers D., et al. (2006): AJ **641**, L145
15. Jorda L., Lamy P., Faury G., et al. (2007): Icarus, **187**, 208
16. Keller H.U., Jorda L., Küppers M., et al. (2005): Science, **310**, 281
17. Keller H.U., Barbieri C., Lamy P., et al. (2007): Space Sci. Rev., **128**, 433–506
18. Keller H.U., Küppers M., Fornasier S., et al. (2007): Icarus, **191**, 241
19. Knight M.M., Walsh K.J., A'Hearn M.F., et al. (2007): Icarus, **191**, 403
20. Küppers M., Bertini I., Fornasier S., et al. (2005): Nature, **437**, 987
21. Lisse C.M., Van Cleve J., Adams A.C., et al. (2006): Science, **313**, 635
22. Mäkinen J.T.T., Combi M.R., Bertaux J.-L., et al. (2007): Icarus, **191**, 263
23. Mason K.O., Chester M., Cucchiara A., et al. (2007): Icarus, **191**, 286
24. Meech K.J., Ageorges N., A'Hearn M.F., et al. (2005): Science, **310**, 265
25. Mumma M.J., DiSanti M.A., Magee-Sauer K., et al. (2005): Science, **310**, 270
26. Rauer H., Weiler M., Sterken C., et al. (2006): A A **459**, 257
27. Rengel M., Küppers M., Keller H.U., et al. (2007): This volume, p. 137
28. Schleicher D.G., Barnes K.L., and Baugh N.F. (2006): Astron. J., **131**, 1130
29. Sugita S., Ootsubo T., Kadono T., et al. (2005): Science, **310**, 274
30. Walker R.G., Weaver W.B., Shane W.W., et al. (2007): Icarus, **191**, 526

References

Comets, Charisma, and Celebrity: Reflections on Their Deep Impact

R. J. M. Olson[1] and J. M. Pasachoff[2]

[1] The New-York Historical Society, 170 Central Park West, New York, NY 10024 U.S.A Olson@NY history.Borg
[2] Williams College, Williamstown, MA 01267 U.S.A.
Jay.M.Pasachoff@williams.edu

Abstract In celebration of the Deep Impact Mission, this essay explores the influence of comets on the arts and sciences since the beginning of recorded time. Through images, ranging from the sublime to the humorous, it probes the reasons why comets are among the most charismatic visual spectacles in the universe and why, even as scientific missions unmask their mysteries, they remain iconic symbols and harbingers of change.

1 Setting the Stage

This paper celebrates the scientific achievement of the Deep Impact Mission to Comet 9P/Tempel in a broader context by reflecting on the influence that comets have exerted on cultural history over the millennia. It also considered comets and the concept of celebrity that superficially have much in common. Like celebrities, comets are visually compelling departures from the normal denizens of the night. They are inherently charismatic and rivet the attention of mere mortals. Once regarded as miraculous, comets, together with meteors, were viewed as errant stars that violated what was perceived as the order of the heavens [1]. But meteors or shooting stars – which after the 18th-century were no longer confused with their comet cousins – are even closer to many celebrities because they burn out after streaking across the firmament. Consider the following definition of celebrity: "Celebrity is a person who is known for his/her well-knownness" [2]. While the concept of celebrity can be shallow, that of comets is profound.

Since ancient times comets and meteors have been ubiquitous symbols in western art and literature. Comets predominate over meteors because of their visibility on the canvas of the night sky for longer periods of time, sometimes for months. Thankfully most of our lives are more like comets, as opposed to those individuals whose lives are meteoric and who make a big splash and then are heard of no longer. Both celestial phenomena caught the imaginations of people in times when there were no electric lights and answers to physical phenomena were only speculative. As the recent appearance of Hale Bopp in 1997

Fig. 1. Anonymous, *The Comet of 1527.* 1527, Thomas le Claire Kunsthandel, Hamburg.

Fig. 2. Obverse and reverse of a Roman coin commemorating Julius Caesar, c. 30 BC–AD 14, Munich, Bayerisches National Museum, Basserman-Jordan Collection.

proved, comets remain mesmerizing and can be spectacular despite modern urban centers. Comets' extraordinary beauty and awesome unpredictability has exerted a powerful appeal across cultures. Early Chinese astronomers kept extensive records of hundreds of apparitions and drew comets quite objectively in their calligraphic style – such as in the Han Dynasty silk comet atlas from tomb number 3 at Mawangdai (before 168 BC). Nevertheless, they regarded them superstitiously as vile apparitions, terming them "long-tailed pheasant stars" or "broom stars", which, as in the West, they associated with different types of disasters, plague and pestilence [3]. In the West, comets, which were frequently compared to swords or daggers, were more imaginatively portrayed and used symbolically. Comets were really two-edged swords, as shown in a mid-15th-century astrological manuscript in the Biblioteca Vaticana (Pal. Lat. 1370) [1]. This tradition continued in illustrations based on the description of 16th-century French physician Ambroise Paré, where the sky rains bloody swords [4]. These bellicose images were fed by the ancient astronomer Ptolemy's classification of comets into different types, some resembling swords [5] that persisted until 1683 in *Description de l'univers* by Alain Mallet [6].

Along with their meteors, comets were unpredictable celestial light shows that either presaged disasters (literally bad stars) or positive occurrences. They captured the public's attention, periodically sweeping up individuals and whole nations in comet fever. A case in point is the recently discovered anonymous watercolor from 1527 representing a personified comet streaking across the nocturnal sky with its arm holding an avenging sword (Fig. 1) [7]. On the one hand, comets and meteors were thought to signal the apocalypse, as in Albrecht Dürer's and Matthias Gerung's woodcuts (and a tapestry after the former). The first illustrates the comet-like form described in Revelation 9:1–13 [8], while the second represents the Star Wormwood from Revelation 8:10–11, which could only have been a spectacular meteor [9]. Destructive comets were also depicted like cannons releasing shot, as in an illustration of two comets in 1472 on fol. 77r of the *Lucerne Chronicles* (1508–13) [10]. Later, they were presciently envisioned as capable of destroying Earth in an amusing French illustration from 1857, a year of many apparitions [11]. Alternatively, comets' positive side put a good spin on events, signaling prosperity, great leadership and celebrity. As such they had a deep impact on art, literature and vernacular culture. Before the advent of advertising and mass media, a comet apparition precipitated the equivalent to a media barrage. It also led to image creation. Take the case of the American writer Mark Twain, who was no stranger to media manipulation and was born in 1835 with the apparition of 1P/Halley. He wanted to live until its next appearance, and his wish was fulfilled in a neat circle, for his life ebbed on 21 April 1910, one day after Halley's Comet reached perihelion. Twain was already a celebrity but the comet's appearance marked him as an even greater one, since his life was bracketed by this cosmic coincidence.

According to the Oxford Classical Dictionary, the word *celebrity* dates from 1600, and meant pomp or solemnity when referring to rites. The word soon became obsolete and morphed into the noun "celebration", exactly the function of the Brussels conference. Triumphing over that usage, the word "celebrity" now refers to persons who are much extolled, celebrated or talked about, who are famous or have certain notoriety, like the true celebrity Edmond Halley, immortalized in the comet named after him [12]. Thus *celebrity* is a word that, like a symbolic comet, is a two-edged sword – it ranges from the sublime to the ridiculous, from Mother Teresa to Paris Hilton. At the very least, in pre-electrical ages comets were dazzling, arresting and riveting. It was only natural that in superstitious times their unpredicted appearances garnered great attention and engendered extreme reactions [1].

Fig. 3. *Left*: Edward Poynter, *Ides of March*. 1883, Manchester City Art Galleries. *Right*: Raphael, detail of *The Madonna of Foligno*. 1512, Pinacoteca Vaticana, Rome (Art Resource, New York).

Fig. 4. Edward Poynter, *The Great Comet of 1882 over Loch Luichart*. 1882, private collection, New York.

2 Synergy Between Comets and Rulers

Comets have been recorded since the dawn of writing on cuneiform tablets in the Near East [13]. Ancient people believed in the power of these potent celestial symbols and developed a tradition linking rulers and comets that was simply negative and positive political propaganda. While Pliny the Elder commented on the horrors brought by the comet of 48 BC, which appeared during the war between Caesar and Pompey, in ancient times not only wars but also kings' or rulers' births, deaths, ascendancies and falls were all augured by comets. If there were no "blazing torches" coinciding with these events, apparitions were invented to accompany significant political developments. Comets also signaled celebrity, singling out individuals as extraordinary, as in a coin commemorating Julius Caesar, whose divine status is noted in the inscription together with a comet on its reverse (Fig. 2). The Roman historian Suetonius in his *Lives* wrote of Caesar: "He was numbered among the gods, not only by a formal decree, but also in the conviction of the vulgar. For at the first of the games which Augustus gave in honor of his apotheosis, a comet shone for seven successive nights [the Great Daylight Comet of 44 B.C.], rising about the eleventh hour, and was believed to be the soul of Caesar, who had been taken to heaven." [14].

The Victorian polymath Edward Poynter took as the source for his *Ides of March* (1883) (Fig. 3) the scene in William Shakespeare's *Julius Caesar* where Calpurnia implores her husband not to go to the Senate, where he will be murdered: "When beggars die there are no comets seen: The heavens themselves blaze forth the death of princes." [15]. Poynter based the comet in this painting on his observations of the Great Comet of 1882 that he had recorded over Loch Luichart in a drawing (Fig. 4) [16]. It was the first comet whose tail was successfully photographed (by Sir David Gill [17], although two people had succeeded in capturing tailless images of Donati's Comet in 1859 [18]). Poynter repeated his rumination on fate and celebrity sparked by the Great Comet of 1882 in several versions.

Since time immemorial comets have been linked to cataclysmic change, especially to political upheavals involving rulers. Take, for example, the 1066 apparition of Halley's Comet depicted in the Bayeux Tapestry (Fig. 5) that celebrates the victory of William the Conqueror at the Battle of Hastings. Its simplified cartoon-like forms, which resemble a primitive rocket spewing forth flames, endow the image with a power worthy of the accompanying inscription: *They are in awe of the star* [19]. Indeed, the 1066 sensational passage of 1P/Halley described in numerous contemporary accounts inflicted mass terror over all of Europe. It was reported to portend all manner of calamities, including war and the death of kings. Conveniently, King Harold II perished at the battle. It seems that in the tapestry the comet has been employed in this negative fashion, because it hangs over Harold. However, the comet was a double-edged sword, for it also augured victory for the Normans, changing the face of England forever. Supposedly a great comet also heralded the 1519 arrival in the New World of Hernando Cortés, the conquistador of Mexico. In his illustrated account the monk Diego Duran relates that Montezuma, ruler of the Aztecs, was so frightened by a comet that on the advice of court astrologers he refused to act while it hung in the sky [20]. Cortés thus easily subdued the Aztec empire and was regarded as the blond-haired god of Aztec myth. Similarly, the Incas of Peru thought comets were an intimation of wrath from their sun god and believed that 1P/Halley's 1531 apparition had heralded Pizarro's invasion. A comet could also signal a ruler's

Fig. 5. Detail of the Bayeux Tapestry with Comet P/Halley of 1066. 1073–1083, Town Hall, Bayeaux (Courtesy of the Town of Bayeux).

Fig. 6. Giotto di Bondone, *The Adoration of the Magi.* 1303–1306, The Scrovegni Chapel, Padua (Art Resource, New York).

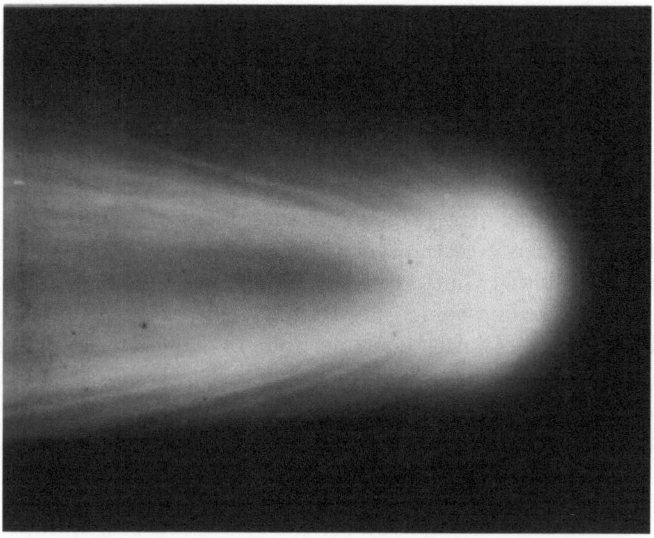

Fig. 7. Comet P/Halley, 8 May 1910, photographed with the sixty-inch reflecting telescope on Mount Wilson (Mount Wilson and Las Campanas Observatories, Carnegie Institution of Washington, Pasadena, California).

death, as in an illuminated manuscript of 1587 where a king's corpse lays on a bier below a comet of the Ptolemaic type *Pertica* [21].

2.1 Meteor Mimics

Like comets, meteors were considered prescient of disasters or doom for rulers. A case in point is the Betrayal of Jesus in *Les Très Riches Heures du Duc de Berry* (1416) that signals the beginning of the end of the so-called "King of the Jews" [22].

2.2 Giotto's Charismatic Portrait of P/Halley

In their capacity as double-edged swords comets also heralded the births of kings and presaged positive reigns. It took the genius of the Florentine painter Giotto di Bondone, in collaboration with his advisors, to link this tradition to the first naturalistic portrait of a comet painted in a large-scale context in the Scrovegni Chapel, Padua [23]. Instead of the standard conventional Star of Bethlehem in the Adoration of the Magi scene, Giotto painted a blazing comet (Fig. 6).

According to contemporary sources, like the eminent historian Giovanni Villani, the comet later identified as 1P/Halley appeared in 1301 as a galvanizing heavenly body (Fig. 7). Villani wrote that the *stella cometa* appeared in the heavens from September 1301 until January 1302 stretching at times across a third of the sky "with great trails of fumes behind". [24] Giotto could have only taken this comet as his model [25], which, not coincidentally, appeared the very year Giotto began his work at the chapel. In each of the chapel's scenes, he painted at least one naturalistic object that not only was a symbolic commentary on the action depicted but also provided his viewers a concrete object with which they could identify to make his narrative more convincing. Giotto's comet was one of those objects, because who could forget that memorable celestial spectacle? His viewers, like the Magi, had witnessed an awesome event, and by including a vivid comet he helped evoke for 14th-century viewers the strong emotions and significance of the scene. An existing patristic tradition existed that theologically allowed Giotto to replace the more customary stylized Star of Bethlehem with a dazzling comet and to marry his naturalistic observations of an historical comet with Church doctrine. John of Damascus and Origen, among other Church Fathers, had described the wondrous Nativity star in language that evoked a comet.

Giotto's comet was so convincing that nearly seven centuries later the European Space Agency named its watershed craft to Comet 1P/Halley in 1985–1996 *Giotto* [26].

2.3 Mere Mortals and More Naturalism

In the Renaissance mere mortals were occasionally singled out as exceptional by comets and meteors. As shown in Raphael's *The Madonna of Foligno* of 1512 (Fig. 3), a fireball actually hit its patron's house in Foligno, not only marking Sigismondo dé Conti as an unusual individual but also presaging his death – the double-edged sword [27]. The specificity of Raphael's fireball suggests that the artist had observed a similar phenomenon to underline the fact that scientific observation and investigation began in earnest long after Giotto, during the 16th and 17th centuries, as in a watercolor of the Comet of 1680 over Beverwijk [28].

Fig. 8. J. Farey and W. Lowry, Cometarium by William Jones. Illustration from Abraham Rees, The Cyclopedia, pl. II, vol. 9, 1819, private collection, New York.

3 Comets as Celebrities of the Century in Science and Art

In 18th-century England during a "Golden Age", when art and science were not as separated as they are today, many links existed between artists and scientists of the late Enlightenment. The resulting popularization of science coincided with the moment when the industrial revolution gained momentum and mass media proliferated. Astronomical lectures and demonstrations provided wildly popular entertainment in polite society, as demonstrated in Joseph Wright of Derby's painting *Philosopher Lecturing with the Aid of An Orrery* (1766) [29]. During lectures cutting-edge devices known as *cometariums or cometaria* demonstrated the orbit of these topical objects around the Sun. They range from the one of 1755 by J.T. Desaguliers to an early 19th-century example by William Jones (Fig. 8) [30].

As scientists, artists and the public became obsessed with comets and meteors, it seemed as though the heavens were literally on fire, precipitating a veritable image explosion of these celestial visitors documenting both their celebrity and their scientific investigation. An example is a steel *tremblant* pin (Fig. 9) that probably once adorned a lady's hat, vibrating and catching the light [31]. With 1P/Halley's 1759 apparition objects like this first began to be produced. One of the marks of 18th-century celebrity was to be "immortalized" in a plaque by Josiah Wedgwood (Fig. 9), as in one of his two portraits of the scientific celebrity Newton that includes a schematic comet (either the Great Comet of 1680 on which he based much of his gravitational theory or 1P/Halley of 1682) [32].

As scientific knowledge challenged traditional notions about them, comets and their symbolism were lampooned – as in William Hogarth's satiric *Marriage*

Fig. 9. *Left*: Anonymous, cut steel comet pin. 1759–1811, private collection, New York. *Right*: Josiah Wedgwood, *Plaque of Isaac Newton*. Before 1780, City Museums and Art Gallery, Birmingham.

Fig. 10. *Left*: R. Hawkins (?), *The Female Philosopher Smelling Out the Comet*. 1790, The Pierpont Morgan Library, Peel Collection, New York. *Right*: Thomas Rowlandson, *Viewing the Comet*. 1821, whereabouts unknown.

à *la Mode* (1745); its first plate contains a portrait of a bewigged ruler holding the thunderbolt of Jupiter (to signify that he is a god) over whom a comet hangs mocking his pretensions [33]. By the early 19th century the sting of superstition, but not the magic, associated with comets had been removed for the scientific community. In fact, the scientific investigation of comets and meteors finally allowed these two phenomena to be differentiated once and for all to check most of the fear and trembling they had engendered, although a certain *frisson* remained for the populous.

4 The Comet-Crazed Century

William Herschel and his sister Caroline observed and Caroline drew the stunning naked-eye comet that appeared early in 1811 in round telescopic fields [34]. This Comet, which before Hale-Bopp held the record as the longest naked-eye apparition in recorded history-visible over ten months – was also one of the first whose structure was analyzed. No longer threatening, comets could be regarded humorously, as in Johann Heinrich Fuseli's 1787 satire of Fortune, or the 1790 etching *The Female Philosopher Smelling Out the Comet* (Fig. 10), wherein Caroline Herschel, amanuensis of her brother who in her 98 years was the discoverer of no fewer than eight comets, receives the barbs in this satire on the popular nature of astronomy and her celebrity [35]! Soon thereafter, comets themselves became celebrities in a wave of cometomania. From Thomas Rowlandson's watercolor of 1821 (Fig. 10) it is apparent that comets captured the urban public's

Fig. 11. Samuel Palmer, *Donati's Comet over Dartmoor*. 1858, private collection, London.

Fig. 12. Charles Williams, *THE COMET!!!*. 1811, © The British Museum, London.

unquenchable curiosity [36]. In this work people use spyglasses and portable telescopes, including one of the most popular, the comet-seeker, to take advantage of the celestial display. Because four versions of Rowlandson's scene still exist, we know that it was in great demand. In another watercolor John J. Chalon recorded the reaction to 1P/Halley's crowd-pleasing apparition in 1835. In this family affair, the group employs a telescope mounted on a tripod [37].

There was no greater comet celebrity in the 19th century than Donati's Comet, which in 1858 mesmerized astronomers and artists alike. Its lengthy apparition was predicted and thoroughly covered in the press. It is no wonder then, that when it passed in front of the bright star Arcturus on the night of 5 October everyone in Europe was watching, including William Turner of Oxford and Samuel Palmer (Fig. 11) [38]. The October second issue of the *London Illustrated News* had even instructed its readers on the comet's path and when it could be seen at its brightest. The fame of Donati's Comet derived in part from its broad, sweeping dust tail, and its two less frequently seen thinner tails (one a gas/ion tail and the other perhaps a synchrone). It was the first comet to receive massive media attention, paving the way for the consumer culture and materialism that would mark Victorian England and our own age, as exemplified by the brass letter opener featuring a comet and a moon (c. 1830–1840)[39].

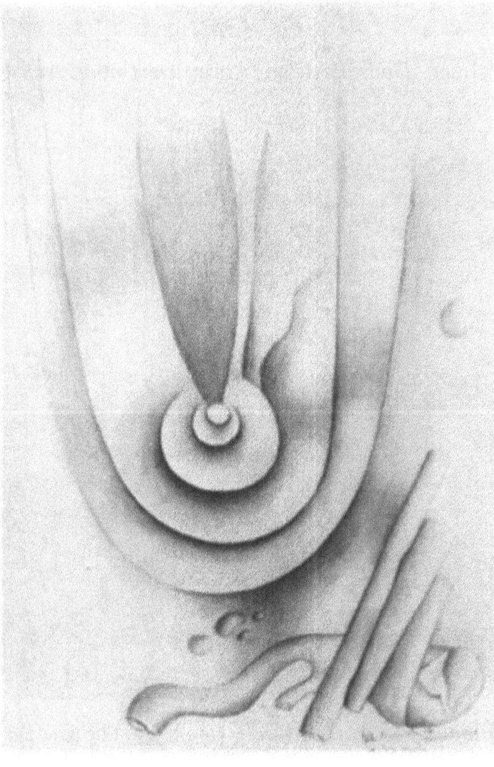

Fig. 13. Ferdinand Léger, *Head of a Comet and Trunk of a Tree*. 1931, private collection, New York.

5 Comets as Personal Symbols

By the late 18th century, comets had morphed into satiric devices for personal attacks – for example in William Dent's caricature of King George III of England, in which the king, who suffered from porphyria, wears his wig backwards signifying his madness, while his alter ego, the Duke of Richmond, regurgitates the comet's tail representing WAR [40]. Since that time comets have functioned as highly personal symbols in Western culture and advertising, as in paintings by the Surrealist Joan Miró or a recruitment postcard from Texas A & M University featuring a comet and claiming: "I Choose, Therefore, I am". The poet Gerald Manley Hopkins immortalized this concept in his line: "I am like a slip of a comet" [41].

Before the initial years of the 19th century, the diminutive Napoleon Bonaparte had cast his shadow over Europe and cannily adopted various comets as protecting *genii*, as shown in a print by Luigi Calamatta of 1834 [42]. He was aware of the legendary tradition wherein comets were associated with great rulers and by adopting comet symbolism, like that of imperial Rome, he lent legitimacy and resonance to his reign. The first historical comet linked with Napoleon was the Great Comet of 1769 (frequently termed "Napoleon's Comet"), which had an unusual red luster. Since portents are multivalent, his enemies later interpreted it as foreshadowing the bloodshed and war he brought. In James Gillray's topical print spoofing the Grand Coronation Procession of Napoleon, the arriviste emperor's entourage is lampooned as they line up behind a figure holding a banner with a huge comet inscribed with the letter N and suspended above a globe ablaze [43]. This comet represents a two-pronged attack: on the ill-fated nature of the French Emperor's reign and on his pompous adoption of comet imagery to legitimize his regime. In the print *The Eight Epochs of Napoleon* (1837–1842) by Charles Bouvier after Steuben, where his signature hat functions as a device to satirize his career, Napoleon's fall at the Battle of Waterloo is followed by a comet to symbolize both war and his fall [44]. His critics, especially the British, had a field day with this comet symbolism; for example, in one hand-colored etching Rowlandson depicted Napoleon as a comet rushing toward the Sun, George III, to reflect a bit of scientific knowledge (that a comet's light is reflected sunlight) and much English pride. English artists repeatedly personified Napoleon as a comet and mocked his pretensions relentlessly [45]. Their inventiveness knew no mercy and was not always in the best taste [46]. In a more seriously ironic treatment of Napoleon's comet fixation, Joseph William Mallord Turner represented the Emperor's abdication at Fontainebleau in his illustration for Sir Walter Scott's *Prose Works*. Exercising artistic license while aware of Scott's admiration of Napoleon, Turner set the event outdoors, where the comet hangs as a cosmic omen of political demise above the heroic but tragic figure, dwarfed by the palace. Turner probably had viewed 1P/Halley in 1835–1836 while he was engaged on this vignette [47] and also painted other generalized sublime light effects in other works with Napoleonic subjects, such as *The Field of Waterloo* (1818) [48].

The floodgates had opened for even more inventive use of comet imagery. In Charles Williams's highly original hand-colored etching *The Comet!!!* (Fig. 12), the comet's head is the visage of the Tory William Pitt, whose hair forms the comet's tail, a conceit derived from etymology of the Greek word *kometes*, meaning long-haired or hairy star. It streams across the design covered with the words: *War* and *Taxes* [49]. A caricatured Charles James Fox (ironically a Whig) floats above the comet's tail with snuffers. Some things don't change, unfortunately (war and taxes)! In a preparatory drawing and its related etching *Passing Events on The Tail of the Comet of 1853* from *George Cruikshank's Magazine* of 1 January 1854, a huge comet functions as a clever synoptic device to sum up the notable and forgettable events of that year – good and bad – a double-edged sword [50]. William Blake, who holds the record for creating the largest number of comets and meteors in his art, used a comet in a relief etching for his own poem "Milton: A Poem in Two Books" to symbolize inspiration. A comet hits the figure of Blake on the foot and calls out his first name as it charges through his spasmodically racked body [51]. The French illustrator Honoré Daumier used a comet in an 1843 lithograph for a different reason: to satirize the unsuccessful opening of a play by Victor Hugo; its rhymed caption asks why the tail of the comet is longer than the ticket lines for Hugo's play [52]. Because of their topical nature, comets and meteors peppered the prints and caricatures of late 18th and 19th century pictorial journalism, further inflaming the public's already strong interest.

One sees a similar fascination in other examples of material culture, for example, in a tankard by Rebecca Emes and Edward Barnard I from 1811 and the pin whose genesis can be linked to the appearance of 1P/Halley in 1759 (Fig. 9) [53]. Afterwards with every major comet, there has been a flurry of activity among jewelers and buyers, especially after the Great Comet of 1811 and the 1835 return of 1P/Halley, which proved to be a commercial watershed for the manufacture of comet items. In 1835 the London theater offered *The Great Comet! A Fearful Tragedy* documented in its poster [54]. Other ephemera of the 19th century testify to the deep fascination with comets and their commercialization. There are card games and board games, almanacs, and even comet valentines, as well as newspapers called "The Comet" or "The Meteor", the latter of which, true to its name, only lasted a short time. Comet wine has always been prized for the magical alchemical qualities of vintages that fermented under the aegis of a comet, and today a comet decorates the label of the Australian Shiraz Yellow Tail. Even gun barrels were decorated with comets and a famous prize English bull, whose appearance is preserved in a chromolithograph, was charismatically named "Comet", but not for his sleekness and swiftness!! The name *Comet* was also frequently assigned to new items of technology – such as flourmills, boats, and many locomotives – as well as in the twentieth century to planes – because the word is magical and connotes sleek speed. To whatever it was attached, the word lent a cachet of glamour, as in Grandville's *Travels of a Comet*, a colored wood engraving from the artist's highly imaginative *Un Autre Monde* (1844) in which the comet is personified as a courtly woman gliding across the firmament, envisioned as a ballroom, the train of her elegant gown decorated with stellar patterns [55]. This passion for comet imagery carried over into the 20th century, as in Wassily Kandinsky's painting of 1900 *Comet (Night Rider)*, where the artist is the rider on horseback, his personal symbol of freedom, in the magical kingdom presided over by a comet, a glamorous signal of success, of making it *ad astra* [56].

Comets were embraced by the glam Art Deco style worldwide, as in a streamlined automobile hood ornament by Renè Lalique [57]. Manufactured in 1927, it is the ultimate chic marriage of form and function that harmonizes with Maurice Guraud-Rivière's sculpture of 1903 *The Comet* [58]. Cecil Beaton photographed his wife Nancy's costume of comets and shooting stars for the Galaxy Ball just before the Stock Market crash in 1929, which it did not presage, while in Mussolini's Rome some sewer grills were decorated with non-utilitarian chic comets. Ferdinand Léger drew his anatomical study of a comet (Fig. 13), which is related to a screen he designed for the villa of the American Gerald Murphy at Antibes in 1931 [59]. Indubitably, he must have seen drawings by astronomers of comet structures, as he rendered a comet's anatomy – its nucleus, center of condensation and envelopes most convincingly. His source remains a mystery, although since 1P/Halley 1835 similar drawings were executed by astronomers, such as John Herschel and Thomas Maclear [60], and with Donati's 1858 by nearly everyone, including George Bond [61]. In 1930 and 1931 at least six comets were observed, but none seem to have been memorable [62]. Nevertheless, those years witnessed an unusual embarrassment of riches of these errant stars to no doubt spark discussion in circles that the artist frequented, or caused him to consult popular books on astronomy that featured anatomical illustrations of comets [63]. For the celebrity couture maven Coco Chanel, Paul Iribe designed the *ne plus ultra* of chic, a comet necklace that snakes around the wearer's neck [64]. Several years ago Karl Lagerfeld, couturier to the stars, began reproducing that necklace, the ultimate in comet jewelry, and marketing it for a cool $ 385,000. Subsequently, the House of Chanel has created a smorgasbord of derivative comet designs in the "Comet Collection", including a ring with double comets that has celebrity appeal. Just as a periodic comet returns to perihelion, designers recycle comet ideas – for example, the Sandwich Comet glass compote of 1835 celebrating 1P/Halley's apparition in that year was re-envisioned for a more recent return by Steuben in its Orbit Bowl. Today, everyone can be an instant celebrity with a Lalique *Jour et Nuit* mirror or high-stepping evening slippers decorated with gold metallic comets. Giotto's Star of Bethlehem is now not only part of the popular culture of the Christmas *presepio*/manger scene but also incites the commercial frenzy of the winter holidays in elite shopping districts all over Italy, including that of Florence. Countering all this materialism is the 1996 charcoal drawing by Vija Celmins *Untitled #13 (Comet)* that was recently acquired by the Museum of Modern Art in New York [65]. She has realized the true charisma of comets – that they are scientifically significant and magically powerful. They inspire people to aim for higher goals whether artistic, scientific or spiritual. Composed of the primeval matter of the universe, comets connect all of us with this higher, continuing order.

6 AD COMETES and Future Celebrations

Astronomers worldwide look forward to the daring scheduled explorations of several comets and with visionary insights, as suggested in Maurits Cornelius

Escher's print *Other World* of 1947 [66], they plan ventures challenging new frontiers. So too, the future promises the advent of new comets – every amateur astronomer's dream – as well as the return of periodic friends that may precipitate reactions similar to those in the lithograph *Une Surprise* (1853) by Daumier [67]. The train of thought we have followed in this meditation on celebrity brings us to the definition of true celebrity in this topsy-turvy, media-saturated world. It does not reinforce Andy Warhol's cynical dictum of the advertising age: "Everyone in the world will be famous for fifteen minutes." Rather, true celebrity results from creative work – like that of the individuals involved with the Deep Impact Mission and its daughters, Deep-Rosetta in 2015 and DIXI (Deep Impact Extended Investigation) – that probes the primordial nature of these enduring natural icons.

Acknowledgements Many thanks to Adam McKay of Williams College and Suzanne G. Valenstein of the Metropolitan Museum of Art, and to other individuals cited below, as well as Chris Sterken, Ulli Käufl and all the people who facilitated the conference at the *Palace of the Royal Academy for Science and the Arts* in Brussels and the publication of its proceedings at Springer-Verlag.

References

1. R. J. M. Olson: Fire and Ice: A History of Comets in Art, Published for the National Air and Space Museum, Smithsonian Institution by Walker and Company, New York (1985)
2. D. J. Boorstin: The Image; Or What Happened to the American Dream, Atheneum, New York, Chap. 1 (1962)
3. F. R. Stephenson and C. B. F. Walker, eds.: Halley's Comet in History, British Museum Publications, London, 41–51 (1985) and J. A. Fernández: Comets: Nature, Dynamics, Origin, and their Cosmological Relevance, Springer, Dordrecht, 3 (2005)
4. S. Schechner Genuth, Comets, Popular Culture, and the Birth of Modern Cosmology, Princeton University Press, Princeton, fig. 2 (1997)
5. R. J. M. Olson, 40–41, fig. 37 (1985)
6. R. J. M. Olson, 53, fig. 48 (1985)
7. Thomas le Claire Kunsthandel, Master Drawings, sale cat. XVII, Hamburg, no. 1 (2005)
8. E. Panofsky: Albrecht Dürer, 2, Princeton University Press, Princeton, fig. 80 (1943)
9. R. J. M. Olson, 59, fig. 33 (1985)
10. R. J. M. Olson, 36, fig. 29 (1985)
11. R. J. M. Olson, 7, fig. 7 (1985)
12. The Oxford Classical Dictionary, The Clarendon Press, Oxford, vol. II, 1019 (1989)
13. F. R. Stephenson, K. K. C. Yau, and H. Hunger, Records of Halley's Comet on Babylonian Tablets, Nature 314 (1985): 587–592; and Stephenson and Walker, 12–40 (1985)
14. R. J. M. Olson, 28–39, fig. 23 (1985)

15. R. J. M. Olson and J. M. Pasachoff, Fire in the Sky: Comets and Meteors, the Decisive Centuries, in British Art and Science, Cambridge University Press, Cambridge, 287–290, fig. 147 (1998)
16. R. J. M. Olson and Pasachoff, fig. 148 (1998)
17. R. J. M. Olson, fig. 67 (1985)
18. J. M. Pasachoff and R. J. M. Olson, The Earliest Comet Photographs: Usherwood, Bond, and Donati 1858. Jour. Hist.Astron. 27, 129–145 (1996)
19. Olson, 14–15, fig. 12 (1985)
20. Olson, 47, fig. 43 (1985)
21. Olson, 40–41, fig. 37 (1985)
22. Olson, 31–32, fig. 26 (1985)
23. R. J. M. Olson, Giotto's Portrait of Halley's Comet: Scientific American 240: 5, 160–170 (1979)
24. G. Villani, Istorie Fiorentine di Giovanni Villani: Cittadino fiorentino fino all'anno MCCCXLVIII, Societá Tipografia, Milan, vol. 4, 71 (1802)
25. R. J. M. Olson, Much Ado About Giotto's Comet. Q.J.R. astr. Soc. 35, 145–148 (1994)
26. R. J. M. Olson and J. M. Pasachoff, New Information on comet P/Halley as depicted by Giotto di Bondone and other Western artists, in the Exploration of Halley's Comet, ed. M. Grewing, F. Praderie, and R. Reinhard, Springer-Verlag, Berlin and Heidelberg, 1–11 (1988)
27. Olson, 46, fig. 44 (1985)
28. Olson, 62, fig. 56 (1985)
29. B. Nicholson, Joseph Wright of Derby, Painter of Light, The Paul Mellon Foundation for British Art, London, vol. 1, 111–112; 2, pl. 54 (1968), and J. Edgerton, Wright of Derby, exh. cat., The Metropolitan Museum of Art, New York, 54–55, no. 18 (1990)
30. Olson and Pasachoff, 44–48, figs. 14–17 (1998)
31. Olson and Pasachoff, 186–187, fig. 102 (1998)
32. Olson and Pasachoff, 39–41, fig. 11 (1998)
33. Olson and Pasachoff, 51–53, fig. 19 (1998)
34. Olson and Pasachoff, 129 and 98–101, figs. 65 and 51, respectively (1998)
35. Olson and Pasachoff, 53–54 and 139, figs. 20 and 72 (1998)
36. Olson, 72, fig. 62 (1985)
37. Olson and Pasachoff, 188, fig. 103 (1998)
38. R. J. M. Olson, A water-colour by Samuel Palmer of Donati's Comet: The Burlington Magazine 132, 95–96 (1990); Olson and Pasachoff, 227–244, figs. 126, 127 and 130 (1998)
39. Olson and Pasachoff, 197, fig. 109 (1998)
40. Olson and Pasachoff, 138–139, fig. 70 (1998).
41. For "I Am Like a Slip of Comet," 1864, see D. H. Levy, Starry Night: Astronomers and Poets Read the Sky, Amherst, NY: Prometheus Books, 122 (2001), and "Poet and Observer: Gerard Manley Hopkins and Some Mid-19th Century Comets" JRASC 75:3, 139–150 (1981), who notes that Hopkins wrote the poem after observing Comet Tempel in 1864 II (C/1864 N1)
42. Olson, 81, fig. 70 (1985)
43. D. Hill, Mr. Gillray, The Caricaturist, A Biography. Phaidon Publishers, London, fig. 73 (1965)
44. Olson, 81, fig. 71 (1985)
45. Olson and Pasachoff, 147–148, fig. 76 (1998)

46. Olson and Pasachoff. 150–151, fig. 80. espec., but also figs. 79 and 81
47. Olson and Pasachoff, 192–196, figs. 107 and 108 (1998)
48. A. Wilton, J. M. W. Turner: His Art and Life, Rizzoli, New York, 267, P138 (1979)
49. Olson and Pasachoff, 152–154, fig, 83 (1998)
50. Olson and Pasachoff, 216–217, figs. 121 and 122 (1998)
51. Olson and Pasachoff, 112–113, fig. 54, as well as figs. 40–50, 52–59
52. Olson, 90, fig. 82 (1985)
53. Olson and Pasachoff, 154–155, fig. 84 (1998)
54. Olson and Pasachoff, 216, fig. 120 (1998)
55. Olson, 103, fig. 93 (1985)
56. Olson, 109–110, fig. 99 (1985)
57. Olson, 116, fig. 102 (1985)
58. Olson, 116, fig. 106 (1985)
59. Olson, 119, fig. 110 (1985). See also G. Bauquier, Fernand Léger: Catalogue raisonné 1929–1931, vol. 2, Maeght diteur, Paris, 236–245, nos. 745–750 [six works with comets including three triptychs] (1995)
60. Olson and Pasachoff, 180–184, figs. 96-98 (1998)
61. G. P. Bond, Account of the Great Comet of 1858, Welch, Bigelow, and Company, Cambridge (1862)
62. Data provided by Dan Green, Harvard-Smithsonian Center for Astrophysics, including reference to B. Marsden's 1993 and 2003 Catalogues of Cometary Orbits, and to L. Kamel's Ph.D. thesis (which contains tabulated data on periodic comets only). See also G. W. Kronk, Comets: A Descriptive Catalog, Enslow Publishers, Hillside, NJ, and Aldershot, Hants, 123–126 (1984). Volume three (1900-1932) of Kronk, Cometography: A Catalog of Comets, Cambridge University Press, forthcoming in July 2007, who lists six in each year.
63. Such as C. Flammarion, Astronomie Populaire: Description Générale du Ciel, C. Marpon et E. Flammarion, Paris, fig. 282 (1880)
64. Olson, 116–117, fig. 107 (1985)
65. J. Elderfield and A. Temkin, Against the Grain: Contemporary Art from the Edward Broida Collection, exh. cat., The Museum of Modern Art, New York, 54, ill. (2006)
66. Olson, 119–120, fig. 111 (1985)
67. Olson, 92, fig. 87 (1985)

The Grain Evolution Model for Icy Grains Ejected from 9P/Tempel 1 by Deep Impact

E. Beer[1], D. H. Wooden[2], and R. Schulz[3]

[1] NPP Research Assoc., NASA Ames Research Center, MS 245-3, Moffett Field, CA 94035-1000 USA ealeal44@yahoo.com
[2] NASA Ames Research Center, MS 245-3, Moffett Field, CA 94035-1000 USA dwooden@mac.com
[3] ESA Research and Scientific Support Dept., ESTEC, Postbus 299, 2200 AG Noordwijk, The Netherlands rschulz@rssd.esa.int

Abstract The GEM (Grain Evolution Model) is a unique model which follows the cometary icy grains from the moment of ejection until complete sublimation. The model takes into accounts the different forces acting on each and every grain from the initial distribution, as they are passing through the coma and sublimating. The GEM is sensitive to the wavelength and composition of the grains and therefore can anticipate which grains can better match the observations at a certain wavelength. The GEM can predict the brightness of the entire coma in steady state, or as in the Deep Impact Mission it can calculate the coma brightness at different cross sections of time while differentiating between the contribution of the nucleus and that of the grains. In this paper we will show that nearly pure ice grains, i.e. 1–5% of Pyroxene, match the observations from the Deep Impact Mission since they can give a reasonable explanation for the UV enhanced rapid and decline after 20–30 minutes. Furthermore nearly pure ice grains explain the sustained brightness in the UV that lasts 7–14 hours [20].

1 Introduction

The GEM code can follow the evolution of the grains and their physical characteristics. Therefore, it can be an important tool in estimating what is being seen by the Deep Impact Mission. On 2005 July 04 UT, the Deep Impact Mission impactor hit Jupiter Family comet 9P/Tempel 1, releasing 19 GJ of kinetic energy and creating and observable gas and dust plume [1].

Indications that icy grains were released from the subsurface nucleus into the coma is confirmed by XMM and Deep Impact Mission spacecraft observations [20, 23] . The icy grains scatter sunlight in the UV band and are quickly disintegrated. These grains contribute to change in color of the brightness (brightness *vs.* wavelength) and are thought to be composed of ice because of their limited coma lifetimes. Two different distributions were tried out: The first one is a full distribution ejected from the nucleus where $a_{min} = 0.01$ μm and a_{max} is the calculated maximal grain size [6] (specifically, a_{max}=0.785 cm; hereafter, this distribution is named "Full"). The second distribution has $a_{min} = 0.01$ µm, but the maximal grain size is smaller than the theoretical one and is set at 5 µm (hereafter this distribution is marked "5MIC"), motivated by the grain sizes inferred in the coma [23]. This 5MIC distribution takes into account that many of the observations done on the Deep Impact mission primarily observed small grains in the plume ejected from the nucleus [10, 11, 14, 7, 19, 1, 15, 12, 24, 18]; the typical sizes from the observations are up to a radius of tens of micron in size. The maximum grain size has been chosen as an indicator, but can easily be changed to any other number in that order of magnitude. The main point in this calculation is to reduce the maximal grain size to the order of microns. The GEM code can supply an estimation of reasonable compositions of icy grains that will match the observed changes in the UV [20]. In Sect. 3.2 we describe the model. In Sect. 3.3 we present the results and discussion.

2 The GEM Model

2.1 Grain Temperatures

The GEM code can be described in 4 steps. The first step is to calculate the grains' temperature from the energy balance equation on the surface of each grain of a given radius a:

$$E_{solar} = E_{rad} + E_{sub} \tag{1}$$

E_{solar} represents the heating due to Solar radiation and is expressed by:

$$E_{solar} = \frac{1}{4d^2} \int_0^\infty Q_{abs} S_\lambda d\lambda \tag{2}$$

Where $S_(\lambda)$ is the monochromatic solar constant at a wavelength λ, Q_{abs} is the efficiency factor of absorption calculated by Mie scattering theory [25] and d is the heliocentric distance. The grains can cool by the two mechanisms listed on the right hand side when E_{rad} is the infrared reradiation expressed as:

$$E_{rad} = \int_0^\infty Q_{emis} F(\lambda, T_g) d\lambda \tag{3}$$

Where Q_{emis} is the efficiency factor of emission, $F(\lambda, T_g)$ is the Planck function at wavelength λ and at the grain temperature T_g. The sublimation mechanism is E_{sub} and is expressed as:

$$E_{sub} = H P_{vap}(T_g \sqrt{\frac{m}{2\pi R_g T_g}}) \tag{4}$$

H is the latent heat of vaporization, P_{vap} is the vapor pressure (both H and P_{vap} are taken for H_2O, since water is considered to be the dominant material of the icy grains). R_g is the gas constant and m is the molar mass of H_2O. For more details see [3]. The energy balance equation is solvable for any mixture of water ice and "dirt". Dirt is used here as a general term for any absorbent material and can take the form of Pyroxene, Olivine, Amorphous carbon, Tholin or any other material that is known to be in cometary grains and has known optical constants in the wavelengths of interest. The grains are spherical, and the mixture rule used is the Symmetrical Bruggeman law [16]. The decision to use this mixing rule has been made after an extensive investigation of many different mixing rules that found that the differences between them is negligible.

2.2 Grain Size Distribution

The grain size distribution is the initial distribution ejected out of the nucleus surface. The grain size distribution is taken to be a power law with index −3.5. The minimal grain size was taken as $a_{min} = 0.01$ μm. The maximal grain size a_{max} was calculated by balancing the gravitational force and the drag force acting on the surface of the nucleus. It was assumed that the vapor pressure driving the grains out into the coma off the nucleus surface is the vapor pressure of H_2O since the Deep Impact Mission encounter happened at a heliocentric distance of 1.51 AU where water is the main gas expected and the dominant gas species observed [17]. The temperature of the grain was calculated independently for grains smaller than 40 μm. For grains larger than 40 μm the temperature was assumed to be equal that the temperature of grains of 40 μm, since the temperature is weekly dependent on grain size and this saves considerable computation time. The maximal grain size that can be ejected into the coma by gas drag can be expressed by:

$$a_{max} = \frac{3R_N^2 m v_0 Z}{4GM_N \rho_d} \tag{5}$$

where v_0 is the velocity of the gas acting on the grains is taken as 0.3 km/s for all grain radii [6], Z is the sublimation rate, G is the gravitation constant, M_N is the mass of the comet nucleus and ρ_d is the composition-dependent grain density ($\sim 1\ g/cm^3$ for nearly pure water ice) [3]. The radius R_N of the nucleus was taken as 3 km in accordance with what has been observed for 9P/Tempel 1 [2]. As mentioned in Sect. 3.1, two different size distributions were applied to the GEM code. The full distribution utilizes a_{max}, the biggest grain size that can be ejected into the coma by the gas drag, while the 5MIC distribution cuts the grain size distribution by setting $a_{max} = 5$ μm. The 5MIC distribution was motivated by the observations of micron to tens of micron size ice grains in the plume [23]. Small icy grains sublime quickly compared to large grains. The question is whether the color evolution of the Full and 5MIC grain size distributions can be distinguished by the observations. The rapid sublimation of small icy grains is a natural explanation [5] for the observed 20–30 minutes decline in the UV [20], and in rise in the H_2O production rate after ∼20 minutes [7].

2.3 Sublimation Step

The sublimation step is calculated by following the change in grain radius due to sublimation as a function of time.

$$|\frac{da}{dt}| = \frac{P_{vap}(T_g)}{\rho_d}\sqrt{\frac{m}{2\pi R_g T_g}} \tag{6}$$

where P_{vap} is the saturated vapor pressure of water ice [9]. At the end of this stage we have a matrix of grains where the first column in the matrix is the initial size distribution and every row in the matrix is the sublimation process acting on this practical grain from the initial size distribution.

2.4 Dynamics

The dynamics in the coma are calculated for each grain in the sublimation matrix independently. The three forces acting on a grain in the coma are assumed to be the drag force (F_{drag}), the nucleus gravity (F_{grav}) and the suns gravity(F_{rad}). The three forces are given by:

$$F_{drag} = \frac{4\pi}{3}a^3\rho_{grain}\frac{dV_{grain}}{dt} = \frac{4\pi}{3}a^2\rho_{gas}V_{grain}(V_{gas} - V_{grain}) \tag{7}$$

$$F_{grav} = -\frac{GM_N m_{grain}}{R^2} \tag{8}$$

$$F_{rad} = -\frac{L_{Sun}\pi a^2[Q_{abs} + (1-g)Q_{scat}]}{4\pi d^2 c} \tag{9}$$

where

$$\rho_{gas} = \frac{R_N^2}{R^2}\frac{P_{vap}}{V_{gas}}\sqrt{\frac{m_{gas}}{2\pi kT}} \tag{10}$$

V_{gas} and T is taken as the thermal velocity of a water molecule and the temperature of a water molecule respectively. ρ_{grain} is the grain density and a is the grain radius. M_N and m_{grain} are the nuclear mass and the grain mass, respectively. R is the distance in the coma. L_{Sun} is the solar luminosity, c is the speed of light Q_{abs}, Q_{scat} and g are the efficiency factors for absorption and scattering and the back scattering coefficients, respectively. Combining all of these three forces together the TFM (Three Forces Model) is formed (for further details see [4]) and the equation below is solved by the Runge-Kutta scheme in the Forth order [13]:

$$\frac{dV_{grain}}{dt} = V_{grain}\frac{dV_{grain}}{dR} = -\frac{\beta(V_{grain} - V_{gas}) + \varsigma}{R^2} - \delta \tag{11}$$

Where $\delta \equiv \frac{3L_{Sun}[Q_{abs}+(1-g)Q_{scat}]}{16\pi d^2 ac\rho_{grain}}$,$\varsigma \equiv GM_n$ and $\beta = \frac{P_{vap}}{a\rho_{grain}}\sqrt{\frac{m_{gas}}{2\pi kT}}R_N^2$.

2.5 Coma Brightness

The total brightness is calculated as the sum of the nucleus and the grains using the following equation:

$$\frac{\Delta^2}{R_\odot^2} 4d^2 B_{N+g} = \frac{1}{J} \sum_{i=1}^{n} \sum_{j=1}^{J} a_i^2 Q_{scat}(1 - g_{i,j}) F(\lambda, \odot) + R_N^2 A_N F(\lambda, \odot) \qquad (12)$$

For the grains (left summand), n is the number of grains in the initial distribution, J is the number of time steps dictated by sublimation and dynamics, i is the indicator for the initial distribution (at $t = 0$), and j designates the evolution through time of the grain's sublimation and dynamics. The geocentric and heliocentric distances to the comet are Δ=0.89 AU and d=1.51 AU. $F(\lambda, \odot)$ is the solar spectrum, with $R_\odot$ the solar radius. We found the Planck function ($F(\lambda, \odot)$ with $T_\odot \approx$6000 K) to be insufficient in assessing the UV from the Sun's corona, so we utilize the measured solar spectrum on 2005 July 04 from the SORCE (Solar Radiation Climate Experiment) satellite. This SORCE UV spectrum is much more accurate and was found to significantly improve the modeling of the UV. Calculations were performed with the SORCE data sets from the 3th and the 5th of July and showed no significant difference. The flux of the nucleus depends on its albedo A_N, taken to be the canonical value of 0.04.

3 Results and Discussion

The results are presented for different compositions, meaning variations in the mixing ratio of ice to refractory material, as well as different refractory materials, specifically Pyroxene $(Mg_{0.6},Fe_{0.4})SiO_3$ and Olivine $(Mg_{0.6},Fe_{0.4})_2SiO_4$ [8, 21, 22]. The GEM code requires the optical constants for water ice [21, 22] to be on the same wavelength grid as the refractory materials. The radiative equilibrium calculations are computed over the wavelength range 0.1–106 μm. The scattering cross sections Q_{scat} and the brightness (Eq. 12) are computed at discrete wavelengths (0.25, 0.30, 0.35, 0.45, 0.57μm). The brightness of the nucleus and the combined brightness of the nucleus and the grains were computed using the measured SORCE solar spectrum on the day of impact (Sect. 3.3). In the GEM code, the brightness of the grains at a specific time after impact is investigated by considering a specific cross section at a particular time, which may correspond to a range of coma distances for grains of different sizes based on the dynamical model [5].

The results of the GEM models are shown in Figs. 1, 2, 3, 4, 5, 6, 7, and 8, which represent the brightness of the nucleus and the grains at different times compared to the brightness of the nucleus alone (*five pointed stars*). Figures 1, 2, 3, and 4 are for a "Full" distribution of grains (a_{max}=0.785 cm), while Figs. 5, 6, 7, and 8 are for the "5MIC" size distribution where the maximal grain size is truncated at 5 μm (see Sect. 3.1).

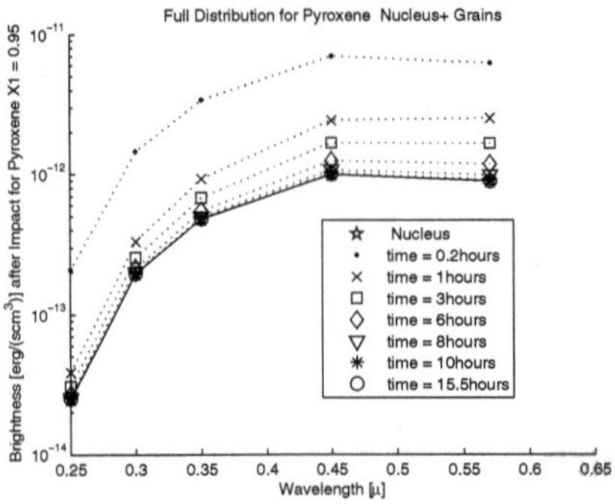

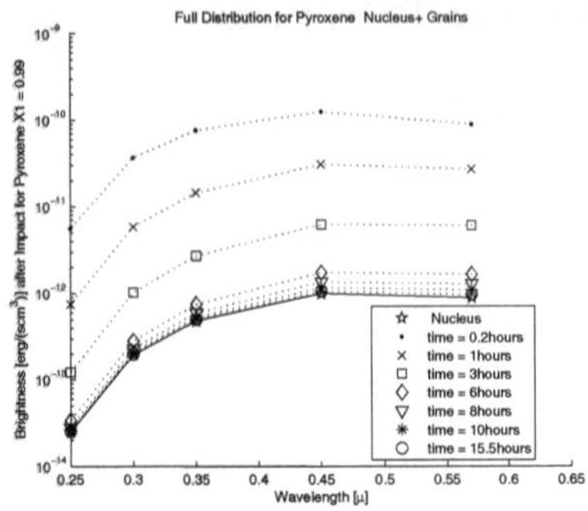

Fig. 1. *Left*: Brightness vs. Wavelength Pyroxene X1=0.95; Full distribution; The brightness of the nucleus is represented by a solid curve and pentagrams while the brightens from the grains and the nucleus is represented by a dotted line and different symbols according to the different times from the impact.
Fig. 2. *Right*: Brightness vs. Wavelength Pyroxene X1=0.99; Full.

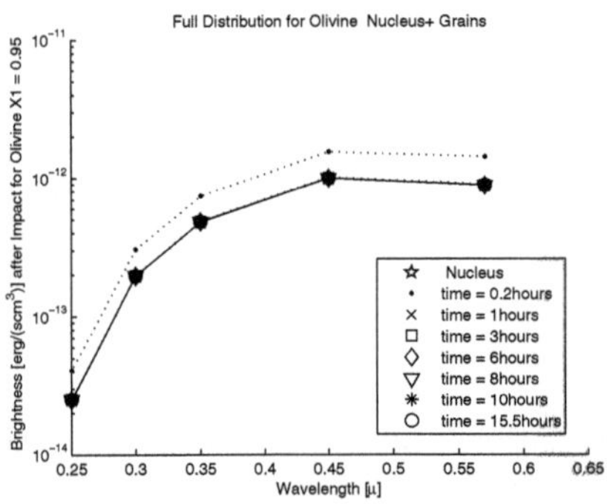

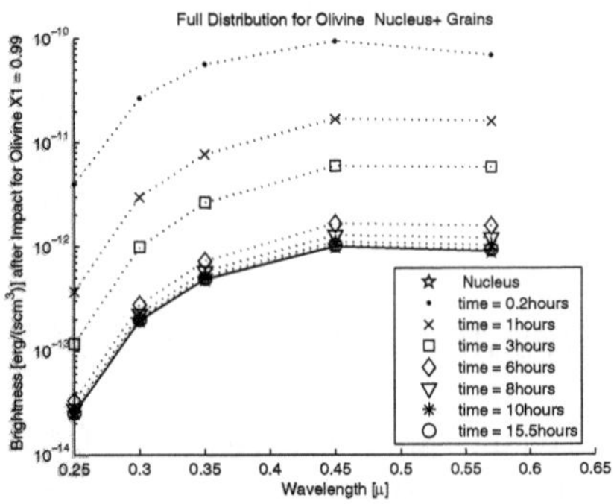

Fig. 3. *Left*: Brightness vs. Wavelength Olivine X1=0.95; Full.
Fig. 4. *Right*: Brightness vs. Wavelength Olivine X1=0.99; Full.

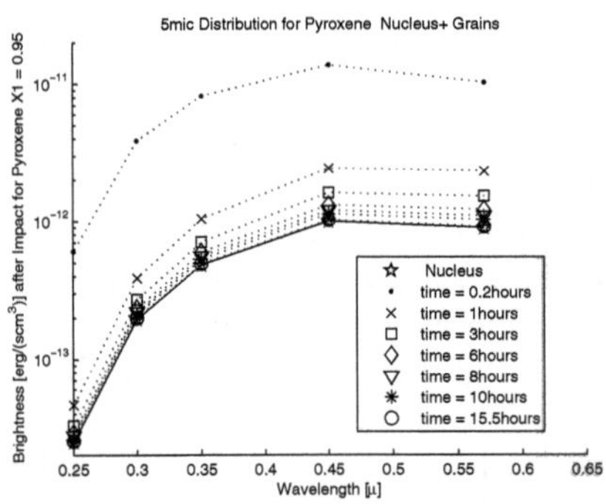

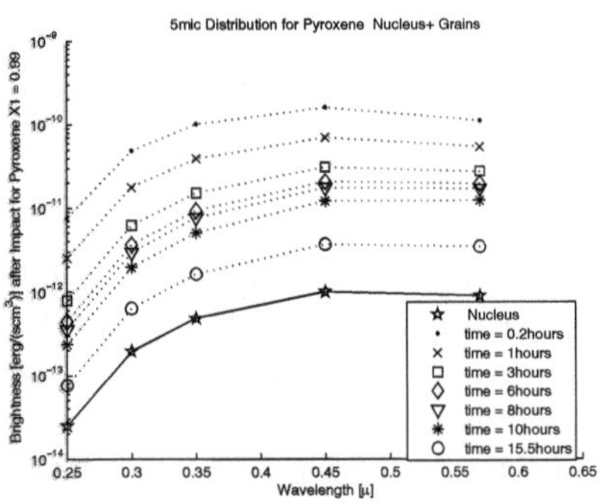

Fig. 5. *Left*: Brightness vs. Wavelength Pyroxene X1=0.95; 5MIC.
Fig. 6. *Right*: Brightness vs. Wavelength Pyroxene X1=0.99; 5MIC.

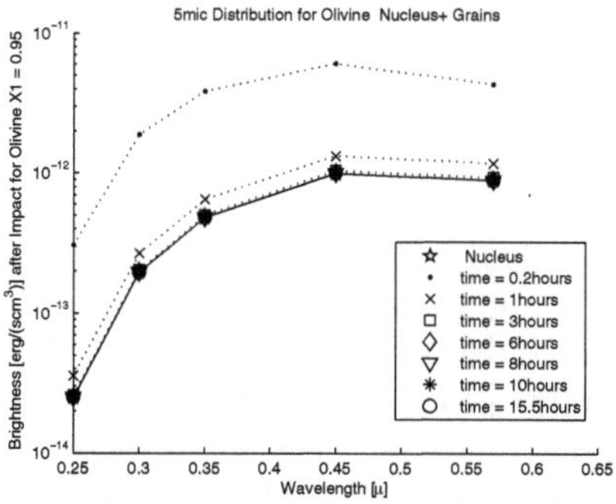

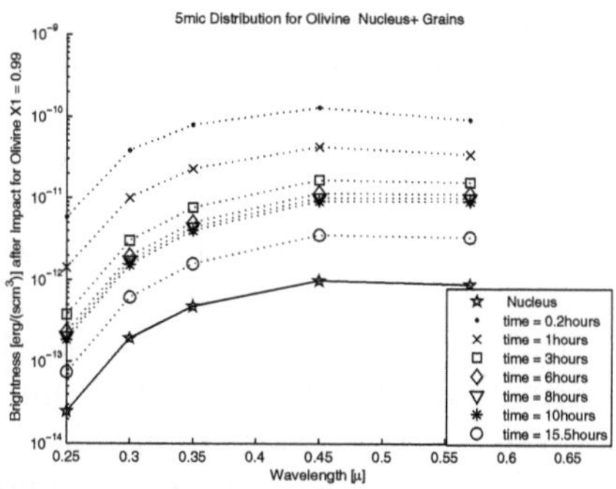

Fig. 7. *Left*: Brightness vs. Wavelength Olivine X1=0.95; 5MIC.
Fig. 8. *Right*: Brightness vs. Wavelength Olivine X1=0.99; 5MIC.

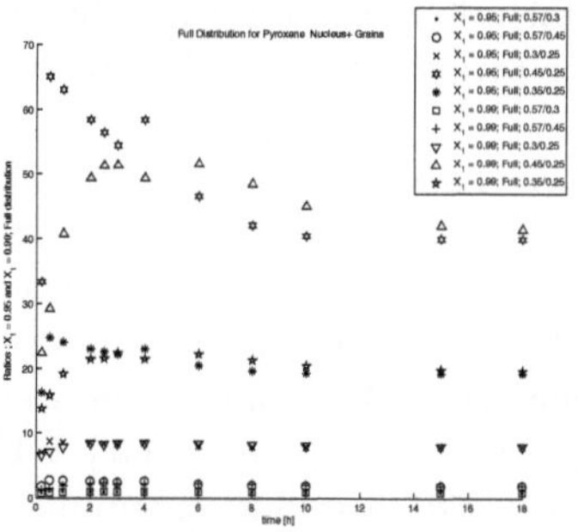

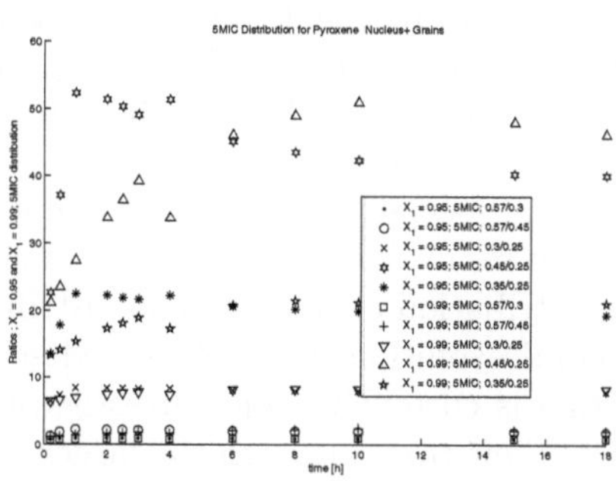

Fig. 9. *Left*: Color Ratios (Ratio of Fluxes at 2 wavelengths) vs. Time for two Pyroxene icy grain compositions: X1=0.95 and 0.99; Full.
Fig. 10. *Right*: Color Ratios vs. Time (same as Fig. 9); 5MIC.

The ice to dust ratio by mass in the figures is indicated as X_1, where $X_1 = 1$ is for pure ice grains, and $X_1 = 0$ is for grains that contain no ice. Figures 1, 2 (Full) and 5, 6 (5MIC) present $X_1 = 0.95$ and $X_1 = 0.99$ for Pyroxene. Figures 3, 4 (Full) and 7, 8 (5MIC) present $X_1 = 0.95$ and $X_1 = 0.99$ for Olivine.

Looking at the evolution of the icy grains at different times and at different compositions, it is clear that the "purer" the icy grains are, the more they contribute to the total scattering brightness of the coma. For example, Pyroxene $X_1 = 0.95$ sustains the grain contribution until 6 hours, i.e., after 6 hours the contribution of the nucleus is the dominant scattering mechanism. Whereas, Pyroxene $X_1 = 0.99$ maintains the contribution from the grains until 10 hours. As expected, the truncated 5MIC distribution is brighter than the full distribution, because it contains more grains; since the same mass is spread into more grains, there are more scattering surfaces. If larger icy grains are present, they do not impact the brightness at the computed wavelengths (colors). Icy grains with Olivine with $X_1 = 0.95$ do not persist long enough to explain the observed UV scattering. In contrast, Olivine $X_1 = 0.99$ has approximately the same coma lifetime as Pyroxene $X_1 = 0.95$. Pyroxene $X_1 = 0.95$ has a slight advantage over Olivine $X_1 = 0.99$ in that at 10 hours there is still enhanced scattering over the nuclear flux at all colors. Comparing the Olivine and Pyroxene, Pyroxene has a larger range of compositions that can fit the temporal evolution of the UV scattering. In summary, in order to have a sustained icy grain contribution to the total scattered brightness we need to have nearly pure ice grains with a small mass fraction of absorbing refractory minerals. This is the most general conclusion from the GEM, since if we will deal with "dirtier" grains they will sublimate too fast to be seen between 7 and 14 hours [20].

Differences in color ratios are subtle (Figs. 9, 10) between the Full and 5MIC distributions. The 0.45 μm/0.25 μm color ratio shows the greatest differences between the size distributions. The main difference in the color evolution is the refractory mineral content and not the size distribution. We only show the evolution of the color ratios for Pyroxene because the trends in the color ratios between Pyroxene X_1 =0.95 and Olivine X_1=0.99 are similar, and Olivine X_1=0.95 does not contribute to the brightness for a long enough time. In summary, the general temporal behavior in the color ratios is very similar between the Full and truncated 5MIC distributions.

The next observational constraint is more detailed and regards the rate of UV decrease, especially in the first 30 minutes after impact. A rapid decrease in the UV may result from the rapid sublimation of submicron ice grains. Whether a Full distribution or a 5MIC-like distribution better fits the early (~30 minutes) temporal evolution of the observed scattered spectrum (UV and longer wavelengths) is currently under investigation with the GEM [5]. Preliminary GEM results show that although the larger portion of the size distribution sublimates more slowly it can not makeup for the rapid loss of the larger number of smallest grains. This is due to the assumption of a power law size distribution: in the beginning we have a few large grains and a lot of small grains. The small grains sublimate rapidly and give rise to the UV in the first 20–30 minutes. After that time, the larger grains that are fewer in number cannot keep up with the loss of the small grains, and the UV decreases.

The final result of the GEM will be a combination of these three limitations: (1) the enhanced and rapidly decreasing UV (20–30 minutes), (2) the sustained UV, and (3) the color evolution. What we can strongly conclude at this time is that we need nearly pure ice grains ($X_1 \geq 0.95$) to sustain the UV scattering (in excess of the nucleus) for 7–14 hours. We consider all icy grains in the distribution to have similar composition, i.e., having the same mixture of ice-to-dust. If there happened to be a large range in the refractory component in the icy grains, say including quite dirty grains ($X_1 = 0.1$), then a full distribution including large grains probably would be required to sustain the UV for as long as ~15 minutes [5].

Many alternatives are possible, both in composition and size distribution of the icy grains. An initial investigation of Tholin and Amorphous Carbon did not result in a good match to the temporal evolution in the color ratios; these calculations were not robust because the optical constants of were interpolated to that of water ice without the effort of preserving the Kramers-Kroning relationship [25]. The exact composition and distribution may have alternatives, but we find that water ice with Pyroxene $(Mg_{0.6},Fe_{0.4})SiO_3$, with a mass ratio of $X_1 \geq 0.95$ fits the observations [20]. The GEM model demonstrates that the UV scattering is a result of submicron nearly pure ice grains in the coma.

References

1. A'Hearn M. F. M, Belton M. J., Delamere W. A. et al. (2005): Science, 310, 258.
2. Basilevsky A. T. and Keller H. U. (2006): Planet. Space. Sci., 54, 808.
3. Beer E. H., Podolak M.and Prialnik D. (2006): Icarus, 180, 473.
4. Beer E., Prialnik D. and Podolak M. (2008): Icarus, 195, 340.
5. Beer E., Wooden, D., Schulz, R., and Weaver, Wm. B. (2007b):Icarus, in preparation.
6. Delsemme A. H. and Miller D. C. (1971): Planet. Space. Sci., 19,1229.
7. DiSanti M. et al. (2006): Icarus accepted
8. Dorschner, J. et al. (1995): Astron. Astrophys., 300, 403.
9. Fanale F. P. and Salvail J. R. (1984): Icarus, 60, 476.
10. Harker D. E., Woodward C. E. and Wooden D. H. (2005): Science, 310, 278.
11. Harker D. et al. (2006): Icarus Accpted
12. Keller H. U., Jorda L., KuppersM. et al. (2005): Science, 310, 281.
13. Kunz S. K., (1957) Numerical Analysis (Mcgraw–Hill Book Company, Inc.).
14. Lisse C. M., VanCleve J., Adams A. C. et al. (2006): Science, 313, 635.
15. Meech K.J., Ageorges N., A'Hearn M. F. et al. (2005): Science, 310, 258.
16. Merrill W. M., Diaze R. E., LoRe M.M. et al. (1999): IEEE Transactions on Antennas and Propagation, 47, 142.
17. Mumma M. J., Disanti M. A., Magee–Sauer K. et al. (2005): Science, 310, 270.
18. Richardson J. E. and Melosh H. J. (2006): Lunar and Planetary Science, 37, 1836.
19. Schleicher D. G., Barnes K. L. and Baugh N. F. (2006): Astrophys. J., 131, 1130.
20. Schulz R., Owens A., Rodrigues–Pascual P. M. et al. (2006): Astron. Astrophys., 448L, 53.
21. Semenov, D. et al. (2003): Astron. Astrophys., 410, 611.
22. Semenov, D. (2005): private communication.
23. Sunshine J. et al. (2006): private comunication.
24. Sugita S., Ootsubo T. , Kadono T. et al. (2005): Science, 310, 274.
25. van de Hulst, Light Scattering by Small Particles (Dover Publications Inc., New York, 1981)

JCMT Observations of the Deep Impact Event

I. M. Coulson[1], H. M. Butner[1], G. Moriarty-Schieven[1], L. M. Woodney[2], S. B. Charnley[3], S. D. Rodgers[3], J. Stüwe[4], R. Schulz[5], K. J. Meech[6], Y. Fernández[6], and P. Vora[7]

[1] Joint Astronomy Centre, Hilo, Hawaii i.coulson@jach.hawaii.edu
[2] California State University, San Bernardino
[3] NASA Ames Research Center, Moffett Field, California
[4] Leiden University, Netherlands
[5] ESTEC, Noordwijk, Netherlands
[6] Institute for Astronomy, University of Hawaii
[7] College of William & Mary, Williamsburg, Virginia

Abstract Spectroscopic observations were made from the James Clerk Maxwell Telescope of comet 9P/Tempel 1 before, during and after the impact by the NASA spacecraft Deep Impact on UT 04 July 2005. Several molecular species (HCN, CH_3OH, CO, CS, HCO^+) were targeted but none was detected. Nonetheless, upper limits on emission line strengths may weakly constrain the chemical composition of the ices in the excavated part of the comet nucleus. Preliminary analysis of the HCN data is presented here.

1 Introduction

The Deep Impact mission offered the first chance to observe newly excavated and newly exposed nuclear material of a Jupiter family comet. Models of comet nuclei – all variations on the theme of Whipple's original [7] dirty snowball – suggested that the excavation of a crater 100 m across and 30 m deep [2], should vapourize pristine nuclear ices which would expand into a coma in the days following the impact. Depending upon the cohesion of the nucleus, the impact might also open up fissures that would outgas similarly over longer periods. The composition of the released gases should be amenable to spectroscopy. The science proposals approved by the JCMT[1] Time Allocation Groups were aimed at determining isotopic ratios and the deuteration fraction in these gases. Observations of several key molecular species would enable the the physical and chemical characteristics of the vapourized gases and hence the parent ices to be determined. Such data

[1] The James Clerk Maxwell Telescope is operated by The Joint Astronomy Centre on behalf of the Particle Physics and Astronomy Research Council of the United Kingdom, the Netherlands Organisation for Scientific Research, and the National Research Council of Canada.

can constrain the original location of the comet within the early solar nebula [3]. The observational challenge seemed merely to establish baseline emission levels prior to the impact, and to measure sufficient numbers of molecular emission lines during and after the impact in order to determine the physical and chemical characteristics of the vapourized gasses.

2 Observations

The present authors established a collaboration and set observational priorities for each phase of the event; with enough flexibility to respond to the results obtained both at JCMT and elsewhere. Several nights of observing in the weeks prior to the impact were used to establish baseline emission levels of the molecules we hoped to follow through impact, and followup observations were scheduled for three weeks afterwards.

Observations during impact were made using the JCMT receiver B tuned to 338.5 GHz. The resolution of 0.625 MHz ($\sim$0.5 km s^{-1}) allows a bandpass of approximately 0.5 GHz, encompassing 20 or more lines of the methanol $J = 7 \rightarrow 6$ ladder. Intensity ratios amongst these lines would allow the determination of the rotational temperature of the gas, a crucial factor in interpreting any detections of other molecular species. The observational plan thereafter was to observe these other species – both the normal and deuterated isotopomers interspersed with methanol observations (there is another ladder at $\sim$241.8 GHz in the A-band) to keep track of the gas temperature. Around the moment of impact and for an hour afterwards we observed with a resolution of 1 minute, relaxing this to 10 minutes once it became clear that there were no strong molecular emissions. While emission lines of several key molecules were detected elsewhere in the optical and infrared [5] no convincing detections of molecular lines were made with JCMT (or any other radio/submm telescopes) so only upper limits to gas temperatures or column densities or production rates can be inferred.

2.1 The HCN Data

Figure 1 shows the HCN $J = 4 \rightarrow 3$ spectra at 3h, 1d, 2d and 3d after impact. HCN is often the strongest submillimetre molecular line detected from comets, and earlier observations elsewhere suggested that there was occasional nuclear activity, unconnected with the impact, releasing small amounts of HCN. The lack of an initial detection at JCMT generated little alarm since the expected gas coma produced by the impact would take $\sim$24 h to expand and fill the JCMT beam. But the results from subsequent days show no detectable change in the amount of HCN released. 3-sigma upper limits to the HCN production are tabulated in Table 1. These data also rule out any possible shielding of localized emission regions due to the 1.7-day rotation period of the nucleus.

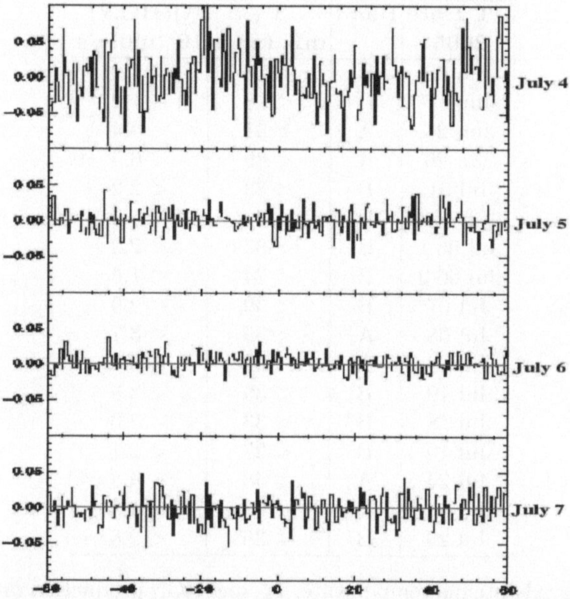

Fig. 1. Spectra of comet 9P/Tempel 1 in the line of HCN $J = 4 \rightarrow 3$ (354.5 GHz). Each day's plot shows antenna temperature T_A^* (K) vs. heliocentric velocity (kms^{-1}) resulting from 1–2 hours of integration. The comet velocity on these days was 0.1, 0.0, +0.1 and +0.2 kms^{-1}, respectively. No detections are made.

3 Conclusions

The impact had no detectable effect upon molecular gas emissions. The post-impact data are consistent with $Q(\mathrm{HCN}) < 2.5 \times 10^{24}$ mols s^{-1}, which is ~4 times smaller than $Q(\mathrm{CN})$ during the same period [6]. This would confirm the assertion by A'Hearn et al. [1] that photodissociation of HCN may be the dominant source of CN in comets, but not the only one. A similar conclusion was reached in the case of C/1995 O1, Hale–Bopp, [8]

References

1. A'Hearn M. et al. (1995): Icarus **118**, 223
2. A'Hearn M. et al. (2003): In *35th COSPAR Scientific Assembly*, p 1667
3. Drouart A., Dubrulle B., Gautier D., Robert F. (1999): Icarus **140**, 129
4. Haser L. (1957): Bull Acad Roy Belg **43**, 740
5. Meech K. et al. (2005): Science, **310**, 265

Table 1. The HCN Data.

UT Date 2005	Band	T^*_A mK.km/s	Q(HCN) 10^{24}mols/s
Jun 09	B	< 18	< 1.2
Jun 17	B	< 18	< 1.2
Jun 25	A	< 51	< 4.4
Jun 29	A	< 89	< 6.0
Jul 01	B	< 33	< 2.2
Jul 04.3	B	< 81	< 5.6
Jul 05.1	B	< 33	< 2.2
Jul 06.2	B	< 24	< 1.6
Jul 07	B	< 39	< 2.9
Jul 08	A	< 39	< 3.7
Jul 09	A	< 39	< 3.8
Jul 10	B	< 39	< 3.8
Jul 18	B	< 33	< 2.6
Jul 19	B	< 27	< 2.1
Jul 24	A	< 48	< 4.8
Jul 24	B	< 63	< 4.8
Jul 25	B	< 36	< 2.8

3σ limits on observed antenna temperature, $T^*_{\rm A}$, and HCN production rate, Q, assuming a Haser [4] model with a spherical, constant outflow of 1 km/s and a scale-length for HCN of 67000 km. JCMT beam sizes at HCN $J = 4 \rightarrow 3$ (354.5 GHz) (B-band) and $J = 3 \rightarrow 2$ (265.9 GHz) (A-band) are 14" and 19", respectively.

6. Schleicher D.G. et al. (2005): private communication
7. Whipple F.L. (1950): ApJ **111**, 375
8. Woodney L.M.. et al. (2002): Icarus **157**, 193

The Visual Brightness Behavior of Comet 9P/Tempel 1 During 1972–2005

V. S. Filonenko[1] and K. I. Churyumov[2]

[1] Astronomical Institute of V.N.Karazin Kharkov National University, Kharkov, Ukraine filonenko@astron.kharkov.ua
[2] Astronomical Observatory of Kiev National University, Kiev, Ukraine klim.churyumov@observ.univ.kiev.ua

Abstract The light curves of integrated visual brightness of the periodic comet 9P/Tempel during 1972, 1983, 1994, and 2005 appearances were constructed and studied. The values of photometric parameters H_0, n, and H_{10} were computed. A secular fading of this comet was studied. The light curve of comet near the impact time was constructed and it was compared with comet's water production rate. The photometric peculiarities of comet and long–term evolution of comet's activity probably will change after impact. Therefore, the results, which are presented in this manuscript, are important.

1 Introduction

The mission Deep Impact was successfully completed. Probably the photometric behavior and a photometric evolution of the target comet will change as a consequence of the artificial impact to the comet nucleus. Therefore, the study of photometric peculiarities and brightness behavior of the target comet in all comet's observed returns is of interest. The results of these investigations are presented here.

2 Observations and Magnitude Analysis

To study the photometric behavior of comet 9P/Tempel 1, we used more than 1,050 visual estimations of its integrated brightness collected from archives of the International Comet Quarterly, of the Comet Section of British Astronomical Association and from some Internet sources. All estimations have been corrected to the standard telescope's aperture [1]. The brightness behavior of comets with respect to heliocentric distance (light curves) is normally represented by equation

$$m_\Delta = H_0 + 2.5n\log(r) \tag{1}$$

where

$$m_\Delta = m_1 - 5\log(\Delta) \tag{2}$$

is heliocentric magnitude of comet, m_1 – comet's observed integrated magnitude, H_0 – absolute magnitude of comet, n is photometric exponent, Δ and r are the geocentric and the heliocentric distances of comet, respectively. An absolute magnitude H_{10}, which had been introduced by Vsekhsvyatsky, is important for the study of the secular evolution of cometary activity. This parameter is represented by the equation

$$H_{10} = m_1 - 5\log(\Delta) - 10\log(r) \tag{3}$$

.

3 Visual Brightness Behavior

The light curve of visual integrated brightness of comets characterizes the change of their activity with time and with heliocentric distance. The comparison of light curves of short-period comets in their different returns gives the information about the evolution of cometary activity and on the relation between cometary brightness behavior, solar activity, and condition in the interplanetary space.

3.1 Light Curves

The brightness behavior of comet 9P/Tempel 1 before perihelion was similar during all four returns (see Fig. 1). As it can be seen, the peak on brightness (associated to comet outbursts) occurs at equal about similar heliocentric distances in all returns.

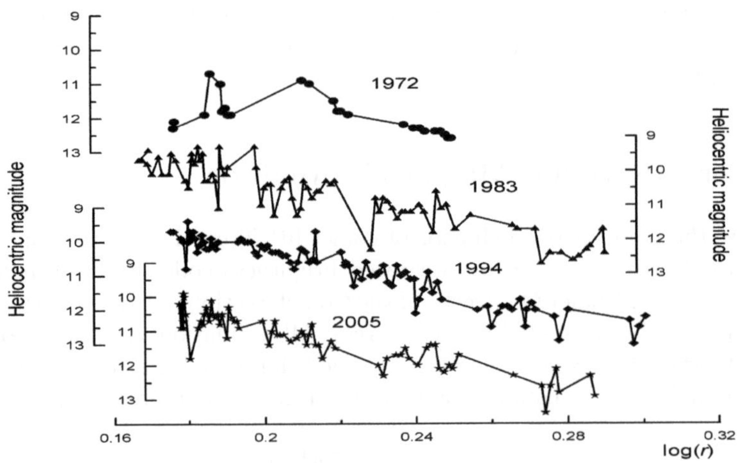

Fig. 1. The pre–perihelion light curves of cornet 9P/Ternpel 1 in four returns.

Especially it can be clearly seen that the outbursts occurred on distances 1.8, 1.76, 1.62, and 1.51 AU from Sun $\log(r) = 0.255$, 0.246, 0.210, and 0.179 accordingly. Authors have found for the first time the analogous phenomenon for comet 10P/Tempel 2 [2, 3]. Later on the authors did show that this phenomenon of cometary outburst activity is typical for all comets [4]. Unfortunately, post-perihelion light curve can be constructed only for comet's returns on 1983 and 2005. For that only 49 visual brightness estimations (24 estimations on 1983 and 25 estimations on 2005) have been used. Due to bad conditions of observations for other comet's returns sufficient observations are absent. Therefore, post–perihelion brightness behavior of comet 9P can be not studied in detail.

3.2 Photometric Parameters

The values of the photometric parameters H_0, n, and H_{10} were collected in Table 1. Authors have determined the values of all photometric parameters for comet's returns on 1972, 1983, 1994, and 2005. For other comet's returns these parameters were taken from [5, 6, 7, 8].

Table 1. Photometrical parameters of comet 9P/Tempel 1 at all apparitions.

Apparition	H_0	n	H_{10}	Notes	Reference
1867 II	–	–	8.4^m	–	5
1873 I	–	–	9.2	–	5
1879 III	–	–	10.4	–	5
1872 V	–	–	12.0 ± 0.1	–	6
	$10.57^m \pm 0.35^m$	3.00 ± 0.64	9.75 ± 0.09	Pre Per	This paper
1978 II	–	–	13.94	–	7
1983 XI	5.36 ± 0.15	–	8.61 ± 0.03	All	8
	5.82 ± 0.26	9.04 ± 0.50	8.47 ± 0.06	All	This paper
	5.30 ± 0.16	10.02 ± 0.28	8.64 ± 0.03	Pre Per	8
	5.96 ± 0.28	8.82 ± 0.51	8.55 ± 0.07	Pre Per	This paper
	7.86 ± 3.06	10.16 ± 0.29	8.25 ± 0.15	Post Per	8
	5.99 ± 1.41	8.36 ± 3.02	8.01 ± 0.14	Post Per	This paper
1994	5.23 ± 0.19	10.41 ± 0.35	8.62 ± 0.06	Pre Per	This paper
2005	7.19 ± 0.42	8.05 ± 0.71	9.37 ± 0.08	Pre Per	This paper

3.3 Secular Brightness Variations

The secular fading of comet 9P/Tempel 1 is important for possible influence of artificial impact to the nucleus of comet and for cometary photometric evolution. The change of the absolute magnitude H_{10} is presented in Fig. 2, which is constructed on the basis of the data, presented in Table 1. From these data the mean fading of comet was calculated to be 0^m0060 per year or about 0^m04 per period. This value well corresponds to the mean value of secular fading for short–period comets belonging to the Jupiter family (0^m05 for period [4]).

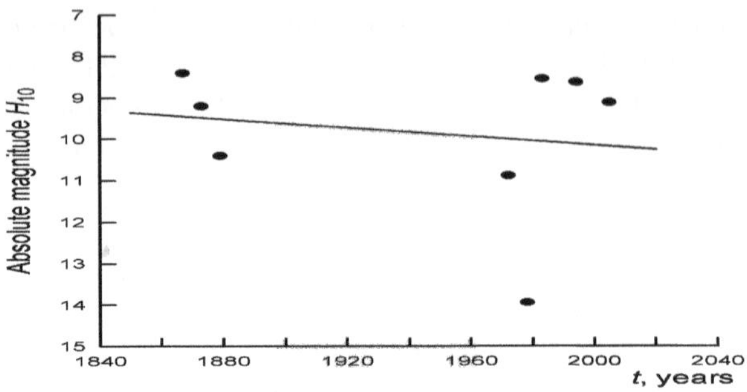

Fig. 2. The secular fading of comet 9P/Tempel 1.

4 Brightness Behavior and Outbursts Before, During, and After Impact

The light curve of comet near impact time is presented on the Fig. 3 (on top). We see that comet had been active before and after impact. The big outburst began on June 22 and reached maximum on June 27. Next outburst began on July 2 and had maximum on July 4, immediately before impact time. The impact created an artificial brightness outburst which began on impact time (as can see in Fig. 3). The amplitude of this outburst is only about 0^m7. But this outburst developed in the time and reached maximum on July 6–7. This photometric behavior good agree with other ground-based and spacecraft observations [9]. In particular, as can see in Fig. 3 (on bottom), light curve constructed by authors very well corresponds to change of water production rate of comet [10].

5 Conclusions

The light curve of comet 9P/Tempel 1 was constructed and studied for all observed comet's returns. The values of photometric parameters H_0, n, and H_{10} were determined. The value of secular fading of comet was calculated. The photometric behavior of comet near impact time very well corresponds to the change of water production rate. The comparison of our results with results of studying of comet's photometric peculiarities after impact will be able to give information on the change of cometary activity as a result of artificial impact to comet nucleus.

References

1. Morris, C. S. (1973): Astron. Soc. Pacific, 33
2. Churyumov K. I., Filonenko V. S. (Liege 1992): Light curve of short–period comet Tempel 2 and new criterion for selection a mechanism of cometary outburst activity. In: *Observations and Physical Properties of Small Solar System Bodies*, pp 295–297

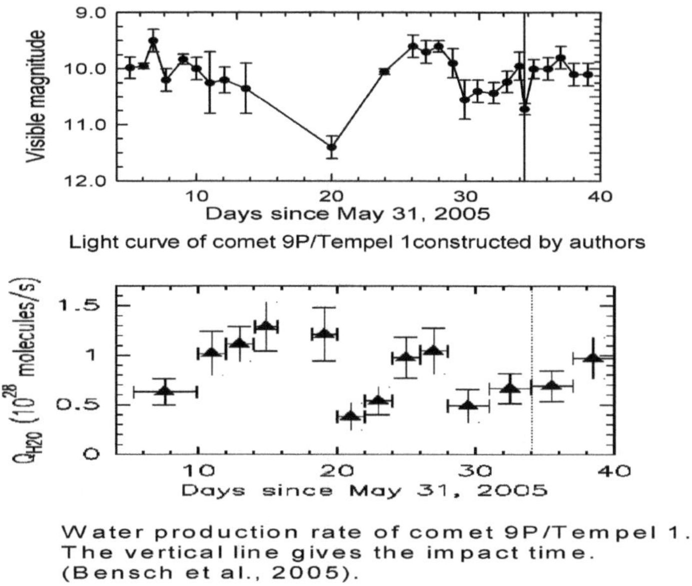

Fig. 3. Light curve of comet 9P/Tempel 1 near impact time (*top*) and change of water production rate (*bottom*).

3. Churyumov K. I., Filonenko V. S. (1992): Pisma Astron. J. **18**, 922
4. Churyumov K. I., Filonenko V. S. (1997): Astron. Vestnik **31**, 43
5. Vsekhsvyatsky S. K. (Moscow 1958): *Physical characteristics of comets.*
6. Vsekhsvyatsky S. K. (Kiev, 1979): *Physical Characteristics of Comets 1971–1975.*
7. Andrienko D. A., Karpenko A. V. (Moscow, 1987): *Physical Characteristics of Comets 1976–1980.*
8. Andrienko D. A., Karpenko A. V. (Kiev, 1993): *Physical Characteristics of Comets 1981–1985.*
9. Bensch F., Melnick G. J., Neufeld D. A. et al. (Cambridge 2005): SWAS observations of comet 9P/Tempel 1 and Deep Impact. In: *Astrochemistry: Recent Successes and Current Challenges, Proceedings of the 231st Symposium of the International Astronomical Union*, pp 489–490
10. Meech K. J., Ageorges N., A'Hearn M. F., et al. (2005): Science, 265.

Imaging Polarimetry of the Dust Coma of the Deep Impact Target Comet Tempel 1

E. Hadamcik and A. C. Levasseur-Regourd

Université Pierre et Marie Curie/Paris 6, UMR 7620, BP3
F-91371 Verrières le Buisson, France
edith.hadamcik@aerov.jussieu.fr
Anny-Chantal.Levasseur@aerov.jussieu.fr

Abstract We have observed comet 9P/Tempel 1 (the target of the Deep Impact mission) by a polarization imaging technique before and after impact with the 80 cm telescope at Haute-Provence observatory (France). The ejected material is well observed in our images and differences in physical properties are characterized.

1 Introduction

The light scattered by dust in a cometary coma is partially linearly polarized. The polarization value depends on the phase angle and on the wavelength. The overall polarization on the whole coma provides information on the dust particles bulk physical properties. Three classes of comets, comets with a low maximum in polarization, comets with a high maximum in polarization and Hale-Bopp have been defined by their dust optical properties [1, 2]. Average physical properties (size distribution, albedo) can be retrieved by such observational results through numerical and laboratory simulations results [3, 4, 5]. The imaging technique gives access to changes (e.g. size differences, fragmentation of particles) of the properties in the different coma regions, such as jets, and circumnucleus halo [6, 7]. Comets with important jet activity generally belong to the high polarization class of comets. Comets with restricted dust emissions or without jet activity generally belong to the low polarization class of comets. A correlation between comets with a high polarization and comets with well structured silicate features in the infrared has been noticed [8]. It provides constraints on the size of the grains.

2 Observational Conditions

The observations were conducted at observatoire de Haute-Provence with the 80 cm telescope in a Cassegrain configuration. Four polarized filters are mounted on a rotating wheel with their fast axis at 45^0 from one another. A red filter centered on 650 nm avoids the contamination by gaseous emissions. A CCD camera with a resolution of 0.84 arcseconds/pixel is used. The comet was observed on 2005, July 2.799, July 3.868, July 5.960, 6.875, respectively about −32 hrs, −7 hrs, +43 hrs and +65 hrs from the impact. The Sun-comet distance was about 1.51 AU and the comet-Earth distance about 0.89 AU. The phase angle was about 41°.

3 Results

For each night, intensity and polarization maps are obtained. A rotational shift on the intensity image allows to emphasize the jets and other features. On Fig. 1, maps of intensity, emphasized intensity and polarization are shown for July 3.87 and July 5.96. The coma is always asymetric and extends on the antisolar direction (close to southeast). Faint jets are present before impact. A large plume extends on more than 15000 km in about the southwest direction after impact. The overall polarization in a 26000 km aperture is about 6.2%, 32 hrs before impact and 6.2%, 7 hrs before impact. It is about 7.3%, 43 hrs after impact and 6.5%, 65 hrs after impact (Fig 2). Near the photometric center the polarization is close to 10% on July 3.87 and 9% on July 5.96. Before impact, the polarization decrease is important in all directions (less than 3% at distances larger than 10000 km) [9]. After impact, the polarization in the extended plume decreases slowly (about 6% at 15000 km) except in a small region at about 4000 km from the nucleus where the polarization is smaller (about 5%).

4 Discussion and Conclusions

The overall polarization value of comet 9P/Tempel 1 was lower by about 1.5% before impact than after impact. At 41° phase angle, the difference between the values for the two cometary classes on the synthetic phase curves (Fig. 2) is about 1%. Before impact the comet presents a limited jet activity and this Jupiter family comet may belong to the low Pmax comets class. Just after impact, the polarization increases [10] and the overall value reaches a value typical of high polarization comets. A good correlation between the increase in polarization and infrared emissions is noticed after impact [11]. From comparison with various data [12], it is suspected that the grains ejected in the plume (and responsible for the increase in polarization) are smaller than in the surrounding coma; they may be included in fluffy aggregates. The narrow lower polarization region in our map can be interpreted by a sorting by grains size, the smallest grains having a greater speed by stronger dynamical coupling to the gas outflow or by fragmentation due to sublimation of ices [13, 14]. The dust ejected from the subsurface seems to be similar to that ejected from more active comets. Particles at the surface may be more processed by the long time in solar vicinity of this short period comet.

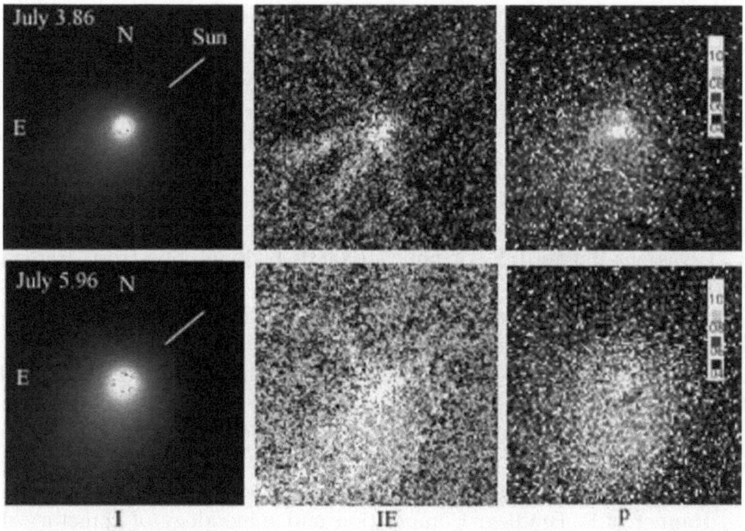

Fig. 1. Maps for July 3.87 (*before impact*), and July 5.96 (*after impact*), intensity (**I**), emphasized intensity (**IE**) and polarization (**P**). Field 70000 km x 70000 km. Scale for P maps: white 10%, dark grey smaller than 4%.

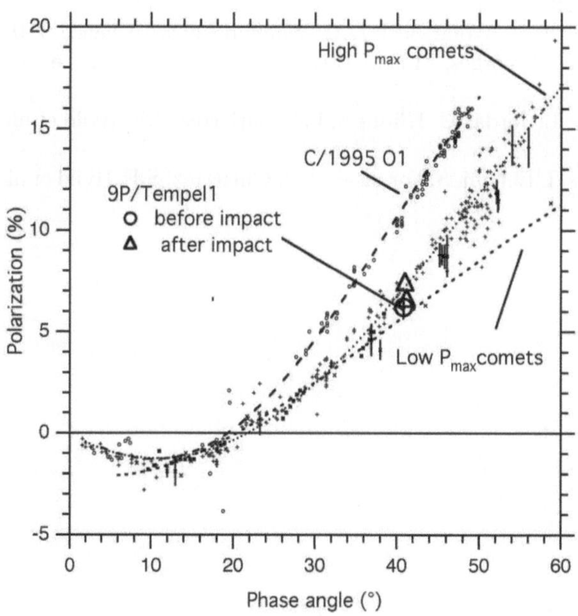

Fig. 2. Overall polarization through a 26000 km aperture. Comparison with other cometary data points and with the fits corresponding to the 3 classes of comets.

References

1. L. Kolokolova, M.S. Hanner, A.C. Levasseur-Regourd, B.A.S. Gustafson: Physical properties of cometary dust, obtained using their light scattering and emission, In: *Comets II*, ed by M.C. Festou, H.U. Keller, H.A. Weaver (Univ. of Arizona press, Tucson 2004) pp 577–604
2. E. Hadamcik, A.C. Levasseur-Regourd (2003a): J. Quant. Spectrosc. Radiat. Transfer **79–80**, 661
3. A.C. Levasseur-Regourd, E. Hadamcik (2003): J. Quant. Spectrosc. Radiat. Transfer **79–80**, 903
4. E. Hadamcik, J.-B. Renard, A.C. Levasseur-Regourd, J. Lasue (2006): J. Quant. Spectrosc. Radiat. Transfer **100**, 143
5. J. Lasue, A.C. Levasseur-Regourd (2006): J. Quant. Spectrosc. Radiat. Transfer **100**, 220
6. G.P. Tozzi, A. Cimatti, S. di Serego Alghieri, A. Cellino (1997): Planet. Space Sci. **45**, 535
7. E. Hadamcik, A.C. Levasseur-Regourd (2003b): Icarus **166**, 188
8. M.S. Hanner, J.P. Bradley: Composition and mineralogy of cometary dust, In: *Comets II*, ed by M.C. Festou, H.U. Keller, H.A. Weaver (Univ. of Arizona press, Tucson 2004) pp 555–564
9. E. Hadamcik, A.C. Levasseur-Regourd, V. Leroi, D. Bardin: Icarus, submitted
10. R. Furusho, Y. Ikeda, D. Kinoshita, W.-H. Ip, H. Kawakita et al. (2006): Icarus, **190**, 454–458
11. D.E. Harker, C.E. Woodward, D.H. Wooden (2005): Science **310**, 278
12. K. Meech, N. Ageorges, M.F. A'Hearn, C. Arpigny, A. Ates et al. (2005): Science **310**, 265
13. H.U. Keller, L. Jorda, M. Küppers, P.J. Gutierrez, S.F. Hvild et al. (2005): Science **310**, 281
14. M. Küppers, I. Bertini, S. Fornasier, P.J. Gutierrez, S.F. Hvild et al. (2005): Nature **437**, 987

Radio Observations from Australia of Comet 9P/Tempel 1 for Deep Impact

P. A. Jones[1], J. M. Sarkissian[2], M. G. Burton[1], M. A. Voronkov[2], and M. D. Filipović[3]

[1] School of Physics, University of New South Wales, Sydney 2052, Australia
Paul.Jones@csiro.au,
[2] Australia Telescope National Facility, PO Box 76, Epping NSW, 1710, Australia
[3] University of Western Sydney, Locked Bag 1797, Penrith South, DC, NSW 1797, Australia

Abstract We present radio observations of comet 9P/Tempel 1, associated with the Deep Impact mission, made with the Australia Telescope National Facility. Weak 18-cm OH emission was detected with the Parkes 64-m telescope, in data averaged over July 4–6, at a level of 12 ± 3 mJy km/s. The OH emission was increasing, but we stopped observing after July 6. We did not detect the HCN 1-0 with the Mopra 22-m telescope over the period July 2–6. The 3 σ limit was 0.06 K km/s for HCN on July 4 after the impact. We did not detect the HCN 1–0 line, 6.7 GHz CH_3OH line or 3.4-mm continuum with the Australia Telescope Compact Array (ATCA) on July 4, giving further limits on any small-scale structure due to an outburst.

1 Introduction

The NASA mission Deep Impact encountered Comet 9P/Tempel 1 on 2005 July 4, with the high velocity collision of the impactor creating a 100 m crater [1, 2]. There was a coordinated international campaign of ground-based and satellite observations [10, 11] to observe the impact event, to complement the observations from the flyby part of the Deep Impact mission, and to follow changes in the comet before and after the impact.

For the radio part of the international campaign, Australia was geographically well-placed to monitor changes in the first few hours after impact (05:44 UT) on July 4. We used all three sites of the Australia Telescope National Facility to monitor the comet around impact choosing OH at 1.67 GHz and HCN and CS in the 3-mm band, as the best tracers that matched the capabilities of these facilities.

The results have been published in MNRAS [9] and a summary is presented here.

2 Parkes OH Observations

Observations with the Parkes 64-m telescope were made of the OH 1667.3590 and 1665.4018 MHz lines simultaneously with 8 MHz BW and frequency switching mode, on July 4, 5 and 6. The correlator configuration gave 8192 channels with 0.95 kHz or 0.176 km/s. After folding the spectra for the frequency switching, spectra around the two lines were extracted and converted onto the velocity scale of the comet (around 9 km/s geocentric, from the JPL Horizons ephemeris).

We expect weak emission, as the inversion parameter was small and positive at around $i = 0.03$ (using [6]) at this time. Integrating the data over a full day of observations (with T_{sys} around 27 K) we obtain sensitivities of a few mK (around that expected), but do not detect the OH lines, except possibly the 1667 MHz line on July 6 (3 σ).

Combining the data for all 3 days, we do get a marginal detection of the 1667 MHz line, as shown in Fig. 1. The fit to this line gives peak 5.5 ± 1.2 mK, centre -0.1 ± 0.2 km/s (relative to the comet ephemeris velocity) and width (FWHM) 2.4 ± 0.7 km/s. The Parkes data (Fig. 1) does have some interference (seen as negative feature here due to the frequency switching folding) and residual baseline effects. However the peak is formally significant above the 3 σ level of fitting uncertainty, at the expected velocity. This peak of 1667 MHz OH is similar to the level of OH (18 mJy km/s integrated) seen with the 100-m Green Bank Telescope [8] over the period July 4–11, although the OH level seen with the GBT is variable.

Rough estimates of the steady-state OH production rate $Q_{OH} = 2.8 \times 10^{28}$ /s) and number of OH molecules (N_{OH}) within the Parkes (7 arcmin or radius 272 000 km) beam were calculated by scaling from the results of [5]. This used the Despois et al. model [6] of inversion (i = 0.03). Note however, that the inversion factor is model-dependent and uncertain close to the zero crossing, so that using the model of Schleicher & A'Hearn [13] (i = 0.08) would decrease the values of Q_{OH} and N_{OH} by factor 3/8 giving $Q_{OH} = 1.0 \times 10^{28}$ /s averaged over the 3 days. For comparison, the pre-impact H_2O production rate was around $Q_{H_2O} = 1.0 \times 10^{28}$ /s [12, 4, 7].

There are tantalising signs of an increase in OH emission over the 3 days, so it is, with hindsight, disappointing that we did not keep observing after July 6.

3 Mopra HCN and CS Observations

Observations with the Mopra 22-m telescope were made of the 88.63 GHz HCN 1-0 triplet (88630.4157, 88631.8473 and 88633.936 MHz) on July 2–6. Observations were also made of the CS 2-1 97.98095 GHz line on July 3. The bandwidth was 32 MHz with 1024 channels giving 31 kHz or 0.106 km/s pixels for HCN and 0.096 kms^{-1} for CS. We did not detect either the HCN or CS lines even after combining several days of HCN observations. The 3 σ limits on HCN were 0.06 K kms^{-1} on July 4 after the impact and 0.04 K kms^{-1} averaged over July 2–6.

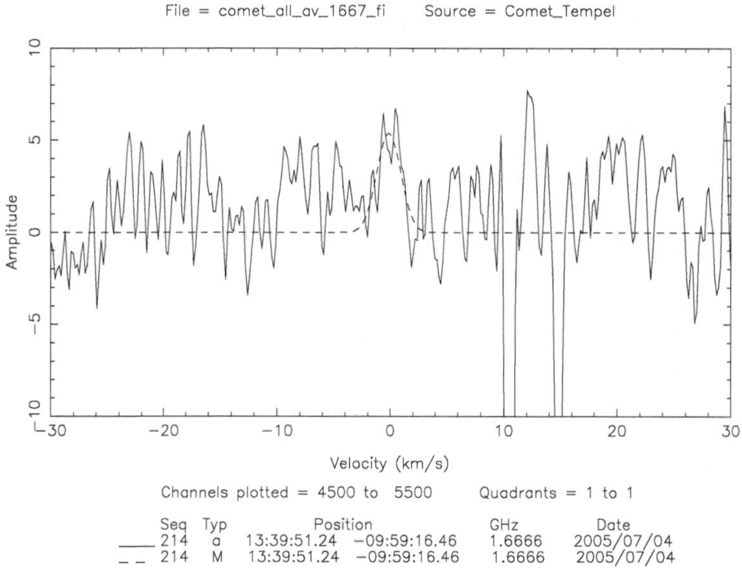

Fig. 1. The Parkes 1667 MHz OH spectrum integrated over July 4–6 showing a weak detection.

The limits on HCN production rate (Q_{HCN}) and number of molecules within the Mopra (35 arcsec or radius 11 300 km) beam (N_{HCN}) were calculated by scaling from the predictions of [10] and correcting for the observed distribution of HCN levels [3]. The 3 σ limits on Q_{HCN} were 1.8×10^{25} s^{-1} on July 4 after impact and 1.1×10^{25} s^{-1} averaged over July 2–6. These limits indicate that the HCN emission did not increase substantially (more than a factor of a few) after the Deep Impact collision above the pre-impact level of around 1×10^{25} s^{-1}.

4 Australia Telescope Compact Array Observations

Observations with the Australia Telescope Compact Array were made of the 88.63 GHz HCN 1–0 triplet on July 4, with the five 22-m telescopes in the H75 configuration, giving resolution 6.7×4.8 arcsec^2. The correlator configuration had one IF of 8 MHz, 257 channels giving spectral resolution 62.5 kHz or 0.21 km/s, and a second IF of 64 MHz BW, 64 channels for calibration and continuum monitoring.

We detected neither the HCN emission, nor 3-mm continuum (to 3 σ limits of 1.2 K kms^{-1} and 4.8 mJy respectively). The limits on HCN production rate (Q_{HCN}) and number of molecules within the ATCA (radius 1800 km) beam (N_{HCN}) were calculated in a similar way as for the Mopra data.

Because of the relatively high resolution of the interferometer the ATCA data probe compact structure in HCN, which we only expected for the first few hours after the Deep Impact collision, in the event of an outburst. Given the low limits on such an outburst (e.g. from the Mopra data) were are not surprised that we did not detect the HCN with the ATCA. Similarly, the limits on 3-mm continuum are much higher than the levels expected from a steep power-law thermal dust emission.

Since the gas from any impact-induced outburst would move out of the ATCA beam in a short time, we use the HCN limit from the ATCA data over the first 2 hours after impact (2.5 K kms^{-1}) to set a 3 σ limit on the number of HCN molecules released by the impact of $< 4 \times 10^{29}$.

We also observed the 6.7 GHz methanol (CH_3OH) transition with antenna 6 of the ATCA as a single dish, simultaneously with the 3-mm observations (as this 6 th dish is not fitted with the 3-mm system). We did not detect this line, but for brevity, the details are not given here (see [9]).

References

1. M.F. A'Hearn, M.J.S. Belton, A. Delamere, W.H. Blume (2005): Space Sci. Rev. **117**, 1
2. M.F. A'Hearn et al. (2005): Science **310**, 258
3. N. Biver, D. Bockelee-Morvan, P. Colom, J. Crovisier, A. Lecacheux, G. Paubert (2005): IAU Circular, **8538**, 1
4. N. Biver, et al. (2005): Div. Planet. Sci, **37**, 710
5. J. Crovisier, P. Colom, E. Gerard, D. Bockelee-Morvan, G. Bourgois (2002): Astron. Astrophys. **393**, 1053
6. D. Despois, E. Gerard, J. Crovisier, I. Kazes (1981): Astron. Astrophys. **99**, 320
7. P.D. Feldman, R.E. Lupu, S.R. McCandliss, H.A. Weaver, M.F.A'Hearn, M.J.S. Belton, K.J. Meech, (2006) Astrophys. J. **647**, L61
8. E.S. Howell, A.J. Lovell, B. Butler, F.P. Schloerb, S.A. Torchinsky, (2005) Bull. Am. Astron. Soc, **37**, 1480
9. P.A. Jones, J.M. Sarkissian, M.A. Burton, M.A. Voronkov, M.D. Filipovic (2006): Mon. Not. RAS **369**, 1995
10. K.J. Meech, M.F. A'Hearn, C.M. Lisse, H.A. Weaver, N. Biver, L.M. Woodney (2005): Space Sci. Rev. **117**, 291
11. K.J. Meech, et al. (2005): Science **310**, 265
12. M.J Mumma, et al. (2005): Science, **310**, 270
13. D.G. Schleicher, M.F. A'Hearn (1988): Astrophys. J. **331**, 1058

The Deep Impact Event as Seen from the University of Nariño Observatory – Colombia

A. Quijano Vodniza, C. Córdoba Barahona, A. J. Quijano Vodniza,
J. Perenguez López, and M. Rojas Pereira

Astronomical Observatory University of Nariño - Pasto, Colombia

Abstract The Astronomical Observatory of the University of Nariño has participated in the program STSP (The Small Telescope Science Program), associate to the project DEEP IMPACT of NASA, by sending photographies and scientific data. We began to study comet 9P/Tempel 1's activity in February 2005, and we have obtained the comet's brightnes curve. The great many data taken during every night have a standard deviation of just 0.03 in magnitude. By means of the curve we can compare the physical behavior before and after the impact. A slight change in the magnitude can be observed as the comet gets closer firstly to the Earth and then to the Sun. Nevertheless, we have detected an abrupt change in the luminous activity after the impact that took place at dawn of July 4th. The magnitude dicreased in 0.76 during the night of that same day. The brightness measurements taken after the impact indicate a greater variation with regard to the average value than those taken before the collision. This could indicate that the system nucleus-coma of the comet acquired a certain instability, an instability which lasted several hours.

1 Methodology

We managed to take a series of images of the comet, owing to the good weather conditions we had in the city of San Juán de Pasto, Department of Nariño, Colombia. The images were taken with a 14" Schmidt-Cassegrain teleswcoep (MEADE LX200GPS), and a commercial CCD camera (ST-7XME from SBIG). Infrared, Red and Visible type Bessell filters supplied by SBIG were used.
Every day we sent to the project STSP[1] a data package made up with BIAS, DARKS, FLAT frames, and crude images of the comet as well as the ones of the standard star. We also sent to the program processed images of Tempel. On Fig. 1 the instrumental magnitudes of 9P/Tempel 1 are shown in function of the date, from March 14 to July 27 of 2005. The perigee took place on May 3 2005 and the perihelion on July 4. The equation which better fits the luminous behavior of the comet from May 18 to June 27 is shown below and it's depicted in Fig. 2. Figure 3 shows the comet's magnitude before and after the impact and how an abrupt transition of brightness occurs.

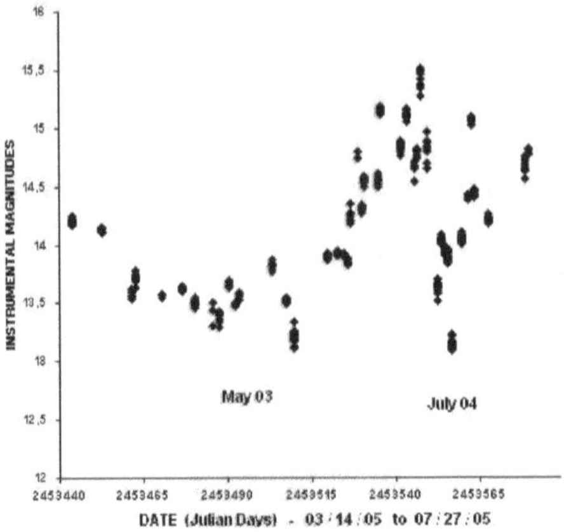

Fig. 1. Instrumental magnitudes as a function of date.

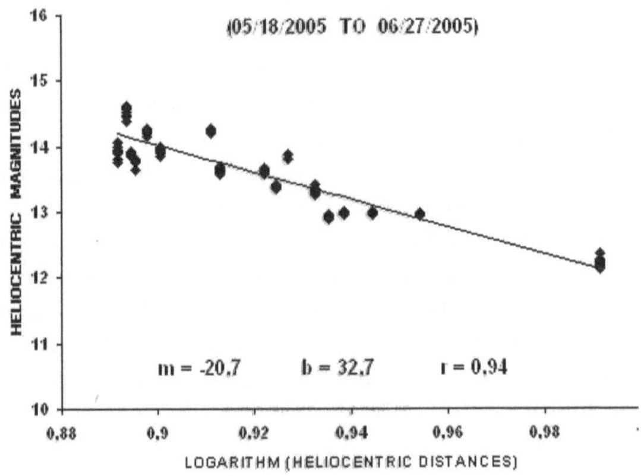

Fig. 2. Heliocentric magnitudes as a function of $\log r$.

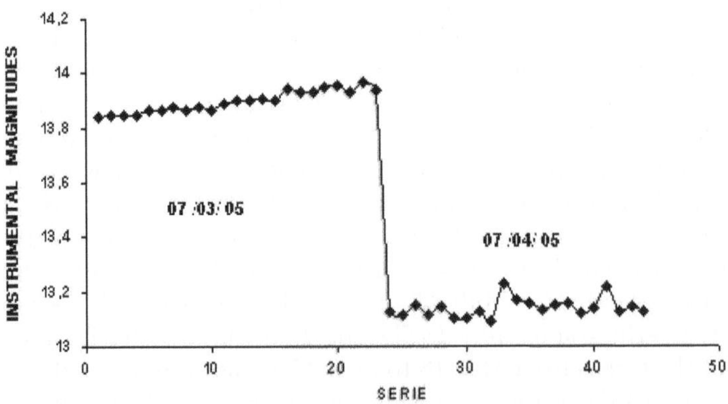

Fig. 3. Comparison between magnitudes (July 3 and 4).

2 Conclusions

A subtle change on the magnitude is detected as the comet approaches first the Earth and then the Sun. However, there is a sudden change in the luminous activity after the impact occurred at dawn on July the 4. The magnitude dropped in 0.76 during night at that same day. Experimental results didn't match up with what was expected theoretically due to the huge amount of dust that was raised during the impact, which substantially brought down the brightness of the comet. In Fig. 3 the magnitude of 9P/Tempel 1 can be seen before and after the impact and how an abrupt transition of brightness occurs. The equation that better matches the luminous behavior of the comet from May 18 to June 27 2005 is shown:

$$M(r) = M_0 + 2.5n \log(r)^{(2)} \quad (1)$$

$$M(r) = 32.7 + 2.5(8.3) \log(r) \quad (2)$$

References

1. `http://deepimpact.umd.edu/stsp/search-action.cfm`

The Earth-Based Deep Impact Observing Program

K. J. Meech[1], J. Pittichová[1], and A. Delsanti[1]

Institute for Astronomy, 2680 Woodlawn Dr., Honolulu HI 96288, USA
meech@ifa.hawaii.edu; Delsanti now at Paris Observatory

Abstract This paper describes the coordinated Earth-based observing program in support of the Deep Impact mission, including a discussion of the types of data available both pre-impact and during encounter. Insights on the interpretation of the Deep Impact results that can be obtained through the comparison of the multi-wavelength ground-based and space based datasets are discussed.

1 Introduction

Comets have long been the subject of investigation because of their potential to yield information about the process of solar system formation. They likely represent the least altered remnants of the era of planetesimal accretion in our solar system. However, since their formation it is likely that they have undergone significant thermal and physical evolution, so it is unclear to what extent observations of their chemistry and other properties can be understood in terms of primordial conditions. Key questions in addressing the connection of today's comet observations to the early solar system include

- Do interstellar ices survive the accretion process?
- To what extent was there chemical processing in the disk?
- Can measurements of volatiles and refractories in comets tell us about the chemistry of the disk?
- What region in the disk best represents the source region(s) for the cometesimals?
- What role did comets play in the delivery of water and organics to the terrestrial planet region?

The extent to which we can address these fundamental questions is limited by our understanding of the extent of radial mixing and dynamical processes in the disk and by how much the formation process during collapse of the protoplanetary disk altered the original interstellar / molecular cloud material. For example, it is expected that shock heating would have caused the sublimation of volatile coatings on interstellar grains during infall to the nebular midplane

[26]. Finally, what do we actually measure in comets, and how representative is it of the nucleus?

Fundamental new insights which address these questions are now emerging from the sample return from the Stardust mission, the results from Deep Impact, and will result from the ESA Rosetta mission rendezvous to comet 67P/Churyumov-Gerasimenko. This paper will describe the coordinated world-observatory that contributed to the success of the Deep Impact mission, and where synergies between data sets are likely to provide insight into the questions listed above.

1.1 Comet Observations and Comparisons

In the absence of an extraordinary event which drives a campaign, comet observations are often obtained with an eclectic mixture of wavelengths, temporal sampling and signal to noise, based in part on the program goals of the observers and in part upon the unpredictable vagaries of the telescope allocation committees, the weather and the comets themselves. The DI Earth-based campaign was the largest coordinated multi-wavelength event ever planned.

Over the past 3 decades, information on the chemical make up of comets has been accumulated, identifying over 80 chemical species, all but one of which has been seen in the interstellar medium (ISM). Some species are identified in only a few comets, although this is often a selection effect caused by different observing sensitivities. Many of the species in the optical and UV wavelengths are dissociation fragments yet few of the chemical pathways have been fully understood [1]. Many species of similar volatilities are seen to be released at the same time and yet for others there are both clear coma sources and changes in abundance with heliocentric distance [5, 6]. These issues make comparison difficult between classes of comets, and with the ISM (see Table 1, from [19, 31, 10, 7]) in order to interpret the chemical processes leading to the birth of our solar system. Thus our knowledge of the chemical ingredients in the coma is good, while the interpretation of linkages to the nucleus and protoplanetary disk is in its infancy [1].

Table 1. Comet and ISM Compositions.

Species	Comets	ISM	Species	Comets	ISM
H_2O	100	100	C_2H_2	0.1–0.3	
CO	1–20	5–50	C_2H_4	0.1–0.6	<0.4
CO_2	3–20	12–37	NH_3	0.6	<6–15
H_2CO	0.1–1	1.7–7	HCN	0.1–0.3	<3
CH_3OH	1–7	0–25	CH_3CN	0.02	
HCOOH	0.05	0.4–3	HC_3N	0.03	
HNCO	0.1		NH_2CHO	0.01	
CH_4	0.1–0.8	1–2			

2 Earth-Based Campaign Observations

The Earth-based campaign observations were obtained during two epochs: the pre-encounter observations which were intended for nucleus characterization and the encounter observations which included several months prior to the impact to establish a baseline of activity, and several months after the impact to watch the development of the activity. For the pre-encounter period, which lasted from 1999 to 2004, data were taken on 14 telescopes at 10 different ground-based observatories over 248 nights, and additional observations were made using the Spitzer and Hubble Space Telescope orbital observatories. At the time of the mission selection in 1999, the comet was moving toward perihelion, but not yet active. The 2000 apparition was not a good one for perihelion observations and no data were taken from July 1999 to April 2000, during the most active phases of the comet. However, the beginnings of activity were observed in January 1999 and when the comet emerged from solar conjunction in 2000, there was still considerable dust present. The comet was then well placed for observations all through aphelion and then inbound toward perihelion as the activity developed. During this period the nucleus size and albedo were determined, we knew the basic rotation state of the comet, and we had information on the phase function [3, 15, 24].

2.1 Encounter Predictions

Using the constraints from the Earth-based observations, Sarid et al. (2005) developed a model to look at the effects of the impact on the activity of the nucleus. Using a quasi 3D porous spherical nucleus (R_N=3.3 km, p_v =0.04 and P_{rot}=41.85 hr – as measured from the campaign) which consisted of both amorphous and crystalline H_2O, CO, CO_2, HCN, NH_3, and C_2H_6, and which had an assumed porosity=0.5 and thermal conductivity of $k = 5.67\times10^2/T$ J m^{-1} s^{-1} K^{-1}, the dust and gas production was matched for the pre-encounter data. The post impact model predicted that the amount of dust that would be ejected would be equivalent to a couple of days of activity, and that the mission of CO and CO_2 could vary by as much as 2 orders of magnitude, and that it would be difficult to distinguish any long-term activity from the event from the normal brightness fluctuations of the comet.

2.2 The Encounter Period

The DI team provided information to assist in the coordination of the Earth-based observing plan by elucidating the prime mission science that would benefit from simultaneous multi-wavelength observations and which science needed key long term monitoring both pre- and post-impact to understand the nature of the impact induced changes. This coordination was accomplished by a series of workshops held prior to proposal submission deadlines to secure the supporting telescope time. The result was a well-coordinated plan where most large facilities were optimized for observations with instrumentation that could provide unique contributions to the collective science. During the few days surrounding encounter, the observers were kept up to date on the discoveries through a multi-channel polycom and telephone conference each day. All together, there were ∼550 whole or partial nights allocated on ground-based facilities worldwide that were reported to the DI team. This represented data obtained on 73 telescopes at 35 different public and private observatories. In addition, nearly all the earth-orbital facilities were being utilized, and the Rosetta spacecraft used its instrumentation to observe the event [28]. A summary of the observing runs is shown in Table 2.

3 Data Set Cross Comparisons

3.1 Subsurface Layering

The imaging cameras on-board the DI spacecraft provided clear evidence of layering on Tempel 1's nucleus [4]. The development of the impact plume as seen from the DI spacecraft produced a light curve which can be interpreted as subsurface layering in the nucleus. Laboratory experiments [32, 33] have shown that the observations were consistent with a highly porous layered target, with the leading edge (high speed component) of the downrange impact plume coinciding with near-surface materials. The first light seen by the spacecraft which was composed of deeper materials, but the earth-based observers would have been primarily looking at the upper few meters of comet material. The long-lasting ejecta above the crater seen in the spacecraft look-back images would have come from larger depths in the later stated of ejection. There were nearly twenty realtime Earth-based program high-speed programs watching the evolution of the light curve during impact and during the days following the impact (see Table 3). All observers reported that there was no immediate impact flash, just a steep increase in brightening during the first 30–60 minutes post-impact. All observers reported a change in the rate of brightening at distinct intervals, followed by a fall off (some of which can be attributed to material moving out of the various sized observing apertures). Preliminary interpretation of the ground-based data has indicated that the changes in slope in the light curve indicate subsurface layering and possible compositional differences [34, 35]. However, a thorough multi-wavelength analysis of all of the light curve data sets to compare with the spacecraft observations to understand the physical interpretation in terms of the cratering excavation has not yet been done. This will be a rich area of investigation since these light curves span wavelengths from the UV⟶thermal IR.

Table 2. Earth-based Observing Circumstances.

PI	UT Date	Telescope	Detector	λ^*	FOV [']	Scale[†]	Res.[‡]	Sky[§]	Pgm.[¶]
1999									
Meech	Jan 19–21, 27–29	UH2.2m	Tek2K	Opt	7.5×7.5	0.219		Clear / Cirrus	Im
Meech	Mar 17–20	UH2.2m	Tek2k	Opt	7.5×7.5	0.219		Clear	Im
Meech	May 8,10,28	UH2.2m	Tek2k	Opt	7.5×7.5	0.219		Clear / Cirrus	Im
Meech	Jul 14–15	UH2.2m	Tek2k	Opt	7.5×7.5	0.219		Clear / Cirrus	Im
2000									
Meech	May 3–7	UH2.2m	Tek2k	Opt	7.5×7.5	0.219		Cirrus	Im
Bauer	May 23–28	UH2.2m	Tek2k	Opt	7.5×7.5	0.219		Clear / Cirrus	Im
Pittichová	Jun 12,28,29	UH2.2m	Tek2k	Opt	7.5×7.5	0.219		Cirrus	Im
Meech	Jul 9, Aug 20–25	UH2.2m	Tek2k	Opt	7.5×7.5	0.219		Varied	Im
Farnham	Aug 26–29	McDonald0.8m	PFC	Opt	46.2×46.2	1.355		Varied	Im
Tozzi	Sep 5–9	TNG3.6m	Dolores	Opt	9.4×9.4	0.275		Varied	Im
Meech	Sep 22–Oct 1	UH2.2m	Tek2k	Opt	7.5×7.5	0.219		Varied	Im
Buie	Sep 28–Oct 1,17,18	Lowell1.8m	Fair2k	Opt	13.3×13.3	0.39		Varied	Im
Sekiguchi	Oct 19–23	Danish1.5m	DFOSC	Opt	13.7×13.7	0.40		Clear	Im
Farnham	Oct 31–Nov 3	McDonald0.8m	PFC	Opt	46.2×46.2	1.355		Clear	Im
Buie	Nov 19–20	Lowell1.8m	Fair2k	Opt	13.3×13.3	0.39		Clear	Im
Hainaut	Nov 26–27	ESO2.2m	WFI	Opt	34×33	0.238		Clear	Im
Bauer	Dec 2–3,18–21	UH2.2m	Tek2k	Opt	7.5×7.5	0.219		Clear	Im
Buie	Dec 16–17	Lowell1.8m	Fair2k	Opt	13.3×13.3	0.39		Cirrus	Im
2001									
Pittichová	Jan 26–28	UH2.2m	Tek2k	Opt	7.5×7.5	0.219		Clear	Im
Meech	Aug 13–22	UH2.2m	Tek2k	Opt	7.5×7.5	0.219		Varied	Im
Tozzi	Aug 17–18,Sep 16–17	TNG3.6m	Dolores	Opt	9.4×9.4	0.275		Clear	Im
Millis	Aug 19–21	CTIO4m	MOSAIC	Opt	35.6×35.6	0.26		Clear	Im
Choi	Sep 19–27	Wise1.1m	CAMERA	Opt	15×30	0.884		Varied	Im

Table 2. Earth-based Observing Circumstances – Contd.

PI	UT Date	Telescope	Detector	λ^*	FOV [']	Scale[†]	Res.[‡]	Sky[§]	Pgm.[¶]
Samarasinha	Sep 21–27	KPNO2.1m	T2KA	Opt	10.4×10.4	0.305		Varied	Im
Meech	Sep 21–27	UH2.2m	Tek2k	Opt	7.5×7.5	0.219		Varied	Im
Buie	Sep 25–26	Lowell1.8m	Fair2k	Opt	13.3×13.3	0.39		Clear	Im
Various	Oct 7,8,29	UH2.2m	Tek2k	Opt	7.5×7.5	0.219		Varied	Im
Various	Nov 9–19	UH2.2m	Tek2k	Opt	7.5×7.5	0.219		Varied	Im
Krisciunas	Nov 10–12	CTIO1.5m	Tek2k	Opt	14.7×14.7	0.432		Clear	Im
Choi	Nov 13,15	Wise1.1m	CAMERA	Opt	15×30	0.884		Clear	Im
Tozzi	Nov 14–15	TNG3.6m	Dolores	Opt	9.4×9.4	0.275		Cirrus	Im
Farnham	Dec 5	McDonald2.7m	IGI	Opt	7×7	0.62		Clear	Im
Meech	Dec 21–24	UH2.2m	Tek2k	Opt	7.5×7.5	0.219		Clear	Im
2002									
Tozzi	Feb 2–4	TNG3.6m	Dolores	Opt	9.4×9.4	0.275		Varied	Im
Fernández	Feb 22–23	UH2.2m	Tek2k	Opt	7.5×7.5	0.219		Clear	Im
Meech	Feb 3–7, Mar 8–9	UH2.2m	Tek2k	Opt	7.5×7.5	0.219		Clear	Im
Meech	Sep 29–30	UH2.2m	Tek2k	Opt	7.5×7.5	0.219		Varied	Im
Mueller	Oct 30–Nov 1	KPNO2.1m	T2KA	Opt	10.4×10.4	0.305		Varied	Im
Hsieh	Nov 5–7	UH2.2m	Tek2k	Opt	7.5×7.5	0.219		Clear	Im
Meech	Nov 30–Dec 2	UH2.2m	Tek2k	Opt	7.5×7.5	0.219		Clear	Im
2003									
Various	Nov 8,9,12,19,20	UH2.2m	Tek2k	Opt	7.5×7.5	0.219		Varied	Im
Various	Dec 4,15,17,18,20	UH2.2m	Tek2k	Opt	7.5×7.5	0.219		Varied	Im
Various	Dec 23–27	UH2.2m	Tek2k	Opt	7.5×7.5	0.219		Varied	Im
Meech	Dec 29	ESONTT	SuSI2	Opt	5.5×5.5	0.081		Clear	Im
Meech	Dec 30–31	CTIO4m	MOSAIC	Opt	35.6×35.6	0.26		Clear	Im
Meech	Dec 29	VLT8m	FORS2	Opt	6.8×6.8	0.25		Clear	Im

Table 2. Earth-based Observing Circumstances – Contd.

PI	UT Date	Telescope	Detector	λ^*	FOV [′]	Scale[†]	Res.[‡]	Sky[§]	Pgm.[¶]
2004									
Meech	Jan 30, Feb 1,24	UH2.2m	Tek2k	Opt	7.5×7.5	0.219		Varied	Im
Meech	Mar 24–25	VLT8m	FORS2	Opt	6.8×6.8	0.25		Clear	Im
Lisse	Mar 23–25	Spitzer	IRS	MIR	1.0×1.3	1.8	Low		Im & Sp
Pittichová	Apr 21–22	UH2.2m	Tek2k	Opt	7.5×7.5	0.219		Clear	Im
Various	May 7–9,11,15	UH2.2m	Tek2k	Opt	7.5×7.5	0.219		Varied	Im
Meech	Nov 18,21, Dec 1	UH2.2m	Tek2k	Opt	7.5×7.5	0.219		Cirrus	Im
Meech	Dec 4	Keck I	LRIS	Opt	7.0×7.0	0.210	Low	Clear	Sp
2005									
Lara	Jan 5–Jun 30[a]	CA2.2m	CAFOS	Opt	9×9	0.53	Medium	Varied	Im & Sp
Lara	Jan 5–Jun 30[b]	OSN1.5m	VersArray	Opt	7.9×7.9	0.464		Varied	Im
A'Hearn	Jan 16–17	KPNO4.0m	MOSAIC	Opt	35.6×35.6	0.26		Cirrus	Im
Meech	Jan 17,18	UH2.2m	Tek2k	Opt	7.5×7.5	0.219		Clear	Im
Buie	Jan 24	Keck I	NIRC	IR	0.64×0.64	0.15		Clear	Im
Farnham	Feb 2–5	KPNO2.1m	CCD	Opt	9.8×3.3	0.19		Cirrus	Im
Crovisier	Mar 4–Jun 8	Nançay	18-cm	Rad	3.5×19		Very high	Clear	Sp
Meech	Mar 10,11	UH2.2m	Tek2k	Opt	7.5×7.5	0.219		Cirrus	Im
Farnham	Mar 11–14	KPNO2.1m	CCD	Opt	9.8×3.3	0.19		Cirrus	Im
Cochran	Mar 12,13	McDonald2.7m	LCS	Opt	0.03×2.5		Low	Clear	Sp
Kidger	Mar 23	ORM	AFOSC	Opt					Sp
Buie	Mar 31–Apr 24	LO-TAM31in	PCCD	Opt	7.7×11.5	1.2		Varied	Im
Meech	Apr 1,4	UH2.2m	Tek2048	Opt	7.5×7.5	0.219		Cloud	Im
Cochran	Apr 6–8	McDonald2.7m	LCS	Opt	0.03×2.5		Low	Cloud	Sp
Howell	Apr 8–9	Arecibo	L-wide	Rad	12×12	200	High	Clear	Sp
Farnham	Apr 9–14	KPNO2.1m	CCD	Opt	9.8×3.3	0.19		Cirrus	Im
Lara	Apr 11–14	CA2.2m	CAFOS	Opt	9×9	0.53	Medium	Clear	Im & Sp

Table 2. Earth-based Observing Circumstances – Contd.

PI	UT Date	Telescope	Detector	λ^*	FOV [']	Scale†	Res.‡	Sky§	Pgm.¶
Hainaut	Apr 19	LSO	EMMI	Opt					Sp
Howell	Apr 18–19	Arecibo	ALFA	Rad	13×13	200	High	Clear	Sp
Buie	Apr 27–May 17	LO–TAM31–in	PCCD	Opt	7.7×11.5	1.2		Varied	Im
Hainaut	May 1	BO	CCD	Opt					Im
Hainaut	May 1	CO	CCD	Opt					Im
Hainaut	May 1	LSO	DFOSC	Opt					Im
Howell	May 2–3,18,29,31	Arecibo	ALFA	Rad	13×13	200	High	Clear	Sp
Biver	May 4–9	IRAM-30m	Rx.A–D	Rad	0.2–0.5	10–27	10^7	Clear	Sp
Cochran	May 10,12	McDonald	LCS	Opt	0.03×2.5		Low	Clear	Sp
Farnham	May 14–17	KPNO2.1m	CCD	Opt	9.8×3.3	0.19		Cirrus	Im
Meech	May 12,15	UH2.2m	Tek2048	Opt	7.5×7.5	0.219		Clear	Im
Howell	May 16	Arecibo	L–wide	Rad	12×12	200	High	Clear	Sp
Harker	May 17	Gemini–N	Michelle	IR	0.53×0.4	0.1005	Low	Clear	Sp
Chaffee	May 24	Keck II	NIRSPEC	IR		0.43×24.0	Med	Cirrus	Sp
Wooden	May 25	Keck II	LWS	MIR	0.18×0.18	0.085	Low	Clear	Sp & Im
Wooden	May 25	Keck II	NIRC	NIR	0.64×0.64	0.15		Clear	Im
Chaffee	May 30	Keck I	HIRESb	Opt		7.0×0.86	Med	Cirrus	Sp
Bauer	May 31–Jun 1	TMO0.6m	CCD	Opt	9.0×9.0	0.52		Clear	Im
Meech	Jun 1,2	UH2.2m	Tek2048	Opt	7.5×7.5	0.219		Cirrus	Im
Buie	Jun 1–3	LO-TAM31in	PCCD	Opt	7.7×11.5	1.2		Varied	Im
Jehin	Jun 2,7,8	VLT UT2	UVES	Opt	0.007×0.17	0.2	80,000	Clear	Sp
Womack	Jun 3	ARO	SIS	Rad					Sp
Farnham	Jun 4–7	KPNO2.1m	CCD	Opt	9.8×3.3	0.19		Cirrus	Im
Bensch	Jun 5–Sep 1	SWAS	Schottky RX	Sub-mm	3.3×4.5		$3.7{\times}10^5$	n/a	Sp
Cochran	Jun 8–9	McDonald	LCS	Opt	0.03×2.5		Low	Clear	Sp

Table 2. Earth-based Observing Circumstances – Contd.

PI	UT Date	Telescope	Detector	λ^*	FOV [′]	Scale†	Res.‡	Sky§	Pgm.¶
Howell	Jun 8–9	Arecibo	L–wide	Rad	12×12	200	High	Clear	Sp
Woodney	Jun 9	JCMT	A3	Rad					Sp
Buie	Jun 12–20	LO-TAM31in	PCCD	Opt	7.7×11.5	1.2		Varied	Im
Birch	Jun 13,14,19,20	PO61cm	CCD	Opt	7.0×5.0	0.60		Variable	Im
Feldman	Jun 13–14	HST	ACS/HRC	Opt	0.5×0.5	0.025		n/a	Im
Bauer	Jun 13–15	TMO0.6m	CCD	Opt	9.0×9.0	0.52		Clear	Im
Buie	Jun 17–24	LO-TAM31in	PCCD	Opt	7.7×11.5	1.2		Varied	Im
Biver	Jun 18	Odin	AOS+AC2	Rad	2.2×2.2	127	10^6	n/a	Sp
Birch	Jun 22,27,28	PO61cm	CCD	Opt	7.0×5.0	0.60		Variable	Im
Osip	Jun 20	LCO	PANIC	IR					Im
Osip	Jun 21	LCO	MagIC	Opt					Im
Osip	Jun 21	LCO	TReCS	IR					Im
Osip	Jun 21	LCO	TReCS	IR					Sp
Biver	Jun 23	Odin	AOS+AC2	Rad	2.2×2.2	127	10^6	n/a	Sp
Kelley	Jun 24–27	IRTF	SpeX	IR	0.008×0.25	0.15	Low	Clear	Sp
Kelley	Jun 25	IRTF	SpeX	IR	1×1	0.12		Clear	Im
Watanabe	Jun 25	NAOJ	CCD	Opt	14×14	0.82			Im
Lisse	Jun 25–Jul 17	IRTF	SPEX	NIR	1.0×1.0	0.3	Low		Im & Sp
Womack	Jun 26–Jul 5	ARO	1.3mm	Rad					Sp
Sekiguchi	Jun 27	IRSF1.4m	SIRIUS	IR	7.7×7.7	0.45	Cloud		Im
Walker	Jun 27–Jul 1	MIRA	CCD	Opt					Im
Walker	Jun 27–Jul 4	MIRA	CCD	Opt					Im
Osip	Jun 27–Jul 9	LCO	WFCCD	Opt					Im
Kelley	Jun 28	MLOF	CCD	Opt	1×2	0.25		Clear	Pol
Keller	Jun 28–Jul 14	Rosetta	OSIRIS-NAC	Opt	132×132	3.9	n/a	n/a	Im

Table 2. Earth-based Observing Circumstances – Contd.

PI	UT Date	Telescope	Detector	λ*	FOV [']	Scale†	Res.‡	Sky§	Pgm.¶
Keller	Jun 28–Jul 14	Rosetta	OSIRIS-WAC	Opt	720×720	20.5	n/a	n/a	Im
Stern	Jun 28–Jul 14	ROSETTA	Alice	UV	6×360	1080	Med	n/a	Sp
Wooden	Jun 29–30	IRTF	HIFOGS	MIR	0.05×0.05		Low	Clear	SpPho
Harrington	Jun 29–Jul 6	AEOS	HiVisS	NIR	0.03×0.12	0.16	Medium	Clear	SpPol
Schleicher	Jun 29–Jul 19	LO–Mesa	KPh	Opt					Pho
Küppers	Jun 30-Jul 15	Rosetta	OSIRIS	Opt			n/a	Im	
Lisse	Jun 30–Jul 24	CXO	ACIS	X–ray	8.3×8.3	0.5	Low		Im & Sp
Gilmore	Jul 1	MtJUO	CCD	Opt	11×11	1.20		Clouds	Im
Gulkis	Jul 1	ROSETTA	MIRO	Rad				n/a	Sp
Rudy	Jul 1–5	LO	NIRIS	IR					Sp
Qi	Jul 1–6	SMA	mm/submm	Rad	0.6–0.9	2–4	10^6	Clear	Im & Sp
Lowry	Jul 1–7	INT	WFC	Opt	34×34	0.33		Clear	Im
Bailey	Jul 1–8	AAO	ES	Opt					Sp
Furushko	Jul 1–9	LOT	PICO	Opt	10.0×5.0	1.874		Cirrus	Pol
Sekiguchi	Jul 1–9	IRSF1.4m	SIRIUS	IR	7.7×7.7	0.45		Cirrus	Im
Crovisier	Jul 1–10	Nançay	18-cm	Rad	3.5×19		Very high	Clear	Sp
Ortiz	Jul 1–10	OSN1.5m	VersArray	Opt	7.9×7.9	0.464		Varaible	Im
Lara	July 1–12	CA2.2m	CAFOS	Opt	9×9	0.53	Medium	Clear	Im & Sp
Birch	Jul 2,4–7	PO61cm	CCD	Opt	7.0×5.0	0.60		Variable	Im
Harker	Jul 2	Gemini-N	Michelle	IR	0.53×0.4	0.1005	Low	Clear	Im
Honda	Jul 2–5	Subaru	COMICS	mid-IR	0.7×0.5	0.13		Clear	Im
Honda	Jul 2–5	Subaru	COMICS	mid-IR	0.7×0.005	0.165	250	Clear	Sp
Christensen	Jul 2–4	SSO	CCD	Opt					Im
Walsh	Jul 2–4	SPM	BChS	Opt	5.3×5.3	0.94	Low	Clear	Sp
Etzel	Jul 2–5	MLO	CCD	Opt	2.7×2.7	0.8		Clear	Im
Feldman	Jul 2–5	HST	ACS/SBC	UV	0.5×0.5	0.025		n/a	Im

Table 2. Earth-based Observing Circumstances – Contd.

PI	UT Date	Telescope	Detector	λ^*	FOV [′]	Scale†	Res.‡	Sky§	Pgm.¶
Feldman	Jul 2–5	GALEX	NUV	UV	72 dia	1.5	Low	n/a	Sp
Fitzsimmons	Jul 2–5	HAO	DilCAM	Opt					Im
Reitsema	Jul 2–5	SPM	BChS	Opt					Sp
Suntzeff	Jul 2–5	CTIO 1.5m	CCD	IR					Im
Walsh	Jul 2–5	SPM	La Ruca	Opt	4.7×4.7	0.274		Clear	Im
Wooden	Jul 2,5,7–9,18	IRTF	MIRSI	MIR	1.42×1.07	0.27	Clear	Pho	
Capria	Jul 2–6,9	ORM	SARG	Opt	0.03×0.13		Med	Clear	Sp
Hadamcik	Jul 2–6	OHP-T80	CCD	Opt	7×7	0.84		Clear	Pol
Jones	Jul 2–6	Mopra	3-mm	Rad	0.6	10^6	Avr	Sp	
Petitpas	Jul 2,4,6	SMA	230GHz Rx	Rad	0.83×0.83	2.5	2.5	Clear	Im & Sp
Hainaut	Jul 2–9	VLT UT1	FORS2	Opt					Im
Buie	Jul 2–7	LO-TAM1.8m	Mimir	IR	9.9×9.9	0.58		Varied	Im
Jones	Jul 2–8	SSO	CCD	Opt					Sp
Snodgrass	Jul 2–8	ORM/LT	RATCAM	Opt	4.6×4.6	0.135		Clear	Im
Snodgrass	Jul 2–8	ORM/LT	SupIRCam	IR	1.7×1.7	0.4		Clear	Im
Boehnhardt	Jul 2–9	LSO3.6m	TIMMI2	IR					Im
Boehnhardt	Jul 2–9	LSO3.6m	TIMMI2	IR					Sp
Boehnhardt	Jul 2–9	ESO/MPI2.2m	WFI	Opt	8.1×16.2	0.238		Clear	Im
Boehnhardt	Jul 2–9	LSONTT	EMMI	Opt					Im
Boehnhardt	Jul 2–9	LSONTT	EMMI	Opt					Sp
Boehnhardt	Jul 2–9	VLT UT1	ISAAC	IR					Sp
Boehnhardt	Jul 2–9	VLT UT1	ISAAC	IR					Im
Boehnhardt	Jul 2–9	VLT UT1	ISAAC	IR					Pol
Boehnhardt	Jul 2–9	VLT UT4	NACO	IR					Im
Boehnhardt	Jul 2–9	VLT UT2	FORS1	Opt					Pol

Table 2. Earth-based Observing Circumstances – Contd.

PI	UT Date	Telescope	Detector	λ^*	FOV [']	Scale[†]	Res.[‡]	Sky[§]	Pgm.[¶]
Boehnhardt	Jul 2–9	VLT UT3	VISIR	MIR	0.32×0.32	0.075	Low		Im
Farnham	Jul 2–9	KPNO4.0m	MOSAIC	Opt	35.6×35.6	0.26		Cirrus	Im
Hodapp	Jul 2–9	UH2.2m	SNIFS	Opt	0.1×0.1	0.4	1000	Clear	Sp
Knight	Jul 2–9	KPNO4.0m	SQIID	IR	5.0×5.3	0.69			Im
Licandro	Jul 2–9	ORM	DOLORES	Opt	9×9	0.275	Low	Clear	Sp & Im
Rauer	Jul 2–10	VLT UT1	FORS2	Opt	6.8×6.8	0.252	Low	Clear	Sp & Im
Biver	Jul 2–10	IRAM-30m	Rx.A–D	Rad	0.2×0.5	10–27	10^7	Average	Sp
Kaeufl	Jul 2–10	LaSO	TIMMI2	IR					Im
Jehin	Jul 2–11	VLT UT2	UVES	Opt	0.007×0.17	0.2	80,000	Clear	Sp
Feldman	Jul 2–16	HST	ACS/HRC	Opt	0.5×0.5	0.025		n/a	Im
Lisse	Jul 2-Aug 24	Spitzer	IRS	MIR	1.0×1.3	1.8	Low		Im & Sp
Van Belle	Jul 3	Keck I	Echelle	Opt				Clear	Sp
Van Belle	Jul 3	Keck II	NIRCCD	IR				Clear	Im
Brosch	Jul 3–4	Wise C18	SBIG CCD	Opt	40.5×27.0	0.9		Clear	Im
Paragi	Jul 3–4	WSRT	DZB	Rad	23.0×23.0	3.0	Med	n/a	Sp
Raju	Jul 3–4	VBO	TK1024	Opt					Im
Biver	Jul 3–5	Odin	AOS+AC2	Rad	2.2×2.2	127	10^6	n/a	Sp
Harker	Jul 3–5	Gemini-N	Michelle	IR	0.53×0.4	0.1005	Low	Clear	Sp
Meech	Jul 3–5	CTIO 1.5m	CPAPIR	IR	35×35	1.03		Variable	Im
Miller	Jul 3–5	UKIRT	CGS4	IR	1.0×0.01	0.8×0.9	Med	Cirrus	Sp
Petitpas	Jul 3–5	SMA	230GHz Rx	Rad	0.58×0.58	1.5	1.5	Clear	Im & Sp
Walker	Jul 3–5	MIRA	BSpe	Opt					Sp
Meech	Jul 3–6	CFHT	MegaCam	Opt	60×60	0.187		Clear	Im
Bauer	Jul 3–7	Palomar	AO	IR	0.67×0.67	0.04		Clear	Im
Mueller	Jul 3–7	WIYN3.5m	DensePak	Opt	0.45×0.72	0.14	Medium	Clear	Sp
Licandro	Jul 3–7	ORM	LIRIS	IR	4×4	0.25	Low	Clear	Im & Sp

Table 2. Earth-based Observing Circumstances – Contd.

PI	UT Date	Telescope	Detector	λ^*	FOV [′]	Scale†	Res.‡	Sky§	Pgm.¶
Weissman	Jul 3–7	TMO	CCD	Opt	8.9×8.9	0.52		Clear	Im
Zhou	Jul 3–8	NAOCh	BATC	Opt					Im
Kidger	Jul 3–8	OdT	CCD	Opt					Im
Licandro	Jul 3–9	ORM	ALFOSC	Opt	7×7	0.19	Low	Clear	Sp & Im
Von Braun	Jul 3–9	LCO Swope1m	SITe3	Opt	14.85×24.3	0.435		Variable	Im
Meech	Jul 3–11	CTIO4m	ISPI	IR	10.25×10.25	0.3		Variable	Im
Chanover	Jul 4	APO3.5m	NIC-FPS	IR	4.66×4.66	0.273		Clear	Im
Chaffee	Jul 4	Keck II	NIRSPEC	IR		0.43×24.0	Med	Clear	Sp
Chaffee	Jul 4	Keck I	HIRESr	Opt		7.0×0.86	Med	Clear	Sp
Drummond	Jul 4	Starfire 3.5m	CCD	Opt				Clear	Im
Fuse	Jul 4	Subaru	HDTV	Opt	1.2×0.7	0.04	n/a	Clear	Im
Jones	Jul 4	ATCA	3-mm	Rad	0.6	3.0	10^6		Sp & Im
Mason	Jul 4	Swift	UVOT	UV				n/a	Im
Seitzer	Jul 4	CTIOSchmidt	CCD	Opt	78×78	2.3		Clear	Im
Slysh	Jul 4	PRO	Spect	Rad					Sp
Sugita	Jul 4	Subaru	COMICS	mid-IR	0.7×0.5	0.13		Clear	Im
Sugita	Jul 4	Subaru	COMICS	mid-IR	0.7×0.005	0.165	250	Clear	Sp
Tucker	Jul 4	SCFS		Opt					Pho
Ehrenfreund	Jun 30	VLT UT3	VISIR	MIR	0.32×0.32	0.075	Low		Sp & Im
Boehnhardt	Jul 4–5	VLT UT3	VISIR	MIR	0.32×0.32	0.075	Low		Sp & Im
Brosch	Jul 4–5	Wise C18	SBI CCD	Opt	40.5×27.0	0.9		Clear	Im
Lis	Jul 4–5	CSO-10m	Rx.345GHz	Rad	0.4×0.4	25	3×10^6	Marginal	Sp
Mumma	Jul 4–5	Keck II	NIRSPEC	IR		0.144	$2.4{\times}10^5$	Clear	Sp
Ries	Jul 4–6	McDonald	PFC	Opt	40×40	1.350		Clear	Im
Dundon	Jul 4–6	CTIO 0.9m	CCD	Opt	13.5×13.5	0.396		Variable	Im
Jones	Jul 4–6	Parkes	18-cm	Rad	7.0		10^6		Sp

Table 2. Earth-based Observing Circumstances – Contd.

PI	UT Date	Telescope	Detector	λ^*	FOV [′]	Scale†	Res.‡	Sky§	Pgm.¶
Slysh	Jul 4–8	KRO	Spect	Rad					Sp
Kidger	Jul 4–8	OdT	CAIN-II	IR					Im
Lovell	Jul 4–12	GBT	SP	Rad	7.0×7.0	442	High	Clear	Sp
Cochran	Jul 5–6	Keck I	Echelle	Opt	0.01×0.12		High	Clear	Sp
Wooden	Jul 5,10	IRTF	HIFOGS	MIR	0.05×0.05		Low	Clear	SpPho
Biver	Jul 7,9,10,12	Odin	AOS+AC2	Rad	2.2×2.2	127	10^6	n/a	Sp
Kelley	Jul 10	IRTF	SpeX	IR	1×1	0.12		Clear	Im
Rauer	Jul 10–12	VLT UT2	FORS1	Opt	6.8×6.8	0.2	Low	Clear	Sp
Buie	Jul 10–14	LO-TAM1.8m	Mimir	IR	9.9×9.9	0.58			Im
Sekiguchi	Jul 11,13,16	IRSF1.4m	SIRIUS	IR	7.7×7.7	0.45	Clear		Im
Boehnhardt	Jul 12–Aug 9	ESO/MPI2.2m	WFI	Opt	8.1×16.2	0.238		Clear	Im
Bauer	Jun 15	Palomar 5.0m	AO	IR	0.67×0.67	0.04		Clear	Im
Biver	Jul 16,25,31	Odin	AOS+AC2	Rad	2.2×2.2	127	10^6	n/a	Sp
Harker	Jul 16,28	Gemini-N	Michelle	IR	0.53×0.4	0.1005	Low	Clear	Sp
Boehnhardt	Jul 17–Aug 17	ESO/MPI2.2m	WFI	Opt	8.1×16.2	0.238		Clear	Im
Buie	Jul 20–22	LO-TAM1.8m	Mimir	IR	9.9×9.9	0.58		Varied	Im
Sekiguchi	Jul 22,25,28	IRSF1.4m	SIRIUS	IR	7.7×7.7	0.45	Clear		Im
Reach	Jul 31	Spitzer	MIPS	IR	22×42	2.5		n/a	Im
Biver	Aug 07	Odin	AOS+AC2	Rad	2.2×2.2	127	10^6	n/a	Sp
Reach	Sep 23	Spitzer	MIPS	IR	22×42	2.5		n/a	Im
Reach	Sep 01	Spitzer	MIPS	IR	8×22	2.5		n/a	Im

Notes: *Wavelength regime: vis=visible, IR=near IR, MIR=mid IR, Rad=Radio/sub-mm; †Detector scale in ″ pix^{-1}; ‡Spectroscopic resolution; §Weather conditions; ¶Observing Program: Im=imaging, Sp=Spectroscopy, SpPho=Spectrophotometry.

Table 3. Light Curve Datasets.

PI	Telescope	Wavelengths
Bauer	Palomar 5m	JHK
Buie	Lowell 1.8m	JHK
Chanover	Apache Point	H
Drummond	Starfire Optical Range 3.5m	
Feldman	HST	F606W
Fernàndez	IRTF 3.0m	near IR
Keller	Rosetta [22]	vis-orange
Knight	KPNO 2.1m	JK
Larson	Mt. Lemmon	H_2O^+
Lisse	Spitzer	mid-IR
Lynch	Lick 40 inch	BVR
Meech	CTIO 4m, 0.9m, Schmidt	VRIJK
Meech	CFHT 3.6m	BVRI
Miller	UKIRT 3.8m	optical
Sugita/Harker	Subaru 8m, Gemini 8m	mid-IR
Wainscoat	UH 0.6m	R
Walsh	San Pedro Martir	R
Weissman	Table Mountain 0.6m	BVRI
Welsh	San Diego State	

3.2 Surface and Sub-Surface Chemical Differences

One of the primary goals of the Deep Impact mission was to excavate beneath the evolved surface layers of a comet, and examine the less altered material from the sub-surface. Implicit in this experimental design is the expectation that the surface layers having been exposed to cosmic ray processing and diurnal and annual solar thermal fluxes, will be devoid of super volatiles and will be altered chemically (and physically) from their initial primordial formation state. Regarding whether the DI experiment produced clear evidence of a different sub-surface chemistry, there is conflicting evidence (summarized in Table 4). One of the expectations was that the impact would create a new area of sustained activity, and that an increase in supervolatile materials, most notably, CO, would be observed. UV observations from HST compared limits to the pre-impact production of CO to the post-impact production, and found that within the errors there was no increase in CO outgassing [13]. In contrast, optical spectra showed no appreciable change in the abundance of species post-impact [9, 27].

The DI spacecraft saw a large increase in organics relative to water post-impact [2], however the resolution was such that the specific organics were not identified. High resolution near-IR spectra obtained at Keck showed an increase in the abundance ratio of ethane after impact, whereas the ratios for methane and hydrogen cyanide remained unchanged. In the mid-IR, spectra from the Spitzer

Table 4. Chemical Differences.

Wavelength	Observations	Results	Observer (PI)
UV	HST Spectra	No difference	Feldman
Optical	Keck, ESO spectra	No difference	Cochran, Manfroid
Near IR	Keck high res spectra	Some in volatile organics	Mumma
Mid-IR	Spitzer, Subaru, Gemini	Yes, grain size effect?	Lisse, Sugita, Harker
Near IR	DI flyby	Yes, change in organics	A'Hearn

space telescope [25] showed a dramatic change from a featureless spectrum pre-impact to a spectrum rich with organic and other mineral features post-impact. However, much of the change can be attributed to a particle size effect; because of the large aggregate particles it was not possible to see the features in the pre-impact spectra.

3.3 Radial Migration and Coma Chemical Heterogeneity

Pre-Deep Impact Evidence

Prior to the Deep Impact experiment, there was an increasing body of evidence suggesting the presence of chemical gradients in the nebula, and that comets must have accumulated from planetesimals formed in different environments. Variable H_2CO production was observed in comet 1P/Halley as it rotated, and this was attributed to a component cometesimal that was rich in H_2CO [29].

In a large spectroscopic survey of 85 comets, A'Hearn et al. (1995) observed a group of comets with a carbon-chain (C_2, C_3) depletion. These comets were all members of the Jupiter-Family, which likely formed in the vicinity of the Kuiper Belt in the outer solar system, and it was established that the depletion was not an evolutionary effect. Rather, this was taken as evidence for a temperature or energy threshold in the nebula for the production of the C-chain parents in the Kuiper Belt.

More recently, with the advent of sensitive high resolution infrared spectrographs, we have the ability to measure abundances of parent organic molecules relative to water. A summary of a survey of organic parent volatiles [30] is shown in Table 5. This compares the unusual composition of comet C/1999 S4 LINEAR with those of other Oort Cloud comets. In this particular comet, the most volatile organic compounds are depleted, suggesting that its ices condensed in a warmer region, perhaps the Jovian sub-nebula.

The Evidence from Deep Impact

Spectral imaging data taken 10 minutes pre-impact by the DI spacecraft showed a clear asymmetry in the H_2O and CO_2 outgassing [11]. The water production was most prevalent in the sub-solar direction, whereas the CO_2 was more prevalent near the south rotation pole (which was not illuminated by sunlight). This suggests large scale chemical heterogeneity in the nucleus of comet 9P/Tempel 1.

Table 5. Nebular Chemical Gradients.

Species	T_{sub}	Others	1999S4
H_2O	180	100	100
CO	25	1–20	0.4–0.9
CH_4	31	0.1–0.8	0.1–0.15
C_2H_6	44	0.1–0.6	0.13
C_2H_2	57	0.1–0.3	<0.14
HCN	95	0.1–0.3	0.1
CH_3OH	99	1–7	<0.17

Spectral modeling of the IR spectra of the DI ejecta show a mixture of high temperature materials, mid-temperature materials and ices [25]. Likewise the mineralogical evidence from the Stardust mission return of grains from comet 81P/Wild 2 also showed that crystalline and amorphous silicates were abundant in the form of weakly bonded assemblies of heterogeneous fragments, that must have formed in different temperature regimes and were subsequently radially mixed [8, 37].

3.4 Grain Size Evolution

There were a large number of observers who took imaging data, pre-impact, during the encounter, and for 1–2 months post-impact to observe the evolution of the dust coma, and to apply dust-dynamical models to the images to infer properties of the dust (e.g. amount of dust, particle size distribution, and if there was new activity [20] as a result of the impact). A summary of the key data sets and preliminary results is shown in Table 6. There are a number of multi-wavelength data sets represented here, which when modeled, will provide comprehensive additional information about the dust evolution.

Table 6. Grain Size Evolution.

Telescope	Lead Observers	Observations
Subaru, Spitzer, Gemini	Ootsubo, Lisse, Harker [25]	Model Spectra
India, CFHT, KPNO	Farnham, Feldman, Pittichová, Vasundhara [12, 14]	Multi-color coma observations and FP modeling
ESO, IRTF, La Palma, San Pedro, Subaru	Boehnhardt, Fernández, Lara, Licandro, Sujita [16, 21, 23]	Coma color post impact
AEOS, ESO, Lulin	Furushko, Hadamcik, Harrington [18]	Change in polarization slope, arc structure
DI Spacecraft	Sunshine [36]	Near IR spectra show surface ice

3.5 Other Areas for Comparison

The DI science team and the world-network of observers have only begun to fully analyze the enormous multi-wavelength data set of observations. From these data new thermal models can be developed using the order of magnitude lower thermal inertia of the surface [17]. It is of intense interest to try to trace the jets and features seen in the coma by the space craft to the nucleus surface in order to gain an understanding of the physical processes associated with activity and the outbursts. The encounter showed us that very little of the activity is coming from the surface, and moreover, much of the dust production is associated with the CO_2-rich vent.

4 Conclusions

The Deep Impact experiment was an unqualified success from the space mission perspective, and has already yielded many fundamental new results which have changed our paradigm of the comet nucleus. However, the extensive rich data set of Earth-based observations is absolutely essential in the interpretation and understanding of many of the spacecraft results, and the integrated analysis of the data sets will provide many new insights into the physics of comets for many years to come. We have seen:

- Possible evidence for chemical gradients in the protoplanetary nebula - from the C-chain depletion seen in certain classes of short-period comets, from the detailed high resolution spectroscopy of comets that is accumulating and from the mixture of both high temperature and low temperature grains seen in the comae of both comets 81P/Wild 2 and 9P/Tempel 1.
- We have seen evidence that the sub-surface layers of comet 9P/Tempel 1 are different from the surface layers. There was a factor of 2 difference in organics seen relative to water post impact compared to pre-impact, and some organic species were seen to be enriched post impact in the high resolution spectra obtained at Keck. The Rosetta spacecraft saw a change in the CN/H_2O ratio post-impact, and there was clear evidence seen for layering on the comet.
- However, there were other indications that the sub-surface layers were not substantially different from the surface layers. In particular there was no significant change in the CO production pre- to post-impact as observed by HST, and there was no likewise change in the abundance of molecules observed spectroscopically at optical wavelengths.
- There was abundant evidence that the dust properties changed from the ambient coma to the post-impact ejecta, both seen from space and from the ground-based observations. The current interpretation is that the comet dust is made of large aggregates of material which became deaggregated in the impact process, which resulted in a change in the size distribution of the dust (and hence color) seen after impact.

- Finally, there is growing evidence that there are inhomogeneities within comet nuclei. This was first seen in comet 1P/Halley. The DI spacecraft observed a coma inhomogeneity with respect to the CO and H_2O production and observers from ESO documented a change in the CN production with rotation.

Merging all of these conclusions into a comprehensive new paradigm of the comet nucleus and activity will rely on cross cutting examination of the spacecraft and Earth-based data sets contained within this unique "world observatory" experiment.

Acknowledgements Support for this work provided through University of Maryland and University of Hawaii subcontract Z667702, which was awarded under prime contract NASW-00004 from NASA. This material is also based upon work supported by the National Aeronautics and Space Administration through the NASA Astrobiology Institute under Cooperative Agreement No. NNA04CC08A issued through the Office of Space Science. We greatfully acknowledge the support of the many telescope allocation committees and directors worldwide who allocated resources to the Earth observing campaign, and to all the observatory support staff who made extraordinary efforts to ensure that these taime critical events were successful. finally, greatful appreciation is acknowledged for all the observers who provided calibration frames or critical rotation data points from their own allocated time.

References

1. M.F. A'Hearn: *Comets II*, ed by M.C. Festou, H.U. Keller and H.A. Weaver, (2004)(Univ. AZ Press, Tucson, AZ 2004) pp 17–22.
2. M.F. A'Hearn, M.J.S. Belton, W.A. Delamere, J. Kissel, K.P. Klaasen, L.A. McFadden, K.J. Meech, H.J. Melosh, P.H. Schultz, J.M. Sunshine + 23 coauthors (2005): Science **310**, 258.
3. M.J.S. Belton, K.J. Meech, M.F. A'Hearn, O. Groussin, L. McFadden, C. Lisse, Y.R. Fernández, J. Pittichová, H. Hsieh, J. Kissel, K. Klaasen, P. Lamy, D. Prialnik, J. Sunshine, P. Thomas, I. Toth (2005): SSR **117**, 137.
4. M.J.S. Belton, P. Thomas, J. Veverka, P. Schultz, M. F. A'Hearn, L. Feaga, T. Farnham, O. Groussin, J.-Y. Li, C. Lisse, L. McFadden, J. Sunshine, K.J. Meech, A.W. Delamere, J. Kissel (2007): Icarus **187**, 332.
5. N. Biver, D. Bockelée-Morvan, J. Crovisier, P. Colom, F. Henry, R. Moreno, G. Paubert, D. Despois, C.D. Lis (2002a): EM&P **90**, 323.
6. N. Biver, D. Bockelée-Morvan, J. Crovisier, P. Colom, F. Henry, E. Lellouch, A. Winnberg, L.E.B. Johansson, M. Gunnarsson, H. Rickman, F. Rantakyro, J.K. Davies et al. +11 authors (2002b): EM&P **90**, 5.
7. D. Bockelée-Morvan, J. Crovisier, M.J. Mumma and H.A. Weaver: *Comets II*, ed by M.C. Festou, H.U. Keller and H.A. Weaver, (2004)(Univ. AZ Press, Tucson, AZ 2004) pp 391–424.
8. D. Brownlee et al. + 182 coauthors (2006): Science **314**, 1711.
9. A.L. Cochran, W.M. Jackson, K.J. Meech, M. Glaz (2007): Icarus **187**, 156.

10. P. Ehrenfreund, S.B. Charnley and D. Wooden(2004): *Comets II*, ed by M.C. Festou, H.U. Keller and H.A. Weaver (Univ. AZ Press, Tucson, AZ 2004) pp 115–136.
11. L.M. Feaga, M.F. A'Hearn, J.M. Sunshine, O. Groussin, T.L. Farnham (2007): Icarus, **191**, 134–145.
12. T.L. Farnham, D.D. Wellniz, D.L. Hampton, J.-Y. Li, J.M. Sunshine, O. Groussin, L.A. McFadden, C.J. Crockett, M.F. A'Hearn, M.J.S. Belton, P. Schulz, C.M. Lisse (2007): Icarus **187**, 26.
13. P.D. Feldman, R.F. Lupu, S.R. McCandliss, H.A. Weaver, M.F. A'Hearn, M.J.S. Belton, K.J. Meech (2006a): ApJ **647**, L61.
14. P.D. Feldman, S.R. McCandliss, M. Route, H.A. Weaver, M.F. A'Hearn, M.J.S. Belton, K.J. Meech (2006b): Icarus **187**, 113.
15. Y.R. Fernández, K.J. Meech, C.M. Lisse, M.F. A'Hearn, J. Pittichová, M.J.S. Belton(2003): Icarus **164**, 481.
16. Y.R. Fernández, C.M. Lisse, M.S. Kelley, N. Dello Russo, A.T. Tokunaga, C.D. Woodward, D.H. Wooden (2007): Icarus **187**, 220.
17. O. Groussin, M.F. A'Hearn, J.-Y. Li, P.C. Thomas, J.M. Sunshine, C.M. Lisse, K.J. Meech, T.L. Farnham, L.M. Feaga, W.A. Delamere (2007): Icarus **187**, 16.
18. D.M. Harrington, K. Meech, L. Kolokolova, J.R. Kuhn, K. Whitman (2007): Icarus **187**, 177.
19. W.M. Irvine, F.P. Schloerb, J. Crovisier, B. Fegley, and M.J. Mumma (2000): *Protostars and Planets IV*, ed by V. Mannings, A.P. Boss and S.S. Russell (Univ. AZ Prss, Tucson, AZ 2000) pp 1159–1200.
20. H.U. Keller et al. (2007): Icarus **187**, 87.
21. M.M. Knight, K.J. Walsh, M.F. A'Hearn, R.A. Swaters, A.B. Zauderer, N.H. Samarasinha, R. Vazques, H. Reitsema (2007): Icarus **187**, 199.
22. M. Küppers, S. Mottola, S.C. Lowry, M. F. A'Hearn, C. Barbieri, M.A. Barucci, S. Fornasier, O. Groussin, P. Gutiérrez, S.F. Hviid, H.U. Keller, P. Lamy (2007): A&A **462**, L13.
23. L.M. Lara, H. Boehnhardt, R. Gredel, P.J. Gutierrez, R. Rodrigo, M.J. Vidal-Nunez (2007): A&A **465**, 1061.
24. J.-Y. Li, M.F. A'Hearn, M.J.S. Belton, C.J. Crockett, T.L. Farnham, C.M. Lisse, L.A. McFadden, K.J. Meech, J.M. Sunshine, P.C. Thomas, J. Veverka (2007):Icarus **187**, 41.
25. C.M. Lisse, J. VanCleve, A.C. Adama, M.F. A'Hearn, Y.R. Fernández, T.L. Farnham, L. Armus, C.J. Grillmair, J. Ingalls, M.J.S. Belton, O. Groussin, L.A. McFadden, K.J. Meech, P.H. Schultz, C.B. Clark, L.M. Feaga, J.M. Sunshine (2006): Science **313**, 635.
26. J.I. Lunine, S. Engel, B. Rizk, M. Horanyi (1991): Icarus **94**, 333.
27. J. Manfroid, D. Hutsemékers, E. Jehin, A.L. Cochran, C. Arpigny, W.M. Jackson, K.J. Meech, R. Schulz, J.-M. Zucconi (2007): Icarus **187**, 144.
28. K.J. Meech et al. + 208 coauthors (2005): Science **310**, 265.
29. M.J. Mumma, D.C. Reuter (1989): ApJ **344**, 940.
30. M.J. Mumma, N. Dello Russo, M.A. DiSanti, K. Magee-Sauer, R. Novak, S. Brittain, T. Rettig, I.S. McLean, D.C. Reuter, Li-H. Xu (2001): Science **292**, 1334.
31. M.J. Mumma, M.A. DiSanti, K. Magee-Sauer, B.P.Bonev, G.L. Villanueva, H. Kawakita, N. Dello Russo, E.L. Gibb, G.A.Blake, J.E. Lyke, R.D. Campbell, J. Aycock, A. Conrad, G.M. Grank (2005): Science **310**, 270.
32. P.H. Schultz, C. Ernst, M.F. A'Hearn, C. Eberhardy, J.M. Sunshine, the Deep Impact Team(2006): LPI **37**, 2294.

33. P.H. Schultz, C.A. Eberhardy, C.M. Ernst (2007): LPI Contrib. **1338**, 1890.
34. S. Sugita, T. Ootsubo, T. Kadono, M. Honda, S. Sako, T. Miyata, I. Sakon, T. Yamashita, H. Kawakita et al. + 13 coauthors (2005): Science **310**, 274.
35. S. Sugita, T. Kadono, S. Sako, T. Ootsubo, M. Honda, H. Kawakita, R. Furushko, J. Watanabe (2007): LPI **38**, 1911.
36. J.M. Sunshine, M.F. A'Hearn, O.Groussin, J.-Y. Li, M.J.S. Belton, W.A. Delamere,J. Kissel, K.P. Klaasen, L.A. McFadden, K.J. Meech (2006): Science **311**, 1453.
37. M.E. Zolensky et al. (2006): Science **314**, 1735.

Part II

The Cometary Dust

Gemini-N Observations of the Dust Excavated from Comet 9P/Tempel 1 During Deep Impact

D. E. Harker[1], C. E. Woodward[2], and D. H. Wooden[3]

[1] University of California, San Diego, CASS, 9500 Gilman Dr. Dept 0424, La Jolla, CA 92093-0424 dharker@ucsd.edu
[2] Department of Astronomy, School of Physics and Astronomy, 116 Church Street, S.E., University of Minnesota, Minneapolis, MN 55455 chelsea@astro.umn.edu
[3] NASA Ames Research Center, Space Science Division, MS245-1, Moffet Field, CA 94035-1000 dwooden@mac.com

Abstract We present mid-infrared spectra of the GEMINI-N (+MICHELLE) observations of comet 9P/Tempel 1 during its encounter with Deep Impact ($\approx$ 7 min cadence). Based on our modeling of the spectra, we find that the submicron sized dust grains travel away from the nucleus faster than the larger grains. Groups of grains of differing mineral composition travel at different ejection velocities. Amorphous olivine and amorphous carbon grains travel together at one speed, while amorphous pyroxene and crystalline olivine travel together at another speed. The different speed of the two mineral groups implies that Deep Impact excavated a layer of material below an amorphous carbon and amorphous olivine rich surface layer to release an amorphous pyroxene and crystalline olivine rich layer.

1 Introduction

Comets contain primitive dust and ice originating from the early solar nebula. The *Deep Impact* (DI) experiment afforded us the opportunity to observe this primitive dust in Jupiter Family (JF) comet 9P/Tempel 1. JF comets have been subjected to processing mechanisms such as collisions, cosmic ray radiation bombardment, UV irradiation, and loss of volatiles by their frequent perihelion passages. These processing mechanisms may alter the morphology, but most likely not the mineralogy, of the dust grains. We examine the thermal emission of the dust excavated by DI as seen by Gemini-N (+MICHELLE). Our observations allow us to probe temporal and spatial variations in the mineralogy and properties of the dust grains after impact.

2 Observations

We obtained mid-infrared (mid-IR) spectra and images of comet 9P before, during, and after impact using the MICHELLE imaging spectrograph on the

8-m Fredrick C. Gilette (Gemini-N) on Mauna Kea, Hawaii. Our observational cadence on the night of impact ($\sim$ 7 min intervals) was chosen to obtain a $S/N \sim 10$ in each spectral observation at 10 μm. Extraction and calibration of the comet spectra from the two-dimensional images used common mid-IR techniques.

3 Ejecta Mineralogy

To search for mineral species detectable in the 10 μm region, we have coadded spectra obtained on the night of impact (04 July 2005 UT) into four temporal blocks: $t = +10$ to $+36$ min, $t = +61$ to $+94$ min, $t = +111$ to $+124$ min, and $t = +148$ to $+181$ min. To facilitate comparison between the 10 μm Gemini and *Spitzer* post-impact spectra, we adopted the analysis methodology (subtraction and normalization of impact-spectra by non-impact spectra) described by Lisse et al. [10] to create residual emissivity spectra (Fig. 1a, b, c, d). A broad feature is clearly seen in the residual emissivity spectrum. Superposed on this broad feature are narrower solid state features. Mg-rich (Mg content > 90%) crystalline olivine is identified by the solid state features at 11.25 and 11.9 μm. Orthopyroxene (Mg content > 90%) is identified through features at 9.3, 10.0, and 10.5 μm.

The wavelength of PAH species at 8.3 μm ("hot" PAHs [8]) and 8.6 μm are not clearly identified, although they are identified in the *Spitzer* spectrum obtained during the interval $t = +30$ to $+60$ post impact [10]. There are four unidentified features in our spectra, located at 8.9, 10.85, 12.4, and 12.9 μm. The 12.4 and 12.9 μm features are possible PAH emission from ionized PAHs.

We clearly identify emission from silicate species such as amorphous olivine, amorphous pyroxene, crystalline olivine and orthopyroxene in the Gemini residual emissivity spectrum. The same minerals also can be identified in the lower SNR, time resolved spectra. Here, our emphasis is on understanding the temporal evolution of the 10 μm emission spectra. Therefore, in our modeling (Sect. 4) we only include those species that can be identified in the lower SNR spectra: the amorphous and crystalline silicates.

4 Model of Thermal Emission from Dust Grains

The thermal emission dust code used to model the observed dust grain emission in the 9P coma is detailed in Harker et al. [6]. Initial estimates of grain size and mineral parameters for coma dust in 9P were taken from Harker et al. [4]. In short, the model assumes that an optically thin collection of discrete (singular mineralogy) dust particles reside at the heliocentric and geocentric distance of the comet observations. The amorphous grains (pyroxene, olivine, and carbon) are assumed to be spherical (Mie Theory) and possibly fractally porous, and range in size from 0.1 to 100 μm. The indices of refraction used in Mie Theory to calculate the optical efficiencies of the silicate grains are from Dorschner et al. [1], and the carbon grains from Edoh [2]. The optical efficiency of the crystalline

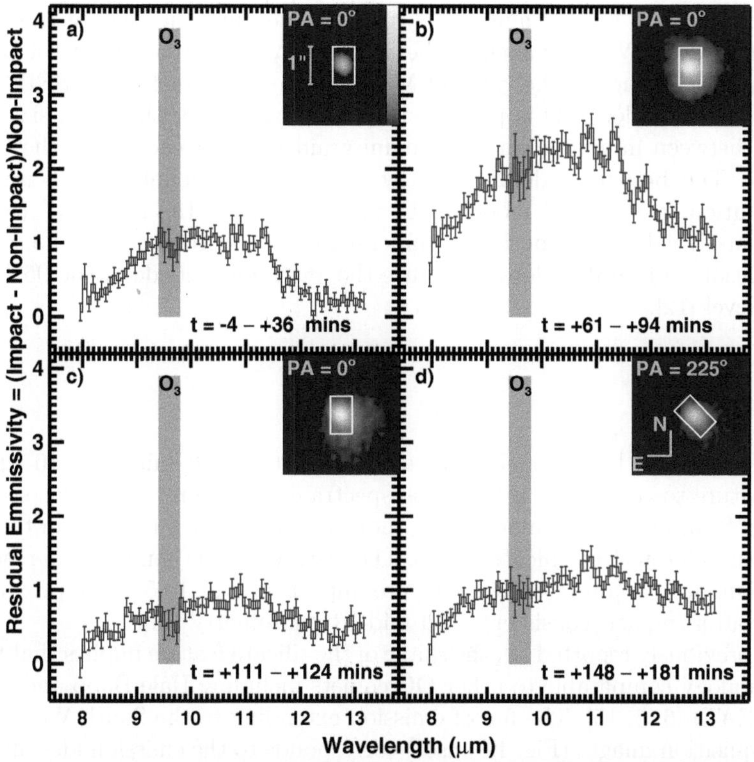

Fig. 1. Coadded spectra of comet 9P/Tempel 1 at four temporal epochs after impact. The images show the comet just prior to the time sequence of the spectra. The rectangle in the images show the size and location of the extraction aperture used to produce the spectra. Telluric ozone is indicated in each graph.

grains are computed as a continuous distribution of ellipsoids (CDE) with one crystallographic axis elongated to a ratio of 10:1:1 to match the wavelength of the emission features with laboratory experiments [3]. The indices of refraction for the crystalline silicates are from Jäger et al. [7]. The radii of the crystalline grains ranges from 0.1 to 1.0 μm; larger crystals do not well match the shape of the observed resonances [11]. The radius of the crystals is an "effective" radius, or the radius a crystal would be if it were spherical.

The temperature of all grains is calculated using a thermal equilibration calculation. The temperature of the crystalline olivine grains are calculated using the "hot crystal model" detailed in [6], and is retained here to compare results between other comets [9].

We use the Hanner modified power law (hereafter, HGSD) as the grain size distribution. The relative amount of each grain size is calculated by: $n(a) = (1 - a_\circ/a)^M (a_\circ/a)^N$, where a is the grain radius, $a_\circ = 0.1$ μm, the minimum grain radius, and M and N are independent parameters. Since we are only modeling the 10 μm region, we set $N = 3.7$, the canonical value determined for comet Hale-Bopp [6, 5]. Values of N and M determine the peak of the HGSD: $a_p = (M + N)/N$. By changing the value of a_p in increments of 0.1 µm, we calculate the change in the value of M used to compute our model SEDs.

The final model SED is produced by integrating over the grain size distribution between minimum and maximum grain radii for each fit to the Gemini spectra. The best-fit is determined by varying the amount of each mineral's contribution to the total model SED, and calculating the lowest χ^2_ν (reduced chi-squared) value. We consider a mineral to be "constrained" by the model if its addition to the SED does not cause the χ^2_ν to fall outside of the 95% confidence level [12].

5 Results

On 4 July 2005 UT, immediately after impact ($\sim$ 10 min) the 10 µm silicate feature-to-continuum ratio in the spectra centered on the nucleus suddenly jumped from unity to a value of > 2, indicating an sudden increase in the number of submicron sized silicate grains. The feature-to-contimuum ratio reached a maximum at 10.5 µm of $\sim$ 2.5 times pre-impact level at 10.5 µm approximately 1 hour after impact, consistent with mid-IR photometry [13].

As previously reported [4], the shape of the silicate feature implies that 9P has a mineralogy comparable to other OC comets including Hale-Bopp and C/2001 Q4 (NEAT) [6, 5, 15]. The fan of emission extending to the South-West seen in our acquisition images (Fig. 1 – inset) corresponds to the emission identified and studied in detail by others [14, 13].

To remove contamination from grains in the coma during quiescence, and the thermal emission from the nucleus, we subtract the non-impact spectrum from the post-impact spectra. We model the resulting non-impact subtracted spectra which represent the emission from only those grains ejected during impact.

Following impact, we see a marked change in the mineralogy and amount of material in the spectra of 9P. We also see an immediate shift in the on source position (centered on the nucleus) in the peak of the grain size distribution (a_p) to the smallest grain sizes ($\sim$ 0.2 µm).

The mass of peak grains for both the center and offset spectra for grain sizes 0.1–1.0 µm and 1.0–10.0 µm are plotted in Fig. 2a, b, c, d. Weighted linear fits to mass values constrained by the model between $t = +61$ and 124 min show the trend of the depletion of grains as they travel through the extraction beams. From Fig. 2 it is evident that between 61 and 124 min after impact, in both the center position and offset positions, the dominant submicron sized mineral is amorphous pyroxene.

The mass of the submicron sized amorphous pyroxene grains changes with time as does the mass of the submicron sized crystalline olivine grains. The rate of mass change of these two species closely track one another, suggesting that the two minerals originated from a similar reservoir of material. A similar correlation is observed between submicron sized amorphous carbon and amorphous olivine grains, towards the center position (albeit the correlation is somewhat weaker).

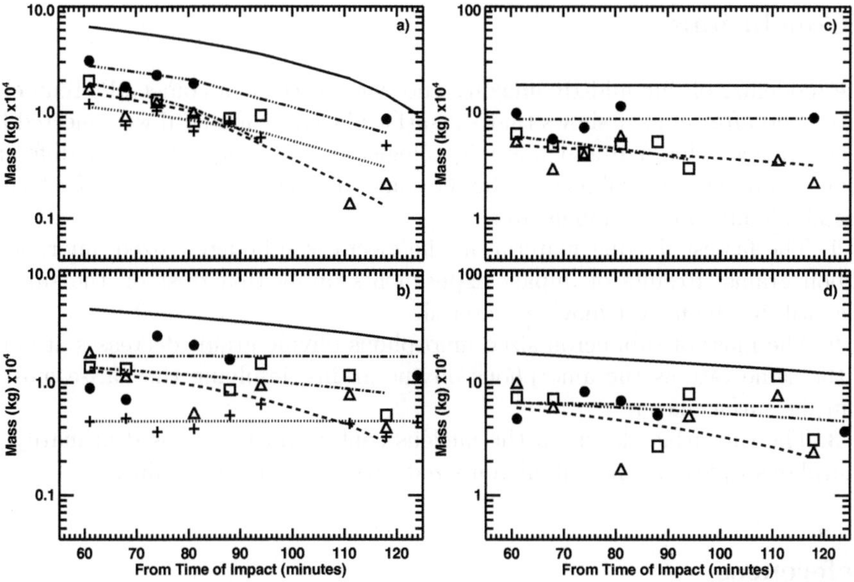

Fig. 2. Panels (**a**) and (**b**) are the mass of grains 0.1 – 1.0 μm in size for spectra extracted centered on the nucleus and 1″ off the nucleus in the direction of the ejected plume, respectively; panels (**c**) and (**d**) are the mass of grains 1.0 – 10.0 μm in size for spectra extracted centered on the nucleus and 1″ off the nucleus in the direction of the ejected plume, respectively. Plotted along with weighted linear fits to the constrained data points is amorphous pyroxene *circles; dashed line*, amorphous olivine *squares; dot-dashed line*, amorphous carbon *triangles; dot-dot-dot-dashed line* and crystalline olivine *"plus" signs; dotted line.* The *solid line* shows the total mass of the grains.

The faster decrease of submicron sized amorphous carbon grain mass in the offset position suggests that the carbon grains are traveling at a greater speed than the amorphous olivine grains. Overall, the amorphous carbon grains seem to be traveling at the highest velocity compared to all other mineral species.

Apparently, grains of different sizes have disperate velocities (Fig. 2a, b, c, d) This figure shows that the mass of the submicron sized grains decreases faster than the mass of the larger grains, in both the center and offset spectra. The observed decrease of the 10 μm silicate feature (caused by the smallest grains in the aggregate population) over the night of impact follows the same trend as depicted in Fig. 2a, b, c, d. This trend suggests that the overall population of the smallest grains are decreasing over time faster than the larger grains. One explanation for this behavior is that the smaller dust grains may be more strongly coupled to material in the temporary volatile gas geyser released as a result of the impact [14].

6 Conclusions

The modeling of our mid-IR imaging and spectroscopy of comet 9P/Tempel 1 before and after impact shows that subsurface layers of material were ejected by Deep Impact. This material was mineralogically and morphologically different from the surface material released by the comet during normal periods of activity. The highlights of our findings are:

1. The fastest moving material in the ejecta is submicron sized amorphous carbon grains. Results of impact experiments imply that dust on the surface layer will be the fastest moving material.

2. The mass of submicron sized amorphous olivine grains decreases at close to the same rate as the amorphous olivine grains, implying they arise mostly from the surface layer.

3. The subsurface layers of the nucleus contain dust composed primarily of amorphous pyroxene and submicron sized crystalline olivine grains.

References

1. Dorschner, J., Begemann, B., Henning, T., Jaeger, C., Mutschke, H. (1995): A&A, **300**, 503
2. Edoh, O. (1983), Ph.D. Thesis, University of Arizona
3. Fabian, D., Henning, T., Jäger, C., Mutschke, H., Dorschner, J., Wehrhan, O. (2001): A&A, **378**, 228–238
4. Harker, D. E., Woodward, C. E., Wooden, D. H. (2005): Science, **310**, 278–280
5. Harker, D. E., Wooden, D. H., Woodward, C. E., Lisse, C. M. (2004): ApJ, **615**, 1081–1081
6. Harker, D. E., Wooden, D. H., Woodward, C. E., Lisse, C. M. (2002): ApJ, **580**, 579
7. Jaeger, C., Molster, F. J., Dorschner, J., Henning, T., Mutschke, H., Waters, L. B. F. M. (1998): A&A, **339**, 904–916
8. Joblin, C., Tielens, A. G. G. M., Geballe, T. R., Wooden, D. H. (1996): ApJ, **460**, L119–L123
9. Kelley, M. S. et al. (2006): ApJ, **651**, 1256
10. Lisse, C. M., et al. (2006): Science, **313**, 635–640
11. Min, M., Hovenier, J. W., de Koter, A., Waters, L. B. F. M., Dominik, C. (2005): Icarus, **179**, 158–173
12. Press, W. H., Teukolsky, S. A., Vetterling, W. T., & Flannery, B. P., (1992) Numerical Recipes. Cambridge Univ. Press, New York
13. Sugita, S., et al. (2006): LPS XXXVIII, **2437**
14. Sugita, S., et al. (2005): Science, **310**, 274–278
15. Wooden, D. H., Woodward, C. E., Harker, D. E. (2004): ApJ, **612**, L77–L80

Release of C_2 Radicals after the Deep Impact Event

R. Schulz[1], J. A. Stüwe[2], C. Erd[3], D. Martin[3], and H. Smit[3]

[1] European Space Agency / ESTEC, Research and Scientific Support Department, P.O. Box 299, NL-2200AG Noordwijk, The Netherlands, rschulz@rssd.esa.int
[2] Sterrewacht Leiden, P.O. Box 9513, NL-2300RA Leiden, The Netherlands, stuwe@strw.leidenuniv.nl
[3] European Space Agency / ESTEC, Science Payloads and Advamced Concepts Office, P.O. Box 299, NL-2200AG Noordwijk, The Netherlands, cerd, dmartin, hsmit@rssd.esa.int

Abstract We present the results of investigating the coma of comet 9P/Tempel 1 in C_2 and continuum, before and after the Deep Impact event of 4 July 2005. A jet-like coma feature was produced by the impact, and its temporal evolution in morphology and brightness has been compared for the dust and gas coma. It shows that the feature remains visible in the dust coma for several days, whereas in the C_2 coma it is visible only on the image taken 15.8 hours after the impact and has vanished when the coma was observed again in the following night. The observational evidence strongly supports that the C_2 in the feature observed on 4 July 2005 was produced from fresh dust particles released by the outburst and forming an extended source for the production of the C_2 radical. This indicates that the disintegration of C_2-bearing dust species can directly be observed during the non-steady state conditions present immediately after outbursts and/or nucleus splitting.

1 Introduction

The Deep Impact mission to comet 9P/Tempel 1 provided the unique opportunity to directly witness an induced activity outburst of a comet nucleus and its consequences on coma morphology and composition [1]. The impact occurred on 4 July 2005, 5:52 UT and led to the release of fresh material from the nucleus into the coma under non-steady-state conditions. Theoretical considerations suggest that the dust ejected from the impact site was most likely fragmented by the energy of the impact process [2]. Observations at the time of the induced outburst revealed that icy grains were ejected from the comet nucleus and rapidly disappeared due to the evaporation of the icy component [3]. Here we present observations, which provide evidence that C_2 was released from the fresh dust grains ejected as a result of the impact.

2 Observations and Data Reduction

Comet 9P/Tempel 1 was observed between 1 July and 6 July 2005, hence before and after the Deep Impact event on 4 July 2005. The ESA 1-m Optical Ground Station (OGS) telescope on Tenerife equipped with the Comet CCD camera was used to obtain images of the coma (field of view: 13.65x13.65 arcmin; pixel size: 0.4x0.4 arcsec) with special narrow-band comet filters showing the comet dust in blue continuum (BC) and red continuum (RC) and in the C_2 (0–0) emission band. The central wavelengths and FWHM of these filters are as follows: BC: 444.0 nm (2.6 nm), RC: 685.6 nm (7.2 nm), C_2: 512.5 nm (12.5 nm). The pre-impact observations were obtained under non-photometric conditions with a seeing of 1.5 arcsec. All other nights were stable with a seeing around 1 arcsec. Table 1 gives an overview of the available images. The narrow-band filter images were flux calibrated (after bias subtraction and flat field correction) by using observations of spectrophotometric standard stars obtained in these filters on each night of observation. The C_2 images were additionally corrected for the underlying continuum in this filter. To search for impact related morphological changes all images obtained after the impact were divided by an image obtained in the same filter before the impact event. The resulting quotient image will not only show any change in coma morphology, but will also provide an estimate of how the coma brightness changed as a result of the impact in each observed component.

Table 1. CCD imaging observations of comet 9P/Tempel 1.

Date 2005	U.T.	τ_{exp} / s	Filter
Jul 1	21:45:07	300	C_2
Jul 1	22:48:21	300	C_2
Jul 2	21:38:40	300	RC
Jul 2	21:46:41	300	BC
Jul 4	21:43:51	900	C_2
Jul 4	22:00:48	900	BC
Jul 4	22:17:57	900	RC
Jul 5	21:38:19	900	C_2
Jul 5	22:01:41	900	BC
Jul 5	22:18:39	900	RC
Jul 6	21:30:42	900	C_2
Jul 6	21:48:24	900	BC
Jul 6	22:05:35	900	RC

3 Results and Conclusions

A jet-like coma structure was produced as a result of the impact. On 4 July, about 16 hours after the impact, a jet structure is visible in the images obtained

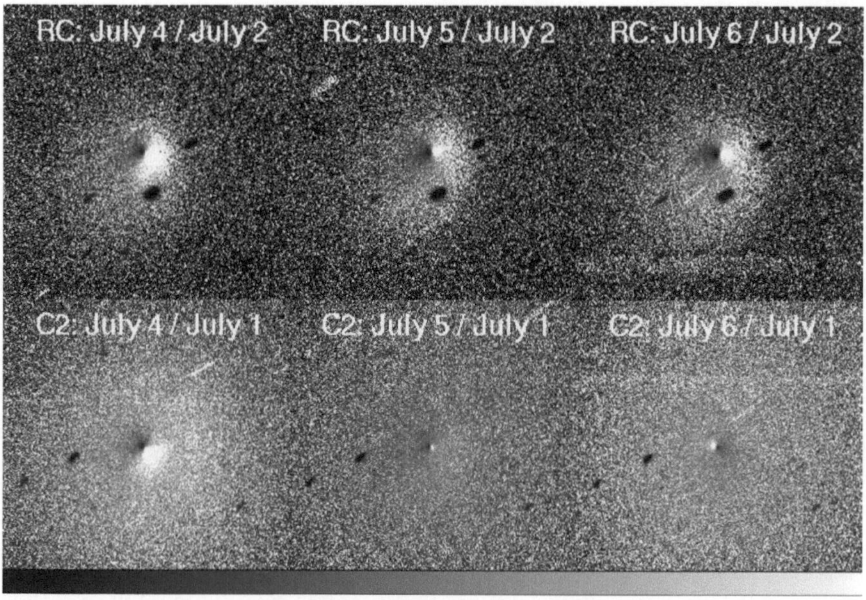

Fig. 1. Quotient images of comet 9P/Tempel 1 in C_2 and red continuum. Images obtained on 3 consecutive nights after the impact are devided by a pre-impact image from 1st July (for C_2) and 2nd July (RC).

in the RC and BC filters (narrow-band red and blue continuum) and in the C_2 filter. The brightness of this structure peaks at a nucleus distance of $\approx 10\,000$ km, but is discernable up to a distance of $\approx 20\,000$ km diluting into the background. Figure 1 shows the evolution of the structure in red continuum and C_2 from 4 to 6 July. It is obvious that in the dust coma, the jet structure is persisting from 4 to 6 July, although it becomes significantly fainter and the peak of brightness moves closer to the nucleus, to a distance of ≈ 7000 km on 6 July. Therefore, dust particles must have been ejected into this jet structure for at least 3 days after the impact occurred. In the C_2 coma the jet structure is only visible on 4 July, indicating that either a C_2 jet was ejected by the outburst which switched off at least 3 hours before the image of 5 July was obtained or the C_2 in the jet visible on 4 July was produced from the fresh material released by the outburst. Although it is not possible to distinguish between these two scenarios from the observational evidence, it seems rather unlikely that C_2 radicals are ejected directly from the nucleus. The C_2 should have come from a parent species, which may be a volatile or dust grains. Consequently, the impact led to the ejection of C_2-bearing volatiles and/or C_2-bearing dust particles which have different properties under steady state conditions. This conclusion is further strengthened by the fact that on 5 and 6 July the dust jet is still visible, while the jet has vanished, meaning that the dust that has been released on these dates does not contain a detectable amount of C_2-bearing species or is not heated to the temperature necessary to release the C_2. The latter would be consistent with the suggestion by Hughes

([2]) that the dust ejected directly after the impact has been broken up by the energetic impact process.

References

1. A'Hearn, M. F. and 32 additional authors (2005): Science, **310**, 258–264
2. Hughes, D. W. (2006): MNRAS, **365**, 673–676
3. Schulz, R., Owens, A., Rodriguez-Pascual, P. M., Lumb, D., Erd, C., and Stüwe, J. A. (2006): A.& A., **448**, L53–L56

A Search for Deep Impact's Large Particle Ejecta

M. S. Kelley,[1] W. T. Reach[2], and C. E. Woodward[1]

[1] Department of Astronomy, University of Minnesota, 116 Church St SE, Minneapolis, MN, 55455 msk@astro.umn.edu
[2] Spitzer Science Center, MS 220-6, California Institute of Technology, Pasadena, CA 91125

Abstract The *Deep Impact* encounter with the nucleus of 9P/Tempel ejected small grains ($a \lesssim 10$ μm) into the comet's coma, evidenced by thermal emission from small dust grains at mid-infrared wavelengths ($\lambda \sim 10$ μm) and dynamical simulations of optical images. Meteor-sized particles ($a \gtrsim 100$ μm) ejected by the impact will likely have the lowest ejection velocities and will weakly interact with solar radiation pressure. Therefore, large particles may remain near the nucleus for weeks or months after ejection by *Deep Impact*. We present initial highlights of our *Spitzer Space Telescope*/MIPS 24 μm camera program to image comet 9P/Tempel at 30, 80, 420, and 560 days after the *Deep Impact* encounter. The MIPS data, combined with our dynamics model, enable detection of large dust grains potentially ejected by *Deep Impact*.

1 Introduction

The *Deep Impact* impactor released an abundance of sub-micron sized particles into the coma of 9P/Tempel. Mid-infrared spectra of the ejecta displayed a strong silicate emission feature, including crystalline silicates, indicating the presence of small particles [3, 12]. Ground-based images of the ejecta showed dust that strongly interacted with radiation pressure, again implying small grains [11]. This small particle ejecta plume was optically thick, apparent from a shadow cast by the ejecta onto the comet nucleus [1]. Though the impact site was available for observation by the *Deep Impact* flyby spacecraft for 800 s after impact, the crater was wholly obscured by the optically thick ejecta. Had the crater been observed, a size estimate of the crater, and therefore, an estimate of the total ejecta mass, may have been directly determined. Instead, we must estimate the total ejecta mass from observations of the ejecta itself. The gas and small particle dust (0.1–10 μm) masses have been constrained to be of order 10^5–10^6 kg [8, 6, 3, 12], but the mass in larger sized dust grains ($\gtrsim 100$ μm) is unknown. If the grains were ejected with typical comet grain size distributions ($dn/da \sim a^{-3.5}$), the large dust mass could dominate the total ejected dust mass.

The particle size distribution may have strongly favored small particles; therefore, the surface brightness of large grains would be faint and difficult to detect in contrast with the smaller, more abundant, grains. Schleicher et al. [11], through dynamic modeling of ground-based optical images, estimated that the effective radii of ejected particles ranged from 0.25 to 1.25 μm with ejection velocities of 0.13–0.23 km s^{-1}. Dynamical simulations of larger particles did not agree with the observations, yet this does not preclude their existence since they may be many times less abundant than the small particles. Moreover, there is some evidence for small and large particle segregation in the ejecta [13], likely due to the ejection of larger particles with lower ejection velocities. If the ejected grain size distribution extends beyond 10–100 μm, then for $v_{ej} < 0.1$ km s^{-1} the largest particles could remain near the nucleus for days, or even weeks after impact. High sensitivity, long temporal baseline observations, beyond the immediate aftermath of *Deep Impact*, are necessary to observe any large dust grains generated by crater formation processes. The slowest particles likely remained gravitationally bound to the nucleus and fell back to the surface, therefore, such observations will only provide a lower limit to the total excavated mass. We present dynamical simulations and *Spitzer*/MIPS images to search for large dust grains in the *Deep Impact* ejecta.

2 Observations and Models

We initiated an observing program with the *Spitzer Space Telescope* [14] to observe comet 9P/Tempel with the MIPS 24 μm camera [10] at 30, 80, 420, and 560 days after the *Deep Impact* encounter. Two observations, obtained on 2005 Jul 31 and on 2005 Sep 23, have been processed and analyzed (see Table 1 and Fig. 1). The images span a 22′ × 42′ area surrounding the nucleus and dust trail. The dust trail is visible in both images as the thin, isolated, linear feature extending from the inner coma to the northwestern edges.

Table 1. *Spitzer*/MIPS 24 μm post-*Deep Impact* observations of comet 9P/Tempel.

Time	ΔT	r_h	Δ_s	Phase Angle	
Date	(UT)	(days)[a]	(AU)[b]	(AU)[a]	(degrees)
2005 Jul 31	04:14	26.96	1.53	0.87	40
2005 Sep 23	21:00	81.66	1.68	1.31	38
2006 Sep 01	08:25	424.11	3.52	3.15	17
2007 Jan 15[c]	00:00	560	4.03	3.55	13

[a]Time elapsed since impact.
[b]Comet-*Spitzer* distance.
[c]Estimated date of observation.

Images of comet 9P/Tempel following the collision with *Deep Impact* showed a short lived plume of high β particles that quickly dispersed from the vicinity of the comet [9, 6, 11]. Schleicher et al. [11] suggest (based on images 6 days after

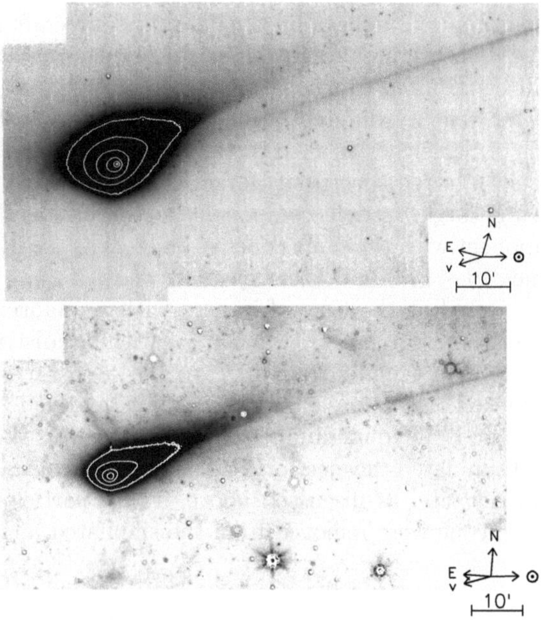

Fig. 1. *Spitzer*/MIPS 24 μm images of comet 9P/Tempel. The lowest contour is placed at 3 MJy sr^{-1}, and the contours are spaced at factor of 3 intervals. Celestial N and E, the projected sunward direction (⊙), and comet 9P's projected velocity vector (v) are provided for each image. The comet's natural dust trail is seen extending from the comet in the negative velocity direction.

impact) that very slow moving particles, $v_{ej} < 0.001$ km s^{-1}, may exist. To predict the location of the large, slow moving, *Deep Impact*-ejected dust grains in the MIPS images, we will work under the assumption that the high β ejecta plume has a low β component, which was ejected in the same manner, but with lower velocities. The simulations will be created with our Monte-Carlo dynamical model [7].

Schleicher et al. [11] reproduced the ejecta plume using particles with $\beta \approx 0.24$–1.2 ($a \approx 0.25$–1.25 μm) and ejection velocities of 0.13–0.23 km s^{-1}. Their best model ejected particles in a cone with an opening angle of 70° and a wall thickness of 30°. The cone was centered at a position angle of 255° and a "phase angle" of 70° (0° is directed toward the Earth, 180° is directly away). We begin modeling Schleicher et al.'s ejection plume with these parameters and verify the parameters by comparison to their images; however, with our model, the ejecta position angle of 225° reported by Sugita et al. [12] produced better agreement with the observations. The simulated images (not shown) exhibit a limb brightened cone at $t_0 + 0.97$ days (t_0 being the time of impact). The limb brightening persists through each of the time steps. Limb brightening is not evident in the images of Schleicher et al. [11]. Although optical images from the *Hubble Space Telescope* (t_0 + 1–2 hrs) [2], mid-IR images from the Subaru Telescope

($t_0 + 3$ hrs) [12], and optical images by the *Rosetta* spacecraft ($t_0 + 21$ hrs) [6] show what may be limb brightening in the ejecta plume. Further modeling and interpretation will be required to resolve this discrepancy.

The evidence for velocity differentiation in the ejected dust is apparent in the optical images of Schleicher et al. [11], the mid-infrared images of Sugita et al. [12], and the mid-infrared spectra of Harker et al. [4, 5]. We posit that there exits a slow moving, large particle component to the ejecta. We modeled the slow moving component with the same cone parameters as the high β grains, but with lower β-values ($\beta \leq 0.1$) and lower ejection velocities ($v_{ej} \leq 0.1$ km s^{-1}). The particles are ejected at the time of impact with a uniform distribution of velocities (between 0 and 0.1 km s^{-1}) and a logarithmic distribution of grain β values ($dn/d \log \beta \propto 1$) ranging from 10^{-4} to 10^{-1} ($a \approx 6 - 6000$ µm). The logarithmic distribution of β-values ensures all grain size decade bins are equally represented (i.e., there is an equal number of 10 µm grains as there are 1000 µm grains). The model simulated the ejection of 10^6 particles in a 180° cone centered on the derived impact site, as discussed above. Those particles outside of the Schleicher et al. [11] cone were removed from the simulated images.

3 Discussion

Synthetic images of the simulations, given a specific observer geometry, predict the location of large particles that may have been ejected as a result of the *Deep Impact* collision. We examined the MIPS images from July and September 2005 (Fig. 1) to determine if slow moving, large dust grains are lingering in the vicinity of the comet. Figure 2 presents sub-sections of the two MIPS images and thermal emission contours (isothermal grains) from the large particle simulations, assuming an ejected particle size distribution of $dn/da \propto a^{-3.5}$. The MIPS images have been unsharp masked to enhance the coma asymmetries. In the 2005 July MIPS image, there is a coma feature extending in the anti-solar direction that is coincident with the contours of the simulation. These particles have β values ranging from 10^{-3} to 10^{-1} (lower β values are found closer to the nucleus). The 2005 September MIPS image shows the same coma feature (in the anti-solar direction), but the simulated *Deep Impact* dust has rotated away from the anti-solar direction. We conclude that the anti-solar feature is a young, normal coma feature and not due to *Deep Impact* liberated dust. The limb brightened cone extending to the lower-right from the nucleus is comprised of high velocity particles with $\beta \lesssim 10^{-3}$ and does not appear to be present in the MIPS images. Possibly, the ambient coma is obscuring the *Deep Impact* ejected dust. Observations and simulations of comet 9P in September 2006 and January 2007 (when the coma emission has diminished significantly) will be required to assess the existence of the largest particles ($\beta = 10^{-4}$–10^{-3}).

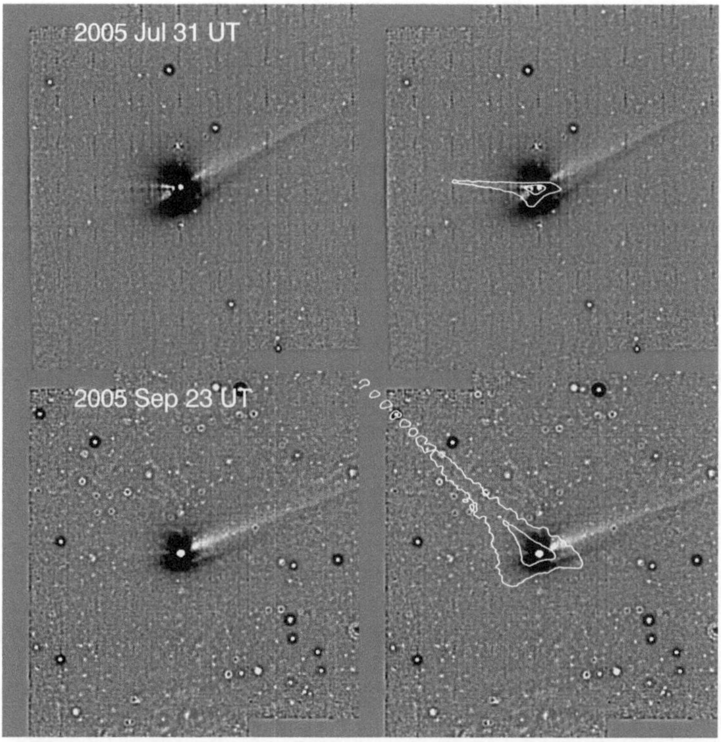

Fig. 2. *Left: Spitzer*/MIPS images of comet 9P/Tempel enhanced with an unsharp mask to show asymmetries in the comet coma (20′ field-of-view). *Right:* Contours from our large particle simulations are overlayed on the enhanced MIPS images to show the possible locations of *Deep Impact* liberated dust at these epochs.

4 Conclusions

The *Deep Impact* collision with comet 9P/Tempel ejected small particles from the comet nucleus, yet the upper size limit, and therefore total ejected dust mass, is observationally not well constrained. Large slow moving grains ($a \gtrsim 100$ μm) in the ejecta plume (if they exist) have yet to be discovered. We simulate observations of large particles ejected by *Deep Impact* to predict the location of grains in *Spitzer*/MIPS images of comet 9P/Tempel at 30, 80, 420, and 560 days after the *Deep Impact* encounter. The larger particles may be observed as an enhancement to comet 9P/Tempel's trail in MIPS images 420 and 560 days after *Deep Impact.*

Acknowledgements This work is based on observations made with the *Spitzer Space Telescope*, which is operated by the Jet Propulsion Laboratory, California Institute of Technology under a contract with NASA. Support for this work was provided by NASA through an award issued by JPL/Caltech.

References

1. A'Hearn, M. F., and 32 colleagues 2005. Deep Impact: Excavating Comet Tempel 1. Science 310, 258–264.
2. Feldman, P. D., McCandliss, S. R., Route, M., Weaver, H. A., A'Hearn, M. F., Belton, M. J. S., Meech, K. J. 2006. Hubble Space Telescope Observations of Comet 9P/Tempel 1 during the Deep Impact Encounter. Icarus, in press (arXiv:astro-ph/0608487)
3. Harker, D. E., Woodward, C. E., Wooden, D. H. 2005. The Dust Grains from 9P/Tempel 1 Before and After the Encounter with Deep Impact. Science 310, 278–280.
4. Harker, D. E., Woodward, C. E., Wooden, D. H., Fisher, S., Trujullio, C. 2006. Gemini-N Mid-IR Observations of the Dust Properties of the Ejecta from Comet 9P/Tempel 1 Excavated During Deep Impact. Icarus, 190, 432–453.
5. Harker, D. E., Woodward, C. E., Wooden, D. H., this volume
6. Keller, H. U., and 11 colleagues 2005. Deep Impact Observations by OSIRIS Onboard the Rosetta Spacecraft. Science 310, 281–283.
7. Kelley, M. S. 2006. The Size, Structure, and Mineralogy of Comet Dust. Ph.D. Thesis.
8. Küppers, M., and 40 colleagues 2005. A large dust/ice ratio in the nucleus of comet 9P/Tempel 1. Nature 437, 987–990.
9. Meech, K. J., and 208 colleagues 2005. Deep Impact: Observations from a Worldwide Earth-Based Campaign. Science 310, 265–269.
10. Rieke, G. H., and 42 colleagues 2004. The Multiband Imaging Photometer for Spitzer (MIPS). Astrophysical Journal Supplement Series 154, 25–29.
11. Schleicher, D. G., Barnes, K. L., Baugh, N. F. 2006. Photometry and Imaging Results for Comet 9P/Tempel 1 and Deep Impact: Gas Production Rates, Postimpact Light Curves, and Ejecta Plume Morphology. Astronomical Journal 131, 1130–1137.
12. Sugita, S., and 22 colleagues 2005. Subaru Telescope Observations of Deep Impact. Science 310, 274–278.
13. Sugita, S., and 13 colleagues 2006. A High-Resolution Mid-IR Observation of the Collision Between Deep Impact Projectile and Comet 9P/Tempel 1. 37th Annual Lunar and Planetary Science Conference 37, 2431.
14. Werner, M. W., and 25 colleagues 2004. The Spitzer Space Telescope Mission. Astrophysical Journal Supplement Series 154, 1–9.

Subaru/COMICS Mid-Infrared Spectroscopic Observations of the Dust Plume from Comet 9P/Tempel

T. Ootsubo[1], S. Sugita[2], J. Watanabe[3], M. Honda[4], H. Kawakita[5], T. Kadono[6], and R. Furusho[7]

[1] Graduate School of Science, Nagoya University, Furo-cho, Chikusa-ku, Nagoya, Aichi 464-8602, Japan, ootsubo@u.phys.nagoya-u.ac.jp
[2] Department of Complexity Science and Engineering, University of Tokyo, Kashiwa, Chiba 277-8561, Japan, sugita@k.u-tokyo.ac.jp
[3] National Astronomical Observatory of Japan, 2-21-1 Osawa, Mitaka, Tokyo 181-8588, Japan, jun.watanabe@nao.ac.jp
[4] Institute of Space and Astronautical Science, Japan Aerospace Exploration Agency, Sagamihara, Kanagawa 229-8510, Japan, hondamt@ir.isas.jaxa.jp
[5] Department of Physics, Kyoto Sangyo University, Kita-ku, Kyoto 603-8555, Japan, kawakthd@cc.kyoto-su.ac.jp
[6] Center for Deep Earth Exploration, Japan Agency for Marine-Earth Science and Technology, Yokohama, Kanagawa 236-0001, Japan, kadono@impact.k.u-tokyo.ac.jp
[7] Waseda University, 1-6-1 Nishiwaseda, Sinjuku, Tokyo 169-8050, Japan, furshork@cc.nao.ac.jp

Abstract We carried out mid-infrared observations of the Deep Impact (DI) collision with a Jupiter-family (JF) comet, 9P/Tempel, using the Cooled Mid-Infrared Camera and Spectrometer (COMICS) on the 8.2 m Subaru Telescope. The N-band spectrum (8–13 μm; $R \sim 250$) of the dust ejecta from the comet nucleus 3.5 hours after the DI event shows a strong crystalline silicate feature. The Subaru/COMICS observation also reveals that small (< 1 μm) amorphous pyroxene and carbon-rich grains flew away in the forefront of the ejecta plume before our spectroscopic observation, leaving behind relatively large submicron (0.8–1 μm) crystalline particles that we subsequently observed. The submicron grains within the subsurface layer of comet 9P/Tempel's nucleus have a higher crystalline-to-amorphous silicate ratio than that in the comae of other JF comets, and even that of Oort cloud comets. Near the comet nucleus 3.5 hours after the impact the crystalline-to-amorphous ratio is similar to that of Hale-Bopp coma dust at 0.97 AU.

1 Introduction

Long-period Oort cloud (OC) comets are thought to have been born in and ejected from the Jupiter-Uranus region, whereas short-period Jupiter-family (JF) comets originated in the trans-Neptunian region [1, 2]. The information on the dust grain properties, such as grain size and mineralogical composition of these two groups, which are believed to have different origins, provides insight into dust grain evolution and planetary formation in the early solar nebula. While there have been many observations of dust in comet comae, we have little information on the dust within a comet nucleus itself. The interior structure of comets are still unknown.

NASA's Deep Impact (DI) mission successfully excavated fresh cometary material (gas, ice, and dust) underneath a presumable refractory crust on the surface of a JF comet, 9P/Tempel [3]. Observations of such fresh and pristine materials from the subsurface of a comet are highly valuable for studying the formation of comet nuclei and the origin of the Solar System. Mid-infrared (mid-IR) observations can be used to constrain fundamental properties of ejecta dust from comet 9P/Tempel after the impact. Since equilibrium thermal radiation from the comet has its peak around 10 μm and silicate grains have a number of key diagnostic spectral features in the mid-IR region, spectroscopic observation in the mid-IR wavelength region is highly valuable. Thus, we carried out mid-IR observations (multi-band imaging and N-band spectroscopy) with the Subaru telescope from the summit of Mauna Kea. The main results of our imaging observations and spectral energy distribution (SED) analyses about the cratering process by the DI impact are already reported [4, 5], and further results are presented by [6] in this volume. Preliminary results of the spectroscopy by Subaru Telescope are given by [4, 7]. The Gemini/MICHELLE team also conducted mid-IR spectroscopic observations at Mauna Kea and derived grain properties of comet 9P/Tempel [8]. Spitzer Space Telescope (SST) also observed the DI event in the mid-IR wavelength region and revealed the compositional properties of the ejecta dust grains [9]. Although SST covers a wider wavelength region (5–35 μm) than Subaru, the higher spatial resolution because of the large (8.2 m) telescope of the Subaru surpasses that of mid-IR space telescopes. Thus, we will study the difference of the ejecta dust properties at different positions from the comet nucleus and derive information on the properties of the pristine subsurface dust grains and the structural properties of the comet nucleus, utilizing the advantages of high spatial resolution.

2 Observation and Results

We used the Cooled Mid-Infrared Camera and Spectrometer (COMICS), which is mounted on the 8.2 m Subaru Telescope on Mauna Kea, to observe the comet before and after the impact. The details of the COMICS instrument are given by [10, 11, 12]. The observation was conducted as a part of the Subaru-Gemini collaboration at the Mauna Kea summit, Hawaii. COMICS has the capability of imaging and long-slit spectroscopy in both of the two atmospheric windows, N- (7.5–13.5 μm) and Q-bands (16–25 μm). N-band low-resolution spectroscopic observations (8–13 μm; $R \sim 250$) with the $0''.33$ wide and $3''$ long slit were conducted 3.5 hours after the impact (4 July 09:00–09:14 UT). We also made spectroscopic observations 24 hours before (3 July UT) and 28 hours after (5 July UT) the DI impact event. Since the first image of comet 9P/Tempel after the impact showed a fan-shaped plume with a symmetry line at 225° position angle (PA) from the comet nucleus [4], we set the slit position along the direction of this symmetry line (225° PA) during the spectroscopic observation on 4 and 5 July UT. We selected κ Vir (HD 124294) [13] as the calibration star for the correction of the atmospheric absorption and the flux standard star for aperture photometry.

For the analysis we extract a $0.33'' \times 4.29''$ (2×26 pixels; $0''.165$/pixel for spectroscopy) region, which has a significant signal from the thermal emission of ejecta dust. We used a narrower slit than Gemini/MICHELLE (i.e., $0.6'' \times 1.0''$ extraction aperture) for the derivation of the spectrum [8]. In order to investigate the difference in the dust properties as a function of the distance from the nucleus, the slit position of the south-west side from the comet nucleus is divided into three regions, SW1 (4 pixels), SW2 (5 pixels) and SW3 (10 pixels) from near the comet nucleus to south-west, and we derived the spectrum for each region (the detailed slit position is shown in the figures of [7, 14]). The spectra at the south-west side of the nucleus (SW1–SW3), which is the side on which the DI impact occurred, exhibits a clear crystalline silicate signature, whereas the north-east side spectrum does not. This result indicates that the crystalline feature in the spectrum can be attributed to the dust ejected by the DI impact (Fig. 1).

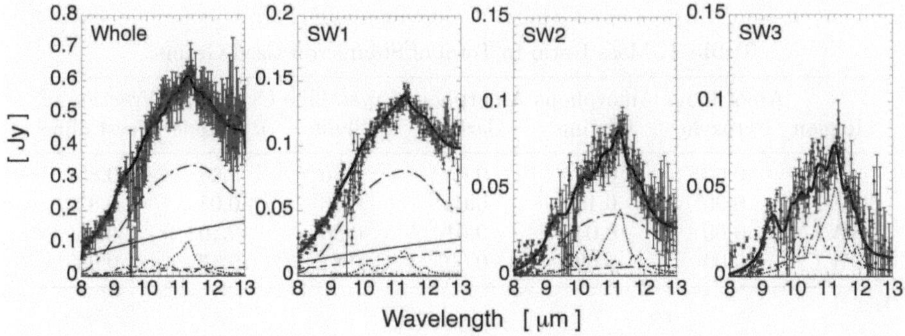

Fig. 1. Observed N-band spectra with the best-fit thermal emission model SEDs. The thick solid line is the resultant best-fit model spectrum, which is the sum of comet nucleus contribution (*thin solid line*), amorphous olivine (*long dashed-dotted line*), amorphous carbon (*dashed line*), crystalline olivine (*dotted line*), and crystalline pyroxene (*dash-dotted line*). Note that there are still a little uncertainties of absolute flux calibrations between the spectra. There is an effect of telluric ozone absorption in 9.4–9.8 μm region.

The observed spectrum in each region is examined using a thermal emission model of the cometary dust grains in order to characterize the grain size and mineralogical composition (including crystallinity) of ejecta dust grains. The method used here is the same as our previous studies [7, 15] and similar to the techniques used by previous authors [8, 16, 17]. In this thermal emission model, the grain size distribution (Hanner size distribution), porosity (fractal, porous dust grains composed of vacuum inclusions in a matrix are considered as cometary aggregate particles), and five grain components (amorphous olivine and pyroxene, amorphous carbon, and crystalline olivine and pyroxene) are taken into account. The χ^2 fitting for this ejecta dust spectrum has been conducted with the Levenberg-Marquardt minimization. More detailed descriptions about this thermal emission model are given by [14, 15].

For the spectrum at the region SW1, the best-fit model spectrum is reproduced by a thermal model with a peak grain size $a_p = 0.8$ µm and slope $N = 3.9$ ($\chi^2_\nu = 1.85$). For SW2 the best-fit parameters are $a_p = 1.0$ µm and $N = 5.0$ ($\chi^2_\nu = 0.891$), and for SW3 $a_p = 0.8$ µm and $N = 6.0$ ($\chi^2_\nu = 1.09$). The fractal dimension for the amorphous materials D is 2.857 (slightly porous) for all regions. For the ejecta dust within the "whole" slit ($0.33'' \times 4.29''$), the best-fit model spectrum is reproduced by a grain size distribution with a peak $a_p = 0.8$ µm and slope $N = 3.7$. We list the mass fraction of submicron grains (0.1–1.0 µm) for each mineral and the fraction of crystalline in the submicron silicate grains ($f_{\rm crystal}$) in Table 1. The main results derived from the modeling analysis are: (1) the depletion of amorphous pyroxene and amorphous carbon grains, and (2) significantly high $f_{\rm crystal}$ values of silicates in the ejecta plume observed by the Subaru/COMICS.

Table 1. Mass Ratio to Total of Submicron-sized Grains.

Region	Amorphous Pyroxene	Amorphous Olivine	Amorphous Carbon	Crystalline Olivine	Crystalline Pyroxene	Fraction of Crystalline
whole	0.00	0.11	0.01	0.80	0.08	0.88
SW1	0.00	0.18	0.03	0.76	0.03	0.82
SW2	0.00	0.04	0.01	0.85	0.10	0.96
SW3	0.00	0.02	0.00	0.81	0.17	0.98

3 Discussion

For our best-fit spectrum, amorphous pyroxene is not needed and amorphous carbon hardly contributed to the total spectrum. The ejected dust grains are slightly fractal porous with $D = 2.857$ and have the peak size around 0.8–1 µm in all the regions. It is likely that subsurface grains in the nucleus of comet 9P/Tempel are porous aggregate particles intrinsically. From the Gemini/MICHELLE observation, it is found that the peak of the size distribution increased from 0.3 µm (1 hour after the impact) to 0.5 µm (1.8 hours after the impact) [8]. The peak of the Hanner size distribution a_p in our model fit is larger than that of Gemini/MICHELLE results. Our spectroscopy was carried out 3.5 hours after the impact, while their spectra were obtained 1.0 and 1.8 hours after the impact. This is indicative of the smaller grains having moved to outside the slit at the time of our spectroscopic observation. In the Gemini spectra, the relative masses of amorphous pyroxene and amorphous carbon have decreased substantially compared with those of amorphous and crystalline olivine between their two observations (0.411–0.106 for amorphous pyroxene and 0.212–0.120 for amorphous carbon). Based on the Gemini results of their analysis, it is natural that the "whole" spectrum of the COMICS observation can be reproduced by a mixture of amorphous olivine and crystalline materials. It is likely that the high-speed ejecta contains much more small amorphous carbon and amorphous pyroxene grains than slower ejecta dust. Polarimetric Imager for COmets (PICO) on the Lulin One-meter Telescope in Taiwan also observed the DI ejecta dust [18]. They report that an arc-shaped higher polarized region was recognized in the PICO polarization map, and dust grains that existed in this higher polarization region had larger expansion velocity than the grains representative of the intensity map. They suggest that small carbonaceous grains might produce such a higher polarization region. The further result by the Gemini/MICHELLE observation also suggests that faster decrease of small grains in the extended ejecta region implies small grains (< 1 μm) are traveling faster through the Gemini aperture than larger grains (> 1 μm) [19]. Combining these observational results together, it can be concluded that small amorphous pyroxene and carbon-rich grains flew away before the COMICS spectroscopic observation, and the relatively large submicron (0.8–1.0 μm) crystalline particles are dominant within the slit at the time of Subaru/COMICS spectroscopic observation.

In SW1–SW3 regions submicron grains in the ejecta have significantly high f_{crystal} values of silicates (see Table 1). This implies the existence of subsurface crystalline silicate grains within the comet nucleus. The fraction of crystalline in the submicron silicate grains (f_{crystal}) derived here is larger than the result by the Gemini observations [8]. The difference in the observation time and in the slit width and direction leads us to infer that our result shows the properties of subsurface dust grains ejected from the deeper interior of the nucleus. More detailed discussion on the difference between surface (amorphous pyroxene and carbon-rich) and subsurface (crystalline silicate-rich) materials of the comet nucleus will be given by [6, 14]. The submicron silicate grains within the subsurface layer have a f_{crystal} value higher than that of OC comets comae dust. In the SW1 region, the slope of the size distribution and the f_{crystal} have similar values to those of the coma of the highly active comet Hale-Bopp at 0.97 AU [20]. In the region SW2 and SW3, the value is larger than in SW1, and most of the dust grains are in crystalline form. These results imply the possibility that JF comets and OC comets have basically the same origin, or that 9P/Tempel is not the classical Kuiper-belt objects but is one of the scattered disk bodies which are thought to have formed in the giant planet region [21, 22].

Acknowledgment This research is based on the data collected at the Subaru Telescope, which is operated by the National Astronomical Observatory of Japan (S05A-042 and S05A-OT11). We are grateful to the staff members of Subaru Telescope. T.O. and H.K. are supported in part by a Grant-in-Aid for Scientific Research (17740119, 18740101) from the Ministry of Education, Culture, Sports, Science and Technology of Japan.

References

1. Stern, S. A., (2003): Nature, **424**, 639
2. H. F. Levison, H. F., & Morbidelli, A., (2003): Nature, **426**, 419
3. A'Hearn, M. F., et al. (2005): Science, **310**, 258
4. Sugita, S., et al. (2005): Science, **310**, 274
5. Meech, K. J., et al. (2005): Science, **310**, 265
6. Kadono, T.,et al.: these proceedings
7. Ootsubo, T., et al., accepted (2006) : In: *Dust in Planetary Systems*
8. D. E. Harker et al. (2005): Science, **310**, 278
9. Lisse, C. M., et al. (2006): Science, **313**, 635
10. Kataza, H., et al. (2000): SPIE, **4008**, 1144
11. Okamoto, Y. K., et al. (2003): SPIE, **4841**, 169
12. Sako, S., et al. (2003): SPIE, **4841**, 1211
13. Cohen, M. et al. (1999): Astron. J., **117**, 1864
14. Ootsubo, T., et al.: Planetary and Space Science, **55**, 1044–1049
15. Ootsubo, T., et al. (2006): Planetary and Space Science, accepted
16. Harker, D. E., et al. (2002): Astrophys. J., **580**, 579
17. Wooden, D. H., et al. (2004): Astrophys. J., **612**, 77
18. Furusho, R., et al. (2007): Icarus, **190**, 454–458
19. Harker, D. E., et al.: these proceedings
20. Harker, D. E., et al. (2004): Astrophys. J., **615**, 1081
21. Gladman, B. (2005): Science, **307**, 71
22. Morbidelli, A. (2005): astro-ph/0512256

Modeling of the Terminal Velocities of the Dust Ejected Material by the Impact

M. Rengel[1], M. Küppers[1], H. U. Keller[1], and P. Gutiérrez[2]

[1] Max-Plack-Institut für Sonnensystemforschung, Max-Planck-Strasse 2, 37191, Katlenburg-Lindau, Germany. rengel@mps.mpg.de
[2] Instituto de Astrofísica de Andalucia-CSIC.C/Camino Bajo de Huétor, 50, 18008, Granada, Spain.

Abstract We compute the distribution of velocities of the particles ejected by the impact of the projectile released from NASA Deep Impact spacecraft on the nucleus of comet 9P/Tempel 1 on the successive 20 h following the collision. This is performed by the development and use of an ill-conditioned inverse problem approach, whose main ingredients are a set of observations taken by the Narrow Angle Camera (NAC) of OSIRIS onboard the Rosetta spacecraft, and a set of simple models of the expansion of the dust ejecta plume for different velocities. Terminal velocities are derived using a maximum likelihood estimator.

We compare our results with published estimates of the expansion velocity of the dust cloud. Our approach and models reproduce well the velocity distribution of the ejected particles. We consider these successful comparisons of the velocities as an evidence for the appropriateness of the approach. This analysis provides a more thorough understanding of the properties of the Deep Impact dust cloud.

1 Introduction

In order to investigate the comet's interior, on 4 July 2004 the NASA mission Deep Impact fired a projectile of 364 kg into the nucleus of 9P/Tempel 1. A crater was formed and material was ejected from the comet [1]. OSIRIS, the scientific camera system on the Rosetta spacecraft [2], was activated five days before the Deep Impact event, and observed 9P/Tempel 1 near-continuously for more than two weeks [3]. Some properties of the cloud (e.g. structure, morphology, number of water molecules produced in the impact, abundance ratio between the CN parent molecules and water) are already reported in the literature.

The main aim of this work is to present the velocity distribution of the ejected dust computed with a discrete linear inverse approach. The Narrow Angle Camera (NAC) of OSIRIS monitored the cometary dust of 9P/Tempel 1, and by aperture photometry on the images, light curves for different circular fields provide the observational input to our approach. A model of the expansion of the dust ejecta plume for different velocities supplies the modeled light curves.

Because of the good temporal parameter sampling that the OSIRIS data offer (into around an image per minute), this approach is very well suitable. The comparison between these results and those derived from other authors not only tests and refines the conclusions derived from other observations and methods, but also provides an evidence for the appropriateness of our models and of our approach, and furthermore, allow to test of additional production of the material.

2 Observational Data and Models

The observational data consist of 73 images of 9P/Tempel 1 taken by the NAC OSIRIS with the clear filter, from the impact time to around 20 h after impact time (for a detailed description of the OSIRIS observations, see [4]). By cometocentric circular aperture photometry on the images of 9P/Tempel 1, the brightness of the comet is integrated over a set of nine apertures of different radii (from 3000 to 15000 km, 1 pixel = 1500 km). An observational data row matrix O is defined, and contains the collection of integrated brightness (73 images $\times$ 9 apertures = 657 data-points):

$$O = [\mathrm{O}_1,\, \mathrm{O}_2,...,\mathrm{O}_{73},...,\mathrm{O}_{657}].$$

The measurement error row matrix E of the observed brightness (or the known individual standard deviation matrix) is defined by:

$$E = [\sigma_1,\, \sigma_2,...,\sigma_{73},...,\sigma_{657}].$$

The models of the expansion of the dust ejecta plume are computed using the decay of the flux when the material leaves the different apertures. For early times (less than 20 h), we assume: (1) that the expanding dust cloud moves with a constant velocity, (2) a neglected effect of the radiation pressure removing the material, (3) and that the dust reaches its final velocity instantly at the time of the impact. We compute the fluxes of the dust ejecta plume for different projected ejecta velocities v, from 1 to 600 ms^{-1}, with a steep i of 50 ms^{-1}. We adopt a maximum velocity of 600 ms^{-1}, although a small fraction of the dust might be actually be ejected at higher velocities [1]. A conservative minimum velocity of 1 ms^{-1} is adopted. The models are normalized in flux by multiplying each modeled flux by a normalization constant. The generated normalized models are stored in a matrix M:

$$M = \begin{bmatrix} M_{1,1} & M_{1,2} & ... & M_{1,657} \\ M_{2,1} & M_{2,2} & ... & M_{2,657} \\ & & & \\ M_{13,1} & M_{13,2} & ... & M_{13,657} \end{bmatrix}$$

where the first sub-index j of each element denotes the velocity bin i and the second sub-index k the modeled brightness point. As an illustration, Fig. 1 shows the lightcurves of the cometary dust for nine circular fields from 2 to 10 pixels, and models with projected velocities of 110, 160 and 300 m s^{-1}.

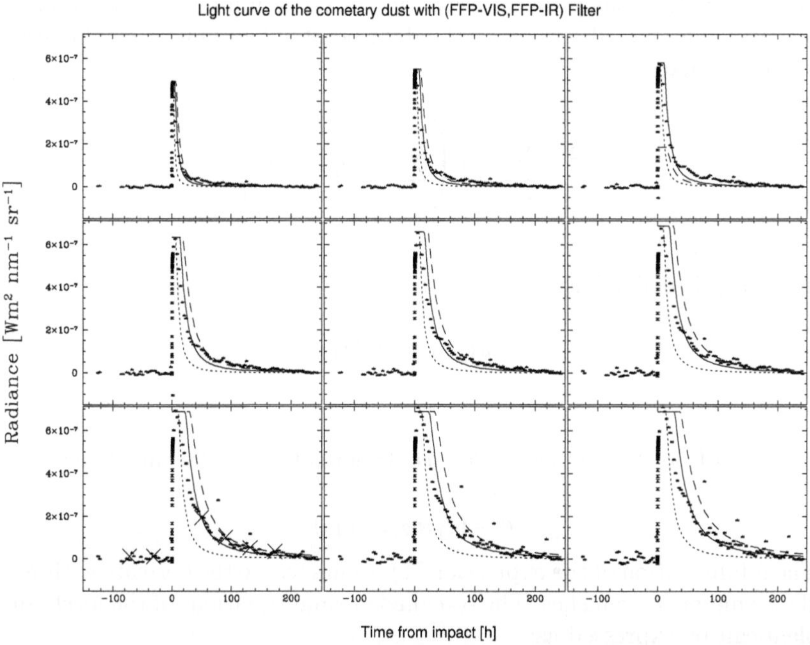

Fig. 1. Light curves of the cometary dust in the clear filter, for nine circular fields from 2 to 10 pixels (or from 3000 to 15000 km radius at the comet) from left to right. The dashed, solid and dotted lines show model profiles for velocities of 110, 160, and 300 ms^{-1}, respectively. The crosses on the bottom left panel are separated by 40.832 h, the rotation period of comet 9P/Tempel 1.

3 Approach

We now consider an approach to retrieve the velocity distribution of particles. Instead to simultaneously fit synthetic images to a selected set of observations (e.g. [5]), we simultaneously fit synthetic light-curves to the observed ones. In the earlier method, in order to clean the images from eventual residuals of cosmic rays and to improve the S/N ratio, new images should be created by stacking a set of successive images. This would decrease the temporal parameter sampling. Because we are extracting the brightness from each image, our method keeps the good OSIRIS temporal parameter sampling. Given the matrices O, E, and M, our focus is to find the curve or the most likely model, among all combination of simple models, that approximately fits the sets of observations, i.e., the observed brightness distribution of the cometary dust. Distribution of terminal velocities reached by the dust are derived using a maximum likelihood estimator and a least squares solution.

We fit simultaneously the observed fluxes O by minimizing the sum of the chi-squared differences between all the data-points O and linear combinations of the modeled fluxes M:

$$\chi^2 = \sum_{k=1}^{657} \frac{1}{\sigma_k^2} \left(\sum_{j=1}^{13} [a(j)]_k - O_k \right)^2 \tag{1}$$

where $[a(j)]_k$ is defined as

$$[a(j)]_k = c_j M_{jk} \tag{2}$$

and c, a set of positive dimensionless coefficients to be determined, as:

$$c = (c_1, c_2, ..., c_{13}). \tag{3}$$

The minimization of the expression (1) occurs where the derivate with respect to all parameters c_j vanishes. The so-called normal equations of the least-squares problem can be expressed as:

$$\frac{\partial \chi^2}{\partial c_j} = 2 \sum_{k=1}^{657} \frac{1}{\sigma_k^2} \left(\sum_{j=1}^{13} c_j M_{jk} - O_k \right) M_{jk} = 0 \tag{4}$$

Equation (4) yields a linear system of 13 equations which are solved using the LAPACK library [6]. The solution corresponds to the deepest minimum of the chi-square function (1). There may be other solutions besides it which corresponds to local minima of the function (1) (with a similar likelihood). This means that our derived parameters c may be affected by large errors.

4 Results

In Fig. 2 we show the velocity distribution . The dust particles have velocities of $\sim$225 ms^{-1}. This velocity is in a good agreement with the projected speed values derived from other authors (e.g. [7] (200 ms^{-1}), [8] (125 ms^{-1}), [9] (130–230 ms^{-1}), [5] (190 ms^{-1})). Nevertheless, [10], from HST images, derive lower velocities (70–80 ms^{-1}).

Dust with a velocity of 200 ms^{-1} reaches the edge of the smallest aperture (2-pixel radius) in slightly more than 4 h. The acceleration phase is not seen, meaning that not more than a few percent of the impact material was produced later than 1 h after the impact.

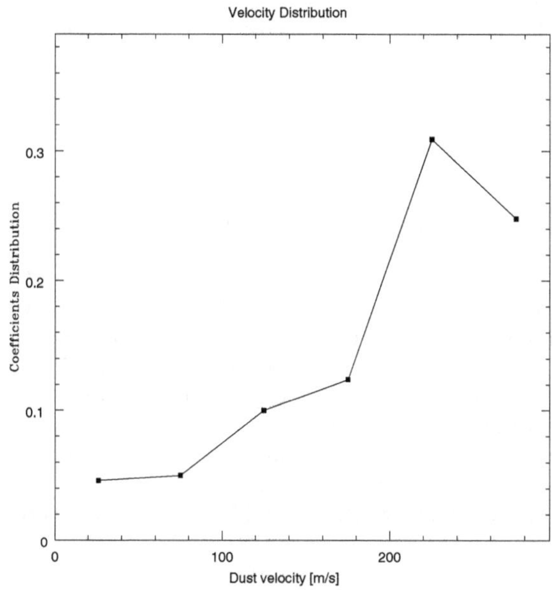

Fig. 2. Parameters c_j versus dust terminal velocity. The distribution peaks at 225 ms^{-1}.

5 Conclusion

We developed a discrete linear inverse problem approach which allows to simultaneously fit modeled light-curves to a set of observational light-curves of cometary dust of 9P/Tempel 1 acquired by NAC aboard Rosetta spacecraft. We derive a broad accurate velocity distribution of the dust particles (between 1 and 600 ms^{-1}). It peaks at 225 ms^{-1}. This value is in good agreement with already published estimates of the expansion velocity of the dust cloud. We consider these successful comparisons as an evidence of the appropriateness of the method.

The velocities of the dust seen in the light curve in the first minutes after the impact suggest early acceleration of the ejected material close to the cometary surface.

Despite the relative success of the models and approach presented in this work, a few points may be raised that need further improvement. In particular, the relative robustness of the method. Furthermore, because of the unstable character of the ill-conditioned solutions, further regularization algorithms for the stabilization of the solution should be investigated. Limits on late acceleration (and therefore in late production of material) also should be enquired, work is in progress.

We acknowledge the funding of the national spaces agencies ASI, CNES, DLR, the Spanish Space Program (Ministerio de Educacion y Ciencia), SNSB, and ESA.

References

1. A'Hearn M. F., Belton M. J. S., Delamere W. A., et al. (2005): Science **310**, 258
2. Keller H. U., Jorda L., Küppers M., et al. (2005): Science, **310**, 281
3. Küppers M., Bertini I., Fornasier S., et al. (2005): Nature, **437**, 987
4. Keller H. U., Küppers M., Fornasier S., et al. (2007): Icarus, **191**, 241
5. Jorda L., Lamy P., Faury G., et al. (2007): Icarus, **187**, 208
6. Anderson E., Bai Z., Bischof C., et al. (1999): *LAPACK User's Guide, Third Edition*, SIAM, Philadelphia, PA, `http://www.netlib.org/lapack`
7. Meech K. J., Ageorges N., A'Hearn M. F., et al. (2005): Science, **310**, 265
8. Sugita S., Ootsubo T., Kadono T., et al. (2005): Science, **310**, 274
9. Schleicher D. G. (2006): Icarus, **181**, 442
10. Feldman P. D., McCandliss S. R., Route M., et al. (2007): Icarus, **191**, 276

The Subsurface Structure of Comet 9P/Tempel 1 Projected into the Dust Plume

T. Kadono[1], S. Sugita[2], S. Sako[3], T. Ootsubo[4], H. Kawakita[5], M. Honda[6], T. Miyata[3], R. Furusho[7], and J. Watanabe[8]

[1] Japan Agency for Marine-Earth Science and Technology
[2] Department of Complexity Science and Engineering, University of Tokyo
[3] Institute of Astronomy, University of Tokyo
[4] Nagoya University
[5] Kyoto Sangyo University
[6] Institute of Space and Astronautical Science, Japan Aerospace Exploration Agency
[7] Waseda University
[8] National Astronomical Observatory of Japan

Abstract Cometary nuclei are believed to contain important information on the condition of the solar nebula, but little observational data is available on their interior structure. Our ground-based observations of NASA's Deep Impact event and the cratering mechanism show that comet 9P/Tempel 1 has a surface layer consisting of small (submicron to micron-sized) carbonaceous grains whose thickness is a few tens of cm.

1 Introduction

The observations of NASA's Deep Impact (DI) event found that crater formation was controlled by gravity [1, 2]. The cratering and ejection mechanism in the gravity regime indicates that higher-speed ejecta are thrown from near the surface around the impact point, while slower ejecta are biased toward material from deeper in the target [3] (Fig. 1). If the nucleus has a surface layer, fast ejecta thrown from near the impact site contain only materials from the surface layer while slow ejecta contain primarily those from the subsurface layer. Hence, the material distribution in the nucleus is reflected in the distribution of ejection velocities of the materials. The ejection velocity is translated to the radial distance from the nucleus in the dust plume; ejecta seen in the outer portion of the plume were ejected at higher velocities than those in the inner portion of the plume. Thus, the materials in the outer portion of the plume originated from the surface layer while those in the inner portion originated primarily from the subsurface. Given this relationship, it is possible to analyze the interior of the comet based on the spatial distributions of materials in the plume.

2 Our Approach

Here, we focus on the spatial distributions of solid grains in the dust plume, analyzing mid-infrared imaging data obtained by the Subaru telescope with three band-pass filters in the N-band with central wavelengths of 8.8, 10.5, and 12.4 μm and in the Q-band at 17.7, 18.8, and 20.5 μm. We denote the flux at each wavelength in the N-band as $I_{8.8}, I_{10.5}$, and $I_{12.4}$, respectively. First, we consider the $I_{feature}/I_{cont}$ ratio, where $I_{feature}$ is $I_{10.5} - I_{cont}$ and I_{cont} is $(I_{8.8} + I_{12.4})/2$ (Fig. 2). This ratio of the intensity of the 10 μm silicate feature to that of the continuum indicates the distribution of small silicate grains in the dust plume. The $I_{feature}/I_{cont}$ ratio is low near the nucleus (~ 1) and increases outward until it reaches maximum ($\sim$ 1.5" from the nucleus), and then decreases toward the edge of the plume. This trend implies that the abundance of small silicate grains is high in the inner portions of the plume, that is, in the subsurface of the comet.

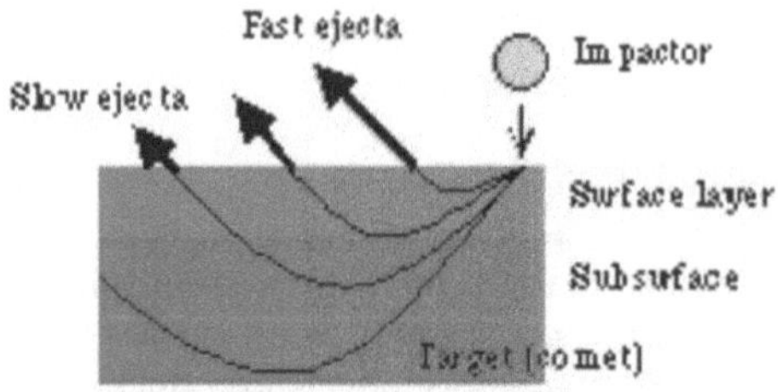

Fig. 1. Excavation flow and ejection velocities in the crater formation.

Next, we consider the spatial distribution of $I_{8.8}/I_{12.4}$, which represents the slope of the continuum. The $I_{8.8}/I_{12.4}$ ratio is controlled by the temperature and composition of the grains in the plume. Figure 2 shows that the $I_{8.8}/I_{12.4}$ ratio increases with the radial distance from the nucleus ($I_{8.8}/I_{12.4} \sim 0.5$) to the leading edge of the plume ($\sim$ 1–2). Annular high $I_{8.8}/I_{12.4}$ regions can be found at 2" from the nucleus. The peak positions of $I_{8.8}/I_{12.4}$ are clearly outside the peaks of $I_{feature}/I_{cont}$ and are associated with low $I_{feature}/I_{cont}$ regions in the outer portions of the plume. High $I_{8.8}/I_{12.4}$ ratios are produced either by very small silicate grains or by small carbonaceous grains, but the low $I_{feature}/I_{cont}$ ratio indicates a small quantity of small silicate grains. Furthermore, the very low surface albedo of the comet ([1]) suggests the dominance of carbonaceous grains in the surface layer. Thus, we conclude that the outer portion (ejected from the surface layer of the nucleus) is dominated by small carbonaceous grains. Assuming that the outer portion consists of only carbonaceous grains, an equilibrium radiation model allows us to estimate the mean grain size from the observed $I_{8.8}/I_{12.4}$ ratio. The observed high $I_{8.8}/I_{12.4}$ ratio (> 1) requires a radiation temperature higher than 350 K. This temperature is the equilibrium temperature of carbon grains with a radius of $\sim$ 0.5 μm at 1.5 AU. Thus, we conclude that the surface layer of the comet is likely to contain a dominant fraction of small (≤ 0.5 μm) carbonaceous grains.

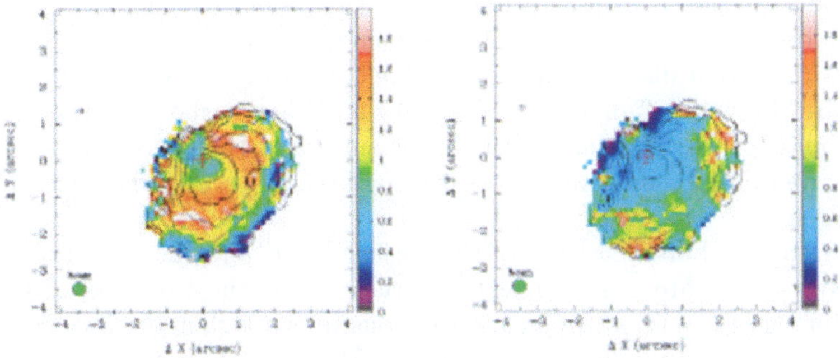

Fig. 2. *Left*: The distribution of small silicate grains in the plume: $I_{feature}/I_{cont}$. Pseudo-color images show $I_{feature}/I_{cont}$ at 2 hours after the impact (7:43–8:08 UT). North is top and east is left. The beam size of the FWHM is shown at the left lower corner. In order to show the spatial extension of the dust plume, $\log(I_{8.8})$ is shown as contours. *Right*: The distribution of the slope of the blackbody continuum: $I_{8.8}/I_{12.4}$. Pseudo-color images show the ratio $I_{8.8}/I_{12.4}$ at 2 hour after the impact. Contours show $\log(I_{8.8})$.

Next, we estimated the thickness of the carbonaceous surface layer using two independent methods. In gravity-controlled cratering in a target with a surface layer (Fig. 1), there is a critical ejection velocity u_c, which corresponds to the excavation streamtube whose maximum depth is equal to the thickness of the surface layer. Ejecta at velocities higher than u_c do not contain material from the subsurface layer of the target. Figure 2 indicate that the outer portion of the plume (the peak of $I_{8.8}/I_{12.4}$) expands with a velocity of 180 m/s while the inner portion (that of $I_{feature}/I_{cont}$) expands with 100 m/s. The critical velocity u_c should be between these values. Assuming the Z-model [3, 4] as the excavation flow, we can derive the maximum depth of the excavation streamtube which corresponds to u_c between 180 and 100 m/s, $\sim$ 40 and 60 cm. Thus, the thickness of the surface carbonaceous layer is expected to be of an order of several tens of cm.

3 Results and Conclusion

We can also obtain the thickness of the surface layer by another method using an estimate for the total mass of carbonaceous grains in the plume based on spectral energy distribution (SED) observations. We fit the SED in the N- and Q-bands [5] observed 2 hours after the DI impact within the 24 pixels ($\sim$2000 km) and the 14 pixels ($\sim$ 1200 km) of aperture, finding the total mass of carbonaceous grains M_c in the plume to be 1.56×10^5 kg and the mass ratio of the silicate to carbonaceous grains in the outer portion of the plume (between 14 and 24 pixels) to be 5.2.

It has been noted that the mass of ejecta observed with a ground-based telescope corresponds to the ejected mass with a velocity higher than the escape velocity ($\sim$ 2 m/s) of the comet [2]. A standard experimental scaling rule [6] predicts the radial distance from the impact point on the surface at which the ejecta are thrown at 2 m/s to be x_{esc} 19 m. When carbonaceous grains with a total mass of M_c uniformly exist on the surface of the comet within a radius of x_{esc}, we obtain the column mass σ of carbonaceous materials per unit area $\sim$141 kg/m^2 (= 14.1 g/cm^2).

Using σ, the bulk density ρ of the surface layer, and the mass ratio of carbonaceous to silicate grains n in the surface layer, the thickness of the layer is written as $(1+n)\sigma/\rho$. Radar observations indicate that the radar albedo of comet surfaces is generally very low, at $\sim$ 0.04–0.1 [7]. Various experiments and theoretical models indicate that the bulk density ρ of porous samples with a radar albedo between 0.04 and 0.1 is between 1.0 and 2.0 g/cm^3 [7]. Here, we assume that the radar albedo of 9P/Tempel 1 is also in this range and set ρ to be 1.0–2.0 g/cm^3. The ratio n is set to be 5.2 obtained from the fitting. Thus, the thickness of the surface layer $(1+n)\sigma/\rho$ is $\sim$ 44–87 cm. This estimated thickness of the surface layer by the second method is consistent with the value obtained by the first method.

Acknowledgment This research is based on the data collected at the Subaru Telescope, which is operated by the National Astronomical Observatory of Japan (S05A-042 and S05A-OT11). This research was supported in part by a Grant-in-Aid from the Japan Society for the Promotion of Science and a grant from The Sumitomo Foundation.

References

1. M. F. A'Hearn et al. (2005): Science **310**, 258
2. S. Sugita et al. (2005): Science **310**, 274
3. Melosh, H.F.: *Impact Cratering: A Geologic Process* , (Oxford University Press, New York, 1989).
4. Maxwell, D.E., In: *Impact and Explosion Cratering* , ed by D. J. Roddy et al., (Pergamon Press, New York, 1977), pp 1003–1008.
5. Ootsubo, T. et al. (2006): Planetary and Space Science, **55**, 1044–1049
6. Housen, K.R., Schmidt, R.M., Holsapple, K.A. (1983): J. Geophys. Res. **88**, 2485
7. Harmon, J.K., et al. In: *Comet II* , ed by M. C. Festou et al., (University of Arizona Press, Tucson, AZ, 2004), pp 265–279.

The Dusty View of DI from ESO Chile

H. Boehnhardt[1], N. Ageorges[2], S. Bagnulo[2]11, L. Barrera[3], T. Bonev[4], O. Hainaut[2], E. Jehin[2], H. U. Käufl[5], F. Kerber[5], G. LoCurto[2], J. Manfroid[6], O. Marco[2], E. Pantin[7], E. Pompei[2], H. Rauer[8], I. Saviane[2], F. Selman[2], C. Sterken[9], G. P. Tozzi[10], and M. Weiler[8]

[1] MPI for Solar System Research, Max-Planck-Str. 2, D-37191 Katlenburg-Lindau, Germany `boehnhardt@mps.mpg.de`
[2] European Southern Observatory, Santiago de Chile
[3] Universidad Metropolitana de Ciencias y de la Educacion, Santiago de Chile
[4] Astronomical Institute of the Bulgarian Academy of Sciences, Sofia, Bulgaria
[5] European Southern Observatory, Garching, Germany
[6] University of Liege, Belgium
[7] CEA Saclay, Paris, France
[8] Deutsche Agentur für Luft- und Raumfahrt, Berlin-Adlershof, Germany
[9] Free University of Brussel, Belgium
[10] INAF Arcetri Observatory, Firenze, Italy
[11] Armagh Observatory, Northern Ireland

Abstract Around the time of the impact of NASA's Deep Impact (DI) mission at comet 9P/Tempel 1, in total 6 telescopes with altogether 7 different instruments, located at the La Silla (LSO) and Paranal (VLT) Observatories of the European Southern Observatory (ESO) in Chile, were used to characterize the dust properties before and after the event. The ejecta cloud expanded at an average speed of about 200 ms^{-1} during the first hours after the event. It reached stagnation distance of 25000 km about 3 days after impact. The pre-impact dust jet and fan activity ('porcupine' pattern) remained undisturbed after impact. In our measurements the jet activity can be traced to a few 100 km nucleus distance. In total 9 comastructures are identified which may originate from at least 4 regions of enhanced dust emission on the nucleus – one of this region may in fact be multiple. No obvious signatures of a new active region created by DI are found. The overall dust production during the impact compares to about 5–10 h of normal activity. The global expansion geometry of the DI cloud is compatible with a majority of dust grains in the micron size range. Indications exist for asymmetric brightness and colour distributions of the dust in the ejecta cloud. The dust temperature rose from about 280–290 K before to 330 K one day after the event and fell to pre-impact level the day thereafter. The dust reflected sunlight was found to be linearly polarized at about 7.5% in the visible and near–IR, at constant level within about 4000 km from the nucleus. No circular polarization of the dust is detected.

1 DI Dust Observations at ESO

The observations of the dust, produced by the DI event at comet 9P/Tempel 1 on 4 July 2005, were performed at ESO's La Silla and Paranal Observatories in Chile. They cover the immediate pre-impact period and extend until several days after the event. However, at the time of the impact (05:53UT) the comet had already set for the ESO sites and could only be re-observed again 14 h (mid-IR) and 16 h(visible and near-IR) thereafter.

All telescopes, operated by ESO at the two sites, together with a number of instruments were 'employed' for the DI dust campaign: at La Silla the 2.2 m ESO/MPG telescope plus the Wide-Field-Imager (WFI), the 3.5 m New Technology Telescope (NTT) plus the EMMI and SOFI instruments and the 3.6 m telescope plus TIMMI2; at Paranal the four 8.2 m unit telescopes (UT1-4) of the Very Large Telescope (VLT) using FORS2 and ISAAC at UT1, FORS1 at UT2, VISIR at UT3 as well as NACO and SINFONI at UT4. The suite of telescope-instrument combinations allowed a comprehensive observing program covering the visible (WFI, EMMI, FORS1, FORS2), near-IR (SOFI, ISAAC, NACO, SINFONI) and mid-IR (TIMMI2, VISIR) wavelength range (0.36 – 20 μm) with different observing techniques (imaging, spectroscopy, polarimetry). Table 1 summarizes the ESO observations obtained for the analysis of the dust component of the Deep Impact event. In parallel, ESO telescopes and instruments were also used for observations of the gas phenomena of the comet and related with DI (for an overview see [3]). The ESO web pages of the La Silla and Paranal observatories provide detailed information on the telescope and instrument equipment used.

2 Results

First results from the ESO DI dust campaign are included in the paper by Meech et al. [7]. Final results from the ESO dust campaign of the DI event – together with more detailed descriptions of the observations, the data processing and analysis methods applied – are found in Boehnhardt et al. [1], Bonev et al. [2], Tozzi et al. [8] and will be released in forthcoming papers (in particular on the polarimetry and adaptive optics observations). Below we summarize our results, focussing on various properties of the ejecta cloud and the 'background' coma produced by normal dust activity of the comet. The instruments that provide the respective observations are indicated as well (usually in parenthesis after the subject title).

Table 1. Overview on the observing schedule of the ESO DI dust campaign in July 2005. The table lists: in the left part – the observing technique (column 'Mode'), the 'field size', the instrument set–up (column 'Setup'); in the central part – indicated by symbol 'X' the usage of the respective configuration for each day during the period 2–11 July 2005; in the right part – the ESO site (column 'Observatory'), the telescopes (column 'Telescopes') and the instruments (column 'Instruments') used for the respective observations. Abbreviations: Small = normal field size of about 5–10 arcmin; Wide = wide field of 32 arcmin; Narrow = very small field size of about 1 arcmin or less; non-AO = seeing limited; AO = adaptive optics; Disp. = dispersion; LSS = longslit spectroscopy; IFU = integral field unit spectroscopy; Spec. = spectroscopy; $BVRIJHKLMNQ$ = broadband astronomical filters; NarrowF = narrowband filters; CometF = narrowband filters for selected cometary molecule and dust continuum bands; Img. linear = linear imaging polarimetry; Spec. linear = linear spectropolarimetry; Spec. circ = circular spectropolarimetry. Column 'Setup' also indicates the wavelength range (in nm or by filter band designations) of the measurements.

Mode	Field Size	Setup	2	3	4	5	6	7	8	9	10	11	Observatory	Telescopes	Instruments
Imaging	Small	NQ	X	X	X	X	X	X	X	X	X		VLT/LSO	UT3/3.6m	VISIR/TIMMI2
	Narrow	LM				X	X	X	X	X			VLT	UT4	NACO
	Narrow	JHK	X	X	(X)	X	X	X	X	X			VLT	UT4	NACO/SINFONI
	Small	JHK	X	X	X		X	X	X	X			LSO/VLT	NTT/UT1	SOFI/ISAAC
	Small	$BVRI$		X	X	X	X	X	X	X	X	X	VLT	UT1/UT2	FORS2/FORS1
	Small	CometF	X	X	X		X	X	X	X			LSO	NTT	EMMI
	Wide	$BVRI$	X	X	X		X	X	X	X			LSO	2.2m	WFI
	Wide	NarrowF	X							X			LSO	2.2m	WFI
Spectroscopy	non-AO LSS	N band	X	X	X	X	X	X	X	X	X		VLT/LSO	UT3/3.6m	VISIR/TIMMI2
	AO/non-AO LSS	L band					X						VLT	UT3/UT1	NACO/ISAAC
	AO LSS/IFU	JHK band	X	X									VLT	UT4	NACO/SINFONI
	non-AO LSS	JHK band	X	X	X		X	X	X	X			LSO	NTT	SOFI
Polarimetry	Img. linear	JHK		X	X		X						LSO	NTT	SOFI
	Img. linear	NarrowF	X							X			VLT	UT2	FORS1
	Spec. linear	400–900nm			X	X							VLT	UT2	FORS1
	Spec. circ.	400–900nm					X						VLT	UT2	FORS1

2.1 Impact Plume

The ejecta cloud (FORS2, EMMI, WFI): The dust cloud produced by the impactor crash into the comet, was detected in broadband and narrowband images from a few hours after the event until at least 8 July 2005. Figure 1 shows in four sub-panels *BVR* composite images of the ejecta cloud at four epochs after the impact. For clarity the coma background from normal activity is removed which allows a better display of the cloud expansion with time. On 4 July 2005, the ejecta cloud had a semi-spherical shape expanding into the southwestern coma quadrant (central position angle PA ∼235 deg; velocity of dust front ∼220 ms^{-1} and of maximum brightness ∼ 120 ms^{-1}). The surface brightness and dust reddening was slightly higher on the sunward side of the cone (compared to the tailward side). Towards Sun direction the cloud expansion got slowed down by radiation pressure during the subsequent days and reached a maximum projected distance of about 25000 km from the nucleus. No indications for 'mini-comets', neither natural nor impact-induced ones, are found in high-resolution imaging data of the inner coma.

Total dust production (FORS2, SOFI): From the flux enhancement of the ejecta cloud over pre-impact level, we estimate a total dust production by the impact that compares to about 5–10 h of 'normal' undisturbed activity of the nucleus at the time of the encounter (this assumes similar dust properties of the ejecta and the normal activity dust).

2.2 Ejecta Dust Grains

β-factor, terminal velocity and reddening slope (FORS2, EMMI, WFI): The sunward apex distances of the cloud, measured every day, allowed to constrain the radiation-pressure-to-gravity parameter β and the initial ejection velocity v_{eject} of the dust grains: the values found decrease from 1.9 and 370 ms^{-1} about 20 h post-impact to 0.2 and 160 ms^{-1} about 3.5 days after the event. The results suggest that the majority of the dust ejecta seen in the visible wavelength range consisted of grains of a few μm in size. In particular, the high β grains may have consisted of absorptive material. The dust reddening changed from a constant level of 9%/100 nm throughout the coma (within 20000 km from the nucleus) before impact to radially increasing values of 17 to 22%/100 nm at 5000–20000 km towards the end of our observing campaign (9 and 10 July 2005). By 6 July 2005 at the latest, the dust started to be expelled into tail direction (PA = 111 deg).

Grain sizes (SOFI, ISAAC, NACO:) Differences in the radial profiles of the *JHK* images of the ejecta cloud suggest that heavier dust was concentrated closer to the nucleus than the lighter one (since the flux profile peaks at 2000 km projected nucleus distance in *K* band compared to a peak distance of 10000 km in *J*). Following the modeling approach of Kolokolova et al. [4], the dust in- and outside of the ejecta cloud was characterized using radial profiles of various dust continuum filters. The slope of the dust size distribution was found to

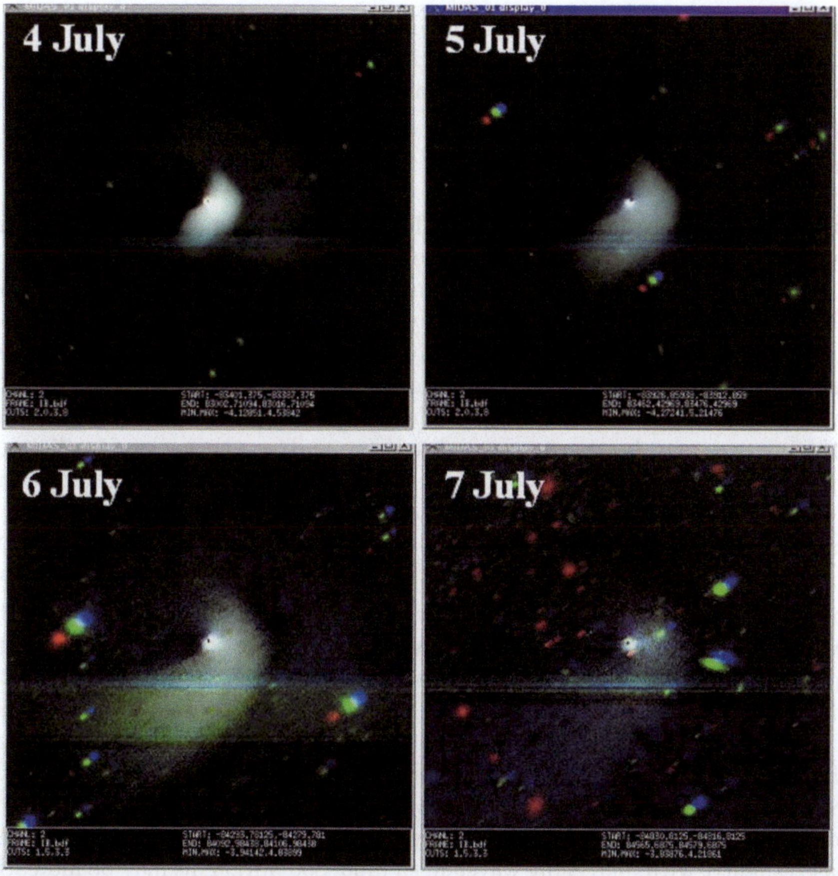

Fig. 1. The expansion of the DI ejecta clouds. The images are composites of BVR exposures taken with FORS2 each night (as noted in the subpanels). The coma background from the normal activity of 9P/Tempel 1 is removed by subtraction of FORS2 pre–impact exposures, obtained in night 3–4 July 2005 through the respective filters. North is up and East is to the left. Field of view is about 150000×150000 km at the distance of the comet. Horizontal stripes result from incomplete removal of the interchip gap of the FORS2 detector array.

be about 2.6 in the overall coma before impact and ∼2 in the impact ejecta alone. The minimum grain size is 0.2 μm in the normal coma and 1 μm in the impact plume. Modeling of the ejecta cloud dynamics indicates a total dust production of 6000–13000 tons for grain sizes between 0.2 and 110 μm and size distribution slopes of 2.9–2.5, respectively. We note that the results on the dust size dsitribution determined from the near-IR and visible colours and the dust dynamics as seen in the visible are not in full agreement and suggest revision of the modeling approaches (in particular, they do not consider dust sublimation and hydrodynamic effects).

Dust sublimation (SOFI, ISAAC, NACO:) The presence of sublimating grains which scattered the light very efficiently in the near-IR, has been detected. The mean lifetime of these grains was found to be of the order of 11 h.

2.3 Dust Material

Dust temperature (VISIR, TIMMI2): We used the mid-IR filter photometry of the cometary coma, namely the filters with central wavelength of 8.7 and 13.2 μm, to estimate the temperature of the very inner dust coma (about 3000 km diameter). During the first day after impact a temperature of about 300 K is measured which is significantly higher than the pre- and later post-impact values (280–290 K). With the exception of the first post-impact day, the variability of the mid-IR flux seen in the near-nucleus region around the days of the impact, may be due to normal activity and/or the nucleus itself (cross section variability by rotation). The N band spectra of the dust reveal a silicate emission feature with different profile pre- and post-impact. Modeling of the spectra indicates the presence of a large amount of absorbing material and an enhancement of amorphous and crystalline silicates in the post-impact dust.

Dust polarization (FORS1): The linear polarization of the dust in the visible wavelength range was found to be the same 7.5% before (3 July 2005) and after impact (5+7+9 July 2005). The polarization degree did not change across the coma within ~5000 km project distance from the nucleus. The anti-correlation between color change and polarization degree throughout the coma is compatiable with a change in the size distribution of the dust grains due to the DI event. Linear spectropolarimetric observations on 5+9 July 2005 support these findings. The cometary dust showed a spatially flat linear polarization that decreased slightly with increasing wavelength (400–800 nm). This may be indicative for a relatively high abundance of organic material in the cometary grains. Measurements of the circular polarization post–impact (on 8 July 2005) showed zero values (within the uncertainties of 0.3%), independant from the nucleus distance (< 5000 km) and wavelength (400–800 nm).

Given the sparse coverage of the DI event by our polarimetry observations and the small coma area (about 5000 km around the nucleus) that provided enough SNR for reasonably accurate polarimetric measurements, we believe that our polarimetry results may be more representative for dust from normal activity of the comet rather than from the impact cloud.

2.4 Localized and Normal Activity of the Nucleus

Localized activity (FORS2, EMMI, WFI, SOFI, ISAAC, NACO): The ejecta plume overlapped in parts with the general dust coma produced by normal activity of the nucleus (see Fig. 2). Image enhancement of the exposures, taken during the nights after impact, reveal the 'porcupine' pattern of localized normal dust activity, also shining through the ejecta cloud (i.e. as of 16 h after DI had produced the plume, it was optically thin). No changes from normal appearance

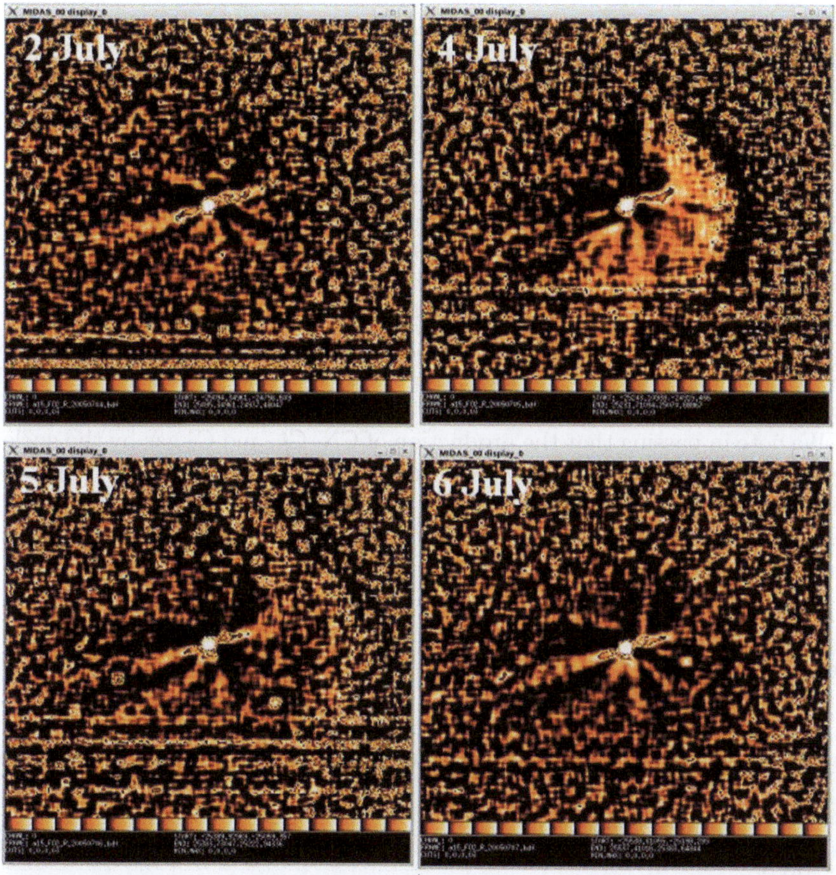

Fig. 2. The 'orcupine' pattern of embedded fans from active regions on the rotating nucleus of 9P/Tempel 1. The subpanels show the coma structures after enhancement by Laplace filtering. The original images were taken with FORS2 through broadband R filter each night (as indicated in the subpanel). The 'procupine' pattern is shining through the DI ejecta cloud on 4 July 2005 (about 16 h after the event). North is up and East is to the left. Field of view is about 85000×85000 km at the distance of the comet. Horizontal stripes result from incomplete removal of the interchip gap of the FORS2 detector array.

in this pattern is found after impact, in particular no jet or fan did evolve that could have been associated with a new, long lasting active area as a DI aftermath. The maximum extension of the 'porcupine' structure varied between – at least – 17000 km in sunward and 58000 km tailward direction. The cometary coma as a whole was back to pre–impact state (as for instance reported by [5, 6]) as of 9 July 2005 and no large-scale signatures of the impact are noticed. The pattern produced by localized activity is compatible with the presence of at least 4 active regions and a rotation axis orientation close to the orbital angular momentum vector of the comet. Coma features seen in normal seeing–limited images, can be traced to near–nucleus distances of a few 100 km.

Normal activity (EMMI, FORS2): The normal dust activity of the comet, seen in the general diffuse coma background and the 'porcupine' pattern, remained rather constant and uniform over the 10 days of our observations around the DI event. This can be concluded from the fact that subtraction of pre- from post/pre-impact images removed - almost - all flux from the general coma background and mostly only impact related signatures remained. The only exception is a marginal diffuse near-nucleus feature in the southwestern coma quadrant that is seen in structure enhanced images taken on 7 July 2005. Radial flux profiles of the dust coma, obtained from pre-impact images by integration over 2π position angle around the nucleus, show slope values close to unity which suggests a rather uniform and homogenuous dust production by the nucleus.

We note in passing that the production of CN, C_3, and C_2 gas by the nucleus, measured from EMMI cometary filter imaging, is at similar level on 3 and 9+10 July 2005. Indications are found that some of the dust jets/fans from enhanced localized activity on the nucleus have counterparts in CN filter images of the coma.

References

1. Boehnhardt, H., Pompei, E., Tozzi, G. P. et al.(2007): A&A, **470**, 1175
2. Bonev, T., Ageorges, N., Boehnhardt, H., et al.(2007): these proceedings and astro-ph **0703554**
3. Käufl, H.U., Ageorges, N., Bagnulo, S., et al.(2005): The Messenger, **121**, 11
4. Kolokolova, L., Jockers, K., Gustafson, B., Lichtenberg, G. (2001): JGR, **106**, 10113
5. Lara, L., Boehnhardt, H., Gredel, R., et al.(2006): A&A, **445**, 1151
6. Lara, L., Boehnhardt, H., Gredel, R., et al.(2007): A&A, **455**, 1061
7. Meech, K., Ageorges, N., A'Hearn, M.K., et al.(2005): Science, **310**, 265
8. Tozzi, G. P., Boehnhardt, H., Kolokolova, L., et al.(2007): A&A, **476**, 979

Spectropolarimetry of the Deep Impact Target Comet 9P/Tempel 1 with HiVIS

D. M. Harrington[1], K. J. Meech[2], L. Kolokolova[2], J. R. Kuhn[1], and K. Whitman[1]

[1] University of Hawaii, Honolulu, HI 96822 dmh@ifa.hawaii.edu
[2] University of Maryland, College Park MD 20742

Abstract Spectropolarimetry of the Deep Impact target, comet 9P/ Tempel 1, was performed during the impact event on July 4th, 2005 with the HiVIS Spectropolarimeter and the AEOS 3.67-m telescope on Haleakala, Maui. We observed atypical polarization spectra that changed significantly in the few hours after the impact. The polarization is sensitive to the geometry, size and composition of the scattering particles. Our first measurement, beginning 8 minutes after impact and centered at 6:30UT, showed a polarization of 4% at 650 nm falling to 3% at 950 nm. The next observation, centered an hour later, showed a polarization of 7% at 650 nm falling to 2% at 950nm. This corresponds to a spectropolarimetric gradient, or slope, of −0.9% per 1000Å 40 minutes after impact, decreasing to a slope of −2.3% per 1000Å 75 minutes after impact. Both are atypical blue polarization slopes. The polarization values of 4 and 7% at 650 nm are typical for comets at this scattering angle, whereas the low polarization of 2 and 3% at 950 nm is not. This, combined with the IR spectroscopy performed by a number of observers during the event, suggests an increase in size, number, and crystallinity of the individual silicate particles (monomers) that are a constituant of the dust particles (aggregates) in the ejecta.

1 Introduction

The angular and spectral dependence of coma polarization are signatures of the size, composition and structure of coma grains [8, 7, 19]. The degree of polarization typically increases with wavelength, called a red polarimetric color for the dust coma, as well as phase angle [19, 18, 22]. A few comets have shown a blue polarimetric color such as comet 21P/Giacobini-Zinner [16]. The polarization and polarimetric color are influenced by the size, composition, and geometry of the individual scattering monomers that make up an aggregate dust particle. On 2005 July 4 at 5:52:02 UT comet 9P/Tempel 1 was impacted by a 364 kg spacecraft as part of the NASA Deep Impact mission [1]. The event presented a unique opportunity to compare spectropolarimetric measurements with other measurements performed simultaneously as part of the Deep Impact mission to further constrain the dust properties.

For comets at phase angle of 40$^\circ\!.$9, average coma polarization in a red filter is 4–10% and perpendicular to the scattering plane plane, usually increasing towards 1 μm [22]. Some exceptions are known however [16, 21, 22]. Some dependence on comet activity was also found and explained as a consequence of the influence of scattering particles of differing size and composition [5, 6, 7]. A change in polarization due to a change of composition and size is also seen in scattering models [18, 19, 15, 20].

The polarization and color of cometary dust scattering has been studied theoretically and experimentally in great depth over a great range of variables such as size, composition, scattering geometry, particle roughness, and aggregation process [3, 19, 15, 20]. Models that fit the typical angle and red wavelength dependence of the polarization best use aggregates of more than 1000 monomers having small radii (~100 nm). Model aggregates that fit the typical polarization measurements have a high index of refraction ($m = n + ik, n = 1.8$–2.0, $k \sim 0.4$–0.6). This corresponds to volume fractions of one-third silicates, two-thirds carbonaceous materials, and a small amount of iron-bearing sulfides [15]. This material is quite dark (high imaginary part of the complex index of refraction) and similar to the composition of 1P/Halley's dust [25]. The carbonaceous material in the model is roughly two thirds amorphous carbon and one third organic-refractory material. The type of aggregation (cluster-cluster vs. particle-cluster) did not play a major role in the polarization models, but it does influence the thermal IR spectra through the porosity [14, 19]. A decrease in polarization with wavelength, called blue polarization gradients or slopes, were not extensively modeled because of their rarity. However, some models demonstrated the tendency of polarimetric color to decrease, *i.e.* get more blue, with increasing size, albedo (*e.g.* icy), and transparency of the material (crystallinity), whereas more absorbing materials typically show an increase in polarzation with wavelength, called a red polarization slope [15]. Multiple scattering in optically thick dust clouds depolarizes the scattered light. With all this information contained in the polarization of scattered light, we expected the polarimetric data to shed light on the change in dust properties resulting from the Deep Impact encounter.

2 The Deep Impact Spectropolarimetry

The AEOS telescope is a 3.67-m, altitude-azimuth telescope. The HiVIS spectrograph is a cross-dispersed echelle spectrograph using the $f/200$ coudé optical path [32]. Since non-normal incidence reflections change the polarization state of the incident light, a careful calibration of the telescope has been performed [10]. We used a 1.5″×7″ slit (970×4530 km at a geocentric distance of Δ=0.89 AU) and the "red" setting (nominally 637.5–968.0 nm) for all comet observations. The spectropolarimetry module for the AEOS spectrograph consists of a rotating achromatic half-wave plate and a calcite Savart Plate. The Savart plate separates an incoming beam into two parallel, orthogonally polarized beams separated in the spatial direction at the focal plane.

A common definition in planetary science is that the degree of polarization is the difference between the intensity polarized parallel and perpendicular to the scattering plane [19]. Since the HiVIS image rotator was not used in order to simplify the polarization calibration, the projection of the slit's position angle onto the sky was not constant, and knowledge of the scattering plane is difficult to extract. In this paper, we will be using an alternative definition that does not require knowledge of the scattering plane, simply calculating the polarization in the instrumental reference frame (see [11] for details).

Comet 9P/Tempel 1 was bright enough to obtain useful spectropolarimetry only on the night of impact. We aquired two complete data sets, from 6–7 and 7–8 UT. The data were reduced using the AEOS-specific reduction pipelines developed for this instrument [10]. Calibration of comet spectropolarimetry is performed by interpolating a sky-map of unpolarized standard star observations to the pointing of the comet to create a telescope-induced polarization calibration. The corrections were typically 2–3% with mild wavelength dependence ($\leq 1\%$) and they did not alter the slope-change seen in the comet 9P/Tempel 1 [11]. In order to present high signal to noise measurements of the polarization, the spectra were averaged 1000:1, giving 19 independent polarization measurements. The spectropolarimetry is plotted in Fig. 1. The 6–7 UT data set, started 8 minutes after impact, shows a slightly anomalous blue-sloped degree of polarization of 4% falling to 3% from 650 to 950 nm ($-0.9\pm0.2\%/10^3$Å). In contrast, the 7–8 UT data set, started 75 minutes after impact, shows a more pronounced blue slope from 7% at 650 nm to 2% at 950 nm ($-2.3\pm0.3\%/10^3$ Å). This is an indication of the change in particle scattering properties. The leading edge of the ejecta plume was moving out from the nuclues at ~200 m s^{-1} [26]. A body moving this speed could move across the slit in 40 minutes, setting the timescale for significant change in scattering by ejected gas and dust, consistent with other observations [26, 13, 30, 28].

For comparison, Comet C/2004 Q2 Machholz was observed on November 27, 2004 at a phase angle of 30°. Two complete data sets (8 images at 1200 s, roughly double the flux) were reduced and calibrated in the same way as the comet 9P/Tempel 1 data. The measured degree of polarization was in the usual range and showed a typical red polarization slope which did not change very signifigantly between the two image sets. This result gives us confidence in our instrument calibration and reduction techniques.

3 Discussion and Conclusions

Our observation of 4 and 7% polarization at 650 nm is typical for comets at these wavelengths and phase angles, 4% being somewhat low. However, the 1 and 3% polarization at 950 nm is not at all typical. The few computer or laboratory simulations that have modeled blue polarization slopes show it can result from larger particles or a predominance of transparent particles (larger crystalline silicates or ices) [4, 19, 14, 15, 20, 16]. At this point, we must discuss other observations of silicates, ices, and particle sizes to fully interpret the polarization.

The infrared observations of comet 9P/Tempel 1 right after the impact showed a strong and complex silicate feature developing by 1 hour after impact, fading 1.8 hours after impact [9, 30, 23]. The size of the silicate particles also evolved strongly. The dust particles were reported to have size distribution peaks (of the aggregates, not monomers) increasing after impact [9, 23, 24]. These models showed that pre impact there was an absence of sub-μm silicates, and the spectrum was mineralogically dominated by larger (0.9 μm) amorphous olivine with no carbon or crystalline silicates. They reported an increase in the number of sub-μm silicates, an emission from relatively transparent Mg-rich crystalline olivine, as well as a doubling of the silicate to carbon ratio and a 4-fold increase in the crystalline to amorphous silicate ratio between the first and second hour after impact. A relative lack of organics was also seen with the amorphous carbon being roughly 20% of the silicates (30–50% is typical) [23].

There was evidence for ice, but in a smaller amount than necessary to produce a blue polarization slope. There was a change in color of the dust in the near-IR post impact interpreted as icy grains, or icy grain mantles being liberated in the impact event [2]. There was direct observation of icy grains in the ejecta plume from the main spacecraft in the form of the 3 μm ice absorbtion from immediately after impact through lookback 46 min later [31]. Evidence for icy grains in the inner 600 km of the coma was seen with the sublimation maximized 1.5 hours after impact [29]. Particle disintegration was suggested from the different spatial evolution of CN, [OI], and dust continuum flux in spectrophotometric measurements [12]. The Spitzer spectra also required ice covering 3% of the dust surface area in their 10" FOV [23]. Thus, ice is also a possibility as a contribution to the transparency of the particles. However, the amount of ice necessary to dominate the scattering is much more than is suggested by the Spitzer and spacecraft data.

Multiple scattering depolarizes light in cometary dust and may be responsible for the low polarization just after impact. An optically thick plume was seen, thinning to $\tau \sim 0.4$, 20–25 minutes after impact [28]. The look-back images from the fly-by spacecraft also showed an optically thick ejecta plume after impact [1]. Since the optical depth was a strong function of time, the influence of multiple scattering is assumed to change strongly as well.

We can explain our observations by a depolarization due to multiple scattering in the first hour and subsequent domination of larger and more transparent monomers (silicates), and possibly ices, in the ejected dust aggregates of comet 9P/Tempel 1. This is consistent with the infrared data on Deep Impact, which indicate a high amount of silicates in the DI dust whereas in situ data for comets 1P/Halley and 81P/Wild 2, typical red polarization comets, show that their dust contained two-thirds of carbonaceous materials and organics [23, 9, 30, 15, 17]. We can speculate that subsurface materials in comet 9P/Tempel 1 had more volatile organics and ices that quickly evaporated or decayed leaving the disrupted and fragmenting dust, rich in silicates, to produce the observed blue polarization slope [2, 27, 1]. Harker et al. have suggested that smaller particles in the ejecta moved faster, size sorting the cloud, leaving larger and more crystalline silicates behind [9]. This could also contribute to the anomalous blue polarization slope we detected for the inner coma.

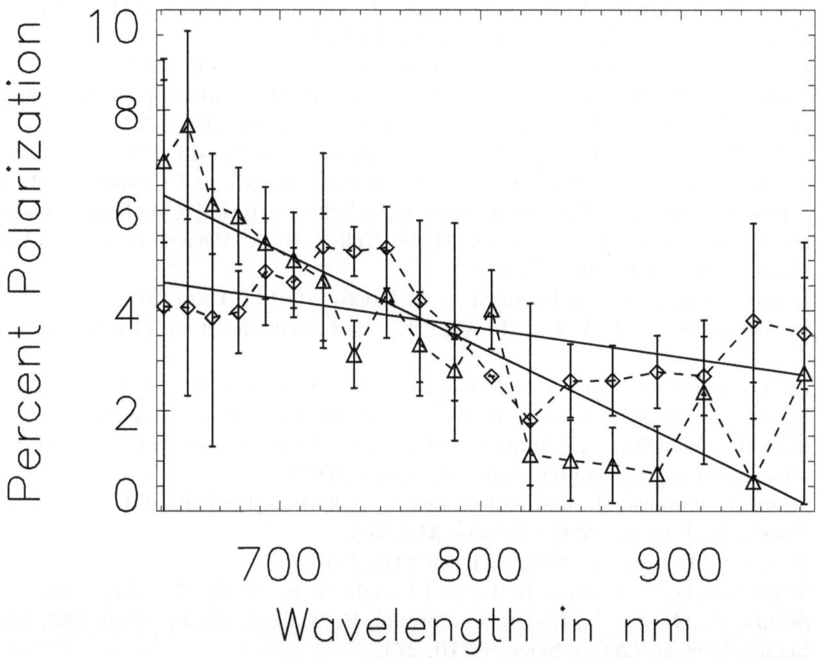

Fig. 1. Polarization for comet 9P/Tempel 1 with 3-σ error bars. The data was taken on impact night from 6 to 7 UT (*diamonds*) and 7–8 UT (*triangles*) at a phase angle of 40.9°. A linear fit is plotted to guide the eye. A single point at 805 nm (order 11) in the 6–7 UT data set has been replaced by the average of neighbor points, and has no error bars. The polarization spectra from 6 to 7 UT curve had a shallow negative slope of $-0.9\pm0.2\%/\ 10^3$ Å. The 7–8 UT curve had a slope of $-2.3\pm0.3\%\ /\ 10^3$ Å.

References

1. A'Hearn, M. F. et al. (2005): Science **310**, 258.
2. Fernández, Y. R., Lisse, C. M., Kelley, M. S. et al. (2006): Icarus, **187**, 220.
3. Gustafson, B. A. S., and Kolokolova, L. (1999): JGR **104**, 31711.
4. Hadamcik, E., Renard, J. B., Worms, J. C. et al. (2002): Icarus **155**, 497.
5. Hadamcik, E., and Levasseur-Regourd, A.-C. (2003a): A&A **403**, 757.
6. Hadamcik, E., and Levasseur-Regourd, A.-C. (2003b): Icarus **166**, 188.
7. Hadamcik, E., and Levasseur-Regourd, A.-C. (2003c): *JQRST* **79–80**, 661.
8. Hanner, M. S.: Comet dust: The view after Hale-Bopp. In: *Proc. IAU 181 Dust in the Solar System and Other Planetary Systems*, S. F. Green, I. P. Williams, J. A. M. McDonnell and N. McBride (Oxford Press 2002) p 239.

9. Harker, D. E., Woodward, C. E., and Wooden, D. H. (2006): Science **310**, 278.
10. Harrington, D. M., Kuhn, J. R., and Whitman, K. (2006): PASP **118**, 845.
11. Harrington, D. M., Meech, K. J., Kolokolova, L. et al. (2007): Icarus, **187**, 177–184
12. Hodapp, K. W., Aldering, G., Meech, K. J. et al. (2006): Icarus, **187**, 185–198
13. Jehin, E., Manfroid, J., Hutsemèkers et al. (2006) ApJ Letters **641**, L145.
14. Kimura, H., and Mann, I. (2004): JQRST **89**, 155.
15. Kimura, H., Kolokolova, L., and Mann, I. (2006): A&A **449**, 1243.
16. Kiselev, N. N., Jockers, K., Rosenbush, et al. (2000): Planet. Space Sci. **48**, 100.
17. Kissel, J., Krueger, F. R., Siln, J. et al. (2004): Science **304**, 1774.
18. Kolokolova, L., and K. Jockers; (1997) Planet. Space Sci. **45**, 1543.
19. L. Kolokolova, M. S. Hanner, A.-C. Levasseur-Regourd, and Gustafson, B. A. S.: Physical properties of cometary dust from light scattering and thermal emission. In: *Comets II*, ed by M. Festou, H. U. Keller, H. A. Weaver (Univ. of Arizona Press, Tucson 2004) pp. 577–604.
20. Lasue J. and Levasseur-Regourd A. C. (2006): JQRST **100**, 220.
21. Levasseur-Regourd, A. C., McBride, N., Hadamcik, E. et al. (1999): A&A **348**, 636.
22. Levasseur-Regourd, A. C. and Hadamcik, E. (2003): JQSRT **79**, 903.
23. Lisse, C., VanCleve, J., Adams, A. C. et al. (2006): Science, **313**, 635.
24. Lisse, C., VanCleve, J., Adams, A. C. et al. (2006): Science: STKE http://www.sciencemag.org/cgi/content/full/sci;1124694/DC1
25. Mann, I., Kimura, H., and Kolokolova, L. (2004): JQRST **89**, 291.
26. Meech, K. J. et al. (2005): Science **310**, 265.
27. Mumma, M. J. et al. (2005): Science **310**, 270.
28. Schleicher, D. G., Barnes, K. L., and Baugh, N. F. (2006): AJ **131**, 1130.
29. Schulz, R., Owens, A., Rodriguez-Pascual, P. M. et al. (2006): A&A **448**, L53.
30. Sugita, S. et al. (2005): Science **310**, 274.
31. Sunshine, J., et al. (2006): LPSC **37**, 1890.
32. Thornton, R. J., Kuhn, J. R., Hodapp, K.-W. et al. (2003): Proceedings of the SPIE **4841**, 1115.

PAHs in Comets: An Overview

A. Li

Department of Physics and Astronomy, University of Missouri, Columbia, MO 65211, USA LiA@missouri.edu

Abstract Polycyclic aromatic hydrocarbon (PAH) molecules, ubiquitously seen in the interstellar medium (ISM) of our own and external galaxies, might have been incorporated into comets if they are formed from relatively unprocessed interstellar matter. The detection of PAHs in comets would be an important link between the ISM and comets. This review compiles our current knowledge on cometary PAHs, based on ground-based and space-borne observations of infrared vibrational and ultraviolet fluorescence spectra of comets, and laboratory analysis of interplanetary dust particles possibly of cometary origin and cometary samples returned to Earth by the *Stardust* spacecraft. The latter provided the most unambiguous evidence for the presence of PAHs in cometary nuclei.

1 Introduction: PAHs as a Link Between the Interstellar Medium and the Solar System

The major goals of cometary science are to determine the chemical composition and physical structure of cometary nuclei and to shed light on the origin of the solar system. It is now widely recognized that comets formed in the cold outer regions of the solar nebula ($\sim$5–55 AU from the Sun, well beyond the "snowline") and have been stored in two distant reservoirs; (i.e. the Oort cloud and the Kuiper Belt) for most of the age of the solar system ($\sim 4.6\times10^9$ yrs). Because of their cold formation and cold storage,[1] it is therefore also widely believed that comets are the most primitive objects in the solar system.

[1] Long-period comets (with orbital periods ranging from 200 yrs up to 10^7 yrs) originate in the Oort Cloud ($\sim$3000–50,000 AU from the Sun) and formed in the giant planets region from Jupiter to Neptune of the pre-planetary nebula ($\sim$5–40 AU from the Sun), where the nebula temperature ranged from >120 K near Jupiter to <30 K near Neptune. The source of Jupiter-family short-period comets (with periods shorter than 20 yrs), formed further out than long-period comets in the trans-Neptune region, is the Kuiper Belt ($\sim$30–50 AU from the Sun). Halley-family short-period comets (with periods of $20\,\text{yrs} < P < 200\,\text{yrs}$) originally came from the Oort cloud and then have been scattered into short-period type orbits by the perturbation of Jupiter and/or Saturn.

However, there is no consensus on to what extent comets preserve the composition of the presolar molecular cloud and the early stages of the protosolar nebula. A compelling theory is that comets are made of unaltered pristine interstellar materials with only the most volatile components partially evaporated [36]. Alternatively, it has also been proposed that cometary materials have been subjected to evaporation, recondensation and other reprocessing in the protosolar nebula and therefore have lost all the records of the presolar molecular cloud out of which they have formed.

Polycyclic aromatic hydrocarbon (PAH) molecules, composed of fused benzene rings (see Fig. 1 for illustration), and a significant constituent of the interstellar medium (ISM) of the Milky Way and external galaxies (see Sect. 2), would also be present in comets if they indeed contain unprocessed interstellar matter. The detection of PAHs in comets would be an important link between the ISM and comets and provides important clues on the processes that occurred during the formation of our solar system.

In this review I attempt to compile all possible evidence for the presence of PAHs in comets (Sect. 4), focusing on ground-based and space-borne spectroscopy of infrared vibrational and ultraviolet (UV) fluorescent emission spectra of comets, and laboratory analysis of stratospherically collected interplanetary dust particles (IDPs) thought to be cometary in origin and cometary samples returned to Earth by the *Stardust* spacecraft, with the latter providing the most unambiguous proof. Using the PAHs in the ISM (Sect. 2) and circumstellar disks (Sect. 3) as a comparison basis, the physical and chemical nature and source of cometary PAHs are discussed in Sect. 5.

2 Ubiquity of PAHs in the Interstellar Medium of the Milky Way and External Galaxies

PAHs reveal their presence in the ISM by emitting a distinctive set of emission features at 3.3, 6.2, 7.7, 8.6, and 11.3 μm (which are also collectively known as the "Unidentified Infrared" [UIR] emission bands).[2] Since their first detection in the planetary nebulae NGC 7027 and BD+30°3639 [35], PAHs have been observed in a wide range of Galactic and extragalactic regions (see Fig. 2 and [24]).

[2] These "UIR" emission features are now generally identified as vibrational modes of PAHs [51, 2]: C–H stretching mode (3.3 μm), C–C stretching modes (6.2, 7.7 μm), C–H in-plane bending mode (8.6 μm), and C–H out-of-plane bending mode (11.3 μm). Other C–H out-of-plane bending modes at 11.9, 12.7 and 13.6 μm have also been detected. The wavelengths of the C–H out-of-plane bending modes depend on the number of neighboring H atoms: 11.3 μm for solo–CH (no adjacent H atom), 11.9 μm for duet–CH (2 adjacent H atoms), 12.7 μm for trio–CH (3 adjacent H atoms), and 13.6 μm for quartet–CH (4 adjacent H atoms). Other prominent features are the C–C–C bending modes at 16.4 μm [63], 17.1, 17.8, 18.9 μm [5, 80, 87, 92].

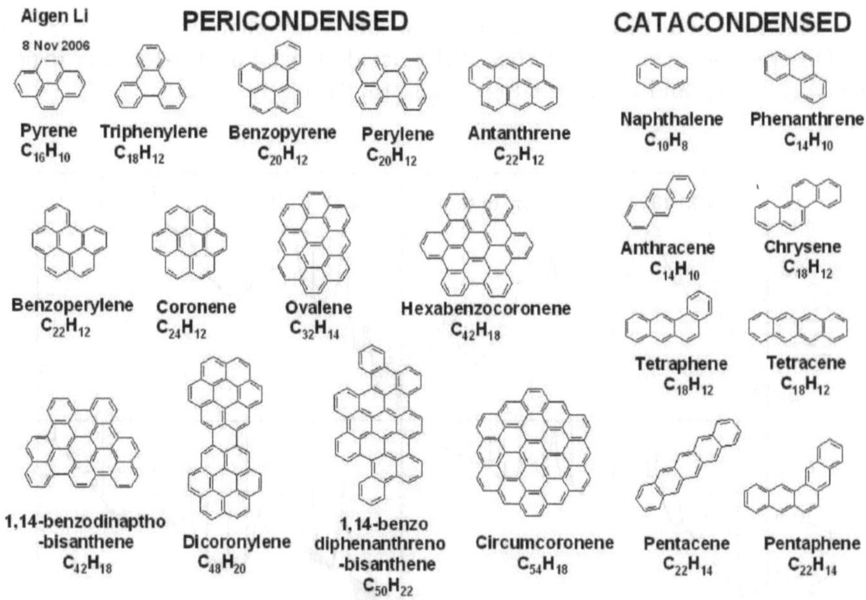

Fig. 1. Structures of 21 *specific* PAH molecules. Both compact pericondensed molecules (*pyrene*, *perylene*) and thermodynamically less favoured catacondensed molecules (*naphthalene*, *phenanthrene*) and their alkylated homologs were identified in the *Stardust* samples [73, 14]. Naphthalene, phenanthrene and their alkylated derivatives were found in IDPs possibly of cometary origin [13]. PAHs with up to 7 rings and their alkyl derivatives are abundant in carbonaceous chondrites [75].

In the Milky Way diffuse ISM, PAHs, containing ~ 45 ppm (parts per million, relative to H) C, account for $\sim 20\%$ of the total power emitted by interstellar dust [53]. The *ISO* (Infrared Space Observatories) and *Spitzer* imaging and spectroscopy have revealed that PAHs are also a ubiquitous feature of external galaxies [84, 81]. Recent discoveries include the detection of PAH emission in a wide range of systems: distant Luminous Infrared Galaxies (LIRGs) with redshift z ranging from 0.1 to 1.2 [27], distant Ultraluminous Infrared Galaxies (ULIRGs) with redshift $z \sim 2$ [94, 95], distant luminous submillimeter galaxies at redshift $z \sim 2.8$ [59], the distant Cloverleaf lensed QSO at redshift $z \sim 2.56$ [60], elliptical galaxies with a hostile environment (containing hot gas of temperature $\sim 10^7$ K) where PAHs can be easily destroyed through sputtering by plasma ions [48], faint tidal dwarf galaxies with metallicity $\sim Z_{\odot}/3$ [39], and galaxy halos (extending ~ 6.5 kpc from the plane of NGC 5907 [43] and >9.5 kpc from the plane of M 82 [29]).

However, the PAH features are weak or even absent in AGNs (as first noticed by Roche et al. [71]) and low-metallicity galaxies [83, 41, 28, 42, 61, 93, 72, 25]. The exact reason for the deficiency or lack of PAHs in low-metallicity galaxies and AGNs is not clear. It is generally interpreted as the destruction of PAHs (1) by hard UV photons (e.g. see [70]) or supernova-driven shocks [67] in metal-poor galaxies,[3] and (2) by extreme UV and soft X-ray photons in AGNs [89, 90, 76]. Observations also suggest that PAHs are destroyed in star-forming regions with very strong and hard radiation fields [15, 31, 6].

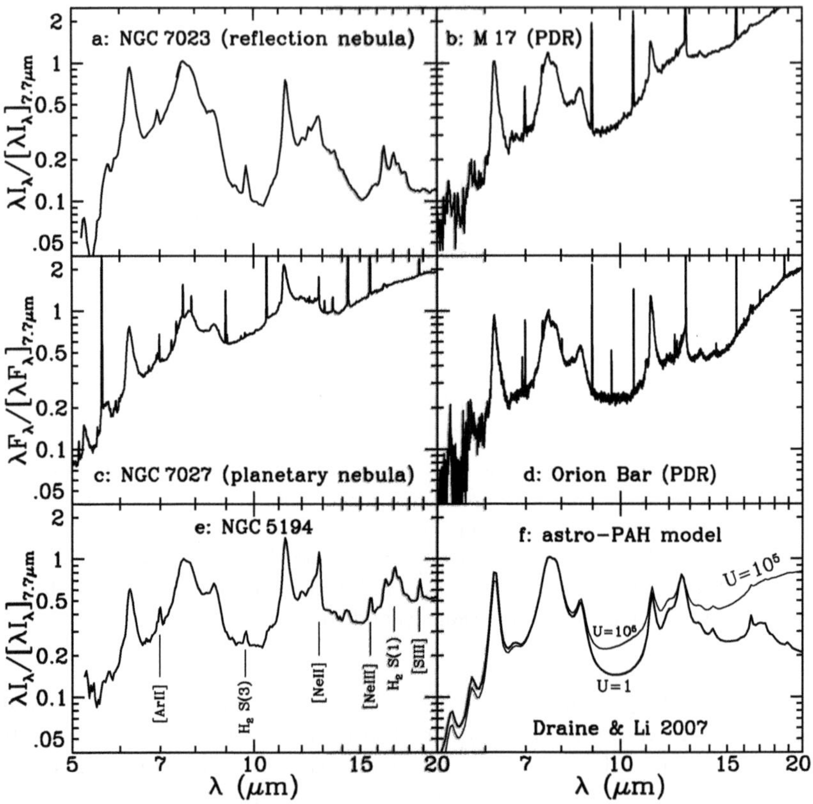

Fig. 2. Observed 5–20 μm spectra for: (**a**) Reflection nebula NGC 7023 [92]; (**b**) Orion Bar photodissociated region (PDR; [88]); (**c**) M17 PDR [69]; (**d**) Planetary nebula NGC 7027 [86]; (**e**) Seyfert Galaxy NGC 5194 [81]. Also shown (**f**) is the emission calculated for the astro-PAH model of Draine & Li [24], illuminated by starlight with an intensity $U = 1$ and 10^5 times of the local interstellar radiation field of Mathis et al. [62]. Taken from Draine & Li [24].

[3] Alternatively, these low-metallicity galaxies may be truly young and PAHs have simply not had time to form due to the delayed injection of carbon molecules from low-mass stars into the ISM [26].

3 PAHs in Circumstellar Dust Disks

Dust disks around young stars, depending on their age, are the source material or the remnants of newly-formed planets, asteroids, and comets. There exists observational evidence for the presence of PAHs in protoplanetary disks around the intermediate-mass (~ 2–$10\,M_\odot$) pre-main-sequence (PMS) Herbig Ae/Be (hereafter HAeBe) stars and their low-mass ($< 2\,M_\odot$) analog T Tauri stars, as well as debris disks around main-sequence (MS) stars (see Fig. 3 for illustration).

- From an analysis of the space-borne and ground-based spectra of 41 HAeBe stars in the 3 μm region, Brooke et al. [8] reported a firm detection of the 3.3 μm PAH C–H stretching emission feature in $\sim 20\%$ of these objects.

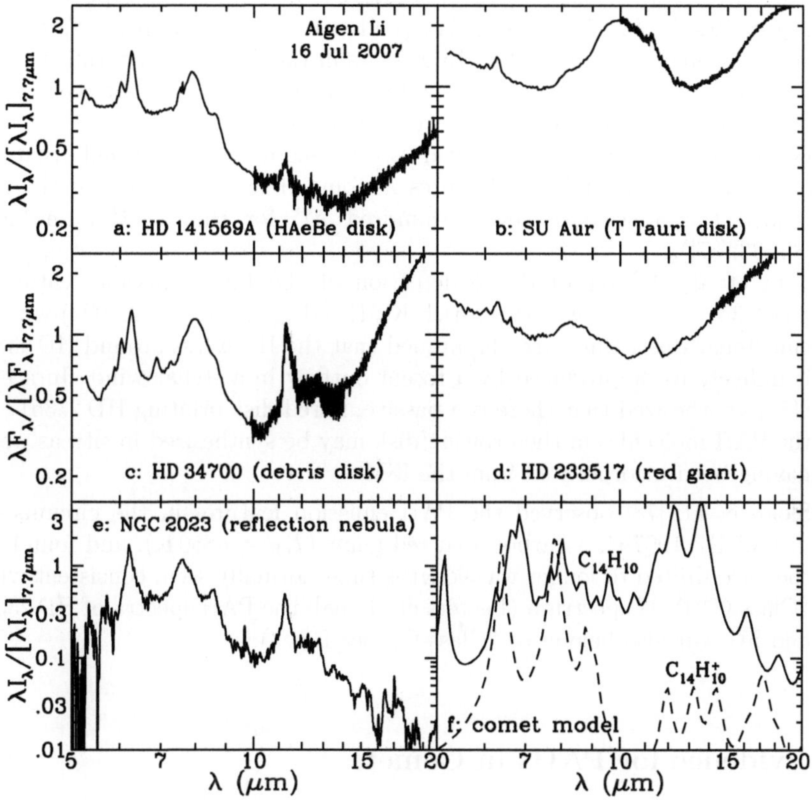

Fig. 3. Observed 5–20 μm spectra for: (**a**) Protoplanetary disk around HAeBe star HD 141569A (B9.5V; $T_{\rm eff} \approx 10{,}000$ K; [77]); (**b**) Protoplanetary disk around T Tauri star SU Aur (G1III; $T_{\rm eff} \approx 5945$ K; [32]); (**c**) Debris disk around HD 34700 (G0V; $T_{\rm eff} \approx 6000$ K; [91]); (**d**) Circumstellar disk around red giant HD 233517 (K2III; $T_{\rm eff} \approx 4390$ K; [45]); (**e**) Reflection nebula NGC 2023 (illuminated by HD 37903 [B1.5V; $T_{\rm eff} \approx 22{,}000$ K]; [88]). Also shown (**f**) is the emission calculated for phenanthrene $C_{14}H_{10}$ and its cation $C_{14}H_{10}^{+}$ at $r_{\rm h} = 1$ AU from the Sun [54].

- Acke & van den Ancker [1] found that the PAH features have been detected in ~57% of the 46 HAeBe stars for which the *ISO* spectroscopic data are available.
- Recent *Spitzer* observations have obtained the PAH spectra of over 20 HAeBe stars [77, 50]. Spectral variations among these stars and their deviations from typical interstellar PAH features were reported (see Fig. 3).
- Geers et al. [34] analyzed the *Spitzer* IRS spectra of 38 T Tauri stars and found PAHs in at least 8% (or probably as much as 45%) of these objects. The PAH spectra of T Tauri stars appear to be quite different from those of HAeBe stars and those typical in the ISM (see Fig. 3).
- The PAH emission features have also been detected in UV-poor dust debris disks around F- and G-type MS stars (e.g. SAO 206462 of spectral type F8V with $T_{\rm eff} \approx 6250$ K [18], and HD 34700 of spectral type G0V with $T_{\rm eff} \approx 5940$ K [82, 79; see Fig.87,84]). However, in an extensive *Spitzer* IRS spectroscopic survey of 111 T Tauri stars in the Taurus star-forming region, Furlan et al. [32] found that the PAH emission bands are not seen in dust disks around T Tauri stars of spectral type later than G1.
- Ground-based spatially resolved spectroscopy has revealed that the 3.3–12.7 μm PAH emission features in some HAeBe disks are spatially extended (on a scale of several hundred AU for the 6.2–12.7 μm bands; [85, 33, 37]).
- Jura et al. [45] reported the detection of the PAH emission features in HD 233517, an evolved oxygen-rich K2III red giant ($T_{\rm eff} \approx 4390$ K) with circumstellar dust. But Jura [44] argued that the IR excess around HD 233517 is unlikely to be produced by a recent outflow in a stellar wind. Jura et al. [45] hypothesized that there is a passive, flared disk orbiting HD 233517 and the PAH molecules in the orbiting disk may be synthesized in situ as well as having been incorporated from the ISM.

  Sloan et al. [78] observed the PAH emission features in the circumstellar disk of HD 100764, a carbon-rich red giant ($T_{\rm eff} \approx 4850$ K), and found that they are shifted to longer wavelengths than normally seen, consistent with a "Class C" PAH spectrum (Peeters et al. [68]; the PAH spectra of HD 233517 and SU Aur also belong to "Class C"; see Fig. 3).

4 Evidence for PAHs in Comets

The presence of PAHs in comets was an open issue in the pre-*Stardust* era. One of the most important discoveries of the *Stardust* Discovery Mission was the first ever most unambiguous detection of PAHs in a comet. Below, I summarize all

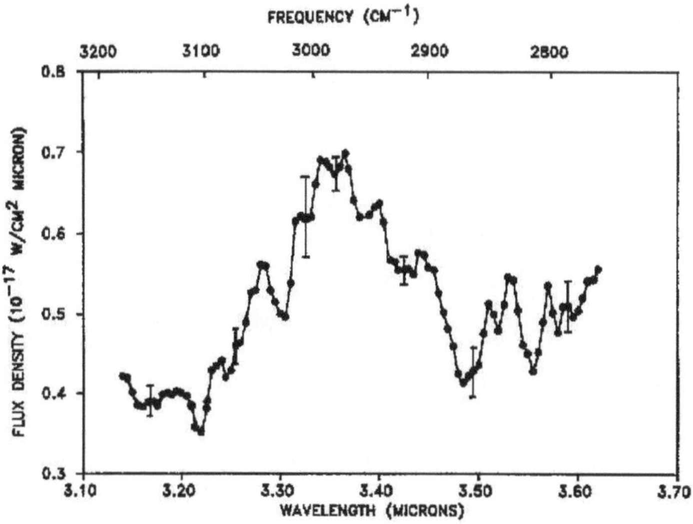

Fig. 4. IR emission spectrum of comet Halley obtained by Baas et al. [4] with the 3.8 m United Kingdom IR Telescope (UKIRT) on Mauna Kea on 1986 April 25 at a heliocentric distance of $r_h = 1.6$ AU. In addition to the major feature at 3.36 μm (often known as the "cometary organic feature"; the 3.33 μm ν_2 band and 3.37 μm ν_9 band of methanol CH_3OH account for about half its total intensity; [7]), there were also subsidiary peaks at 3.28 and 3.52 μm (ν_3 band of methanol; [40]). Taken from Baas et al. [4].

tentative evidence in the pre-*Stardust* era which may suggest the presence of PAHs in comets and finally, the more definite proof from the *Stardust* mission.[4]

- Baas et al. [4] by the first time reported the detection of a discrete emission feature at 3.28 μm in comet Halley after perihelion, at heliocentric distances of $r_h = 1.6$ AU (on 1986 April 25; see Fig. 4) and 2.0 AU (on 1986 May 24). This feature was also seen in comet Levy at $r_h = 1.4$ AU (on 1990 August 27; [20]; see Fig. 5). It was tentatively attributed to the C–H stretching mode of PAHs (e.g. see [7]). However, other species, such as CH_4 and OH prompt emission are also contributing at this wavelength (e.g. see [64]).
  Using a $C_{14}H_{10}i^2$-fitting technique, Lisse et al. [56] found the 6.2, 7.7, 8.6 and 11.3 μm emission features of PAHs in the spectrum of the *Deep Impact* ejecta of comet Tempel 1 (at $r_h = 1.51$ AU), obtained with *Spitzer* 45 minutes after

[4] We should note that PAHs (e.g. naphthalene, phenanthrene, pyrene, chrysene, perylene, benzo[ghi]perylene, and coronene; see Fig. 1) with varying degrees of alkylation have been identified in primitive carbonaceous chondritic meteorites [75, 21]. Although most primitive meteorites are *asteroidal* (originated in the asteroid belt, somewhere between 2–5 AU from the Sun), Campins & Swindle [10] argued that some meteorites may have a *cometary* origin (but so far no known meteorites appear to come from comets).

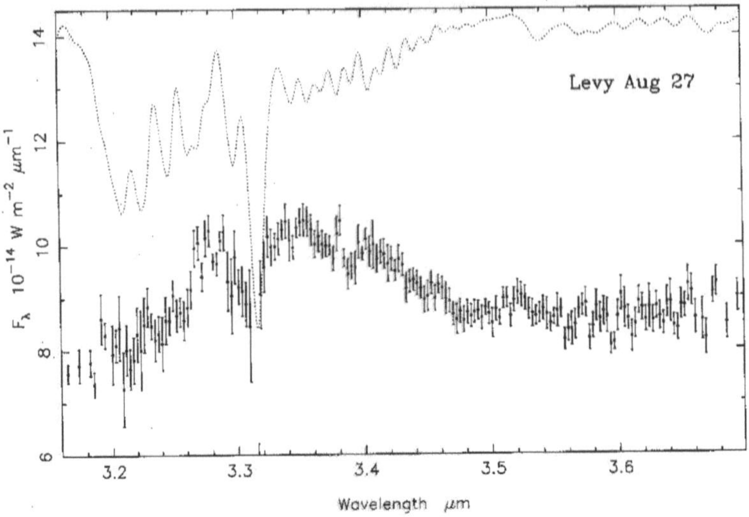

Fig. 5. IR emission spectrum of comet Levy obtained by Davies et al. [20] with UKIRT on 1990 August 27 at $r_h = 1.4$ AU. The 3.28 µm feature was as prominent as the broad 3.4 µm (which was resolved into two components peaking at 3.35 µm and 3.41 µm). The dotted curve was the relative transmission of the atmosphere above Mauna Kea. Comet Levy had the strongest 3.28 µm feature relative to the 3.4 µm feature among all comets. Taken from Davies et al. [20].

impact. This identification did not rely on a search for individual spectral features, but from the decrease of the residual χ^2 after a fit of the observed spectrum by a modeled spectrum.[5] Using the same technique, Lisse et al. [57] re-analyzed the *ISO* spectrum of Hale-Bopp at $r_h = 2.8$ AU taken on 1996 October 6 and found PAH signals in this comet (but see [19, 18]).

- Moreels et al. [65] reported the detection of four emission bands at 347, 356, 364 and 374 nm in the near-UV spectrum of comet Halley at $r_h = 0.83$ AU obtained with the Three-Channel-Spectrometer (TKS) on board the Vega-2 spacecraft (see Fig. 6). They attributed these bands to the fluorescence of phenanthrene ($C_{14}H_{10}$). But the production rate of phenanthrene alone required to account for the observed band intensities ($\sim 1.5 \times 10^{27}$ mol s^{-1}) would be ~ 100 times higher than that of PAHs derived from the 3.28 µm feature [7]. More recently, Clairemidi et al. [12] identified in the TKS-Vega spectrum of the inner coma of comet Halley a broad-band emission feature between 340 and 390 nm with 3 peaks at 371, 376 and 382 nm. They tentatively attributed these bands to pyrene ($C_{16}H_{10}$).

[5] In the ground-based 7.8–13.2 µm mid-IR spectra of comet Tempel 1 obtained with *Gemini-N* 61–94 minutes after impact, Harker et al. [38] possibly detected emission from PAHs at 8.25 and 8.6 µm (perhaps also at 12.4 and 12.9 µm). It is unclear whether the 11.3 µm PAH feature was present or not since it might have been hidden by the much stronger 11.3 µm crystalline olivine feature.

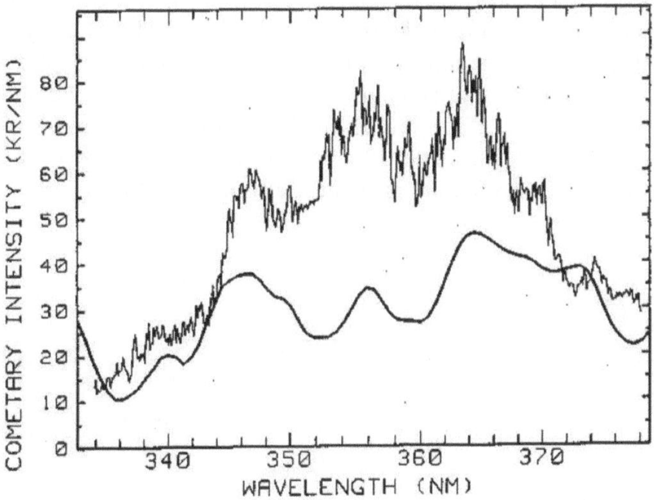

Fig. 6. Comparison of the near-UV emission spectrum of comet Halley measured by TKS-Vega at $r_h = 0.83$ AU (on 1986 March 9) with the experimentally-measured laser-induced fluorescence spectrum of phenanthrene (*upper curve*). Three main peaks coincide at 347, 356, and 364 nm. Taken from Moreels et al. [65].

- Clemett et al. [13] identified small PAH molecules (including naphthalene $C_{10}H_8$ and phenanthrene) and their alkylated derivatives in the microprobe two-step laser desorption laser ionization mass spectrometry (μL^2MS) of IDPs possibly of cometary origin.[6]
- Very recently, Sandford et al. [73], Keller et al. [49], and Clemett et al. [14] analyzed the cometary materials returned to Earth by the *Stardust* spacecraft and clearly demonstrated that PAHs are present in comet Wild 2, the *Stardust* mission target: (1) The L^2MS (two-step laser desorption laser ionization mass spectrometry) spectra of the *Stardust* samples revealed the presence of

[6] Stratospheric IDPs are believed to have originated primarily from asteroids and short-period comets (i.e. collisional debris from main belt asteroids and cometary dust captured in Earth's atmosphere when spiraling toward the Sun due to Poynting-Robertson drag). Atmospheric entry velocities, as determined from the atmospheric entry temperatures measured from the stepped He-release method [66] and the atmospheric entry model of Love & Brownlee [58], have been used to distinguish between IDPs arising from comets and asteroids, based on the marked differences between typical asteroidal and cometary orbits – asteroidal IDPs spiraling in toward the Sun from low inclination, low eccentric asteroidal sources will enter Earth's atmosphere, on average, at relatively lower velocities than cometary IDPs [30, 47]. Very recently, Joswiak et al. [47] analyzed 31 IDPs and distinguished two groups: 12 porous cometary IDPs with atmospheric entry velocities $> 18\,\mathrm{km\,s^{-1}}$, an average density of $\sim 1.0\,\mathrm{g\,cm^{-3}}$, and an anhydrous mineralogy; 4 more compact asteroidal IDPs mainly composed of hydrated minerals with atmospheric entry velocities $< 14\,\mathrm{km\,s^{-1}}$, and an average density of $\sim 3.3\,\mathrm{g\,cm^{-3}}$.

naphthalene, phenanthrene, pyrene, perylene ($C_{20}H_{12}$) and their alkylated homologs extending up to at least C_4-alkyl ([73, 14]; Fig. 7); (2) The Raman spectra of the *Stardust* samples exhibit pronounced "D" ($\sim 1360\,\Delta\,\mathrm{cm}^{-1}$) and "G" ($\sim 1580\,\Delta\,\mathrm{cm}^{-1}$) bands, characteristic of highly disordered sp^2-bonded aromatic carbon ([73]; Fig. 8); and (3) The Fourier transform infrared (FTIR) spectra clearly show an absorption feature at $3050\,\mathrm{cm}^{-1}$, corresponding to the aromatic C–H stretching mode at $3.28\,\mu\mathrm{m}$ ([49]; Fig. 9).

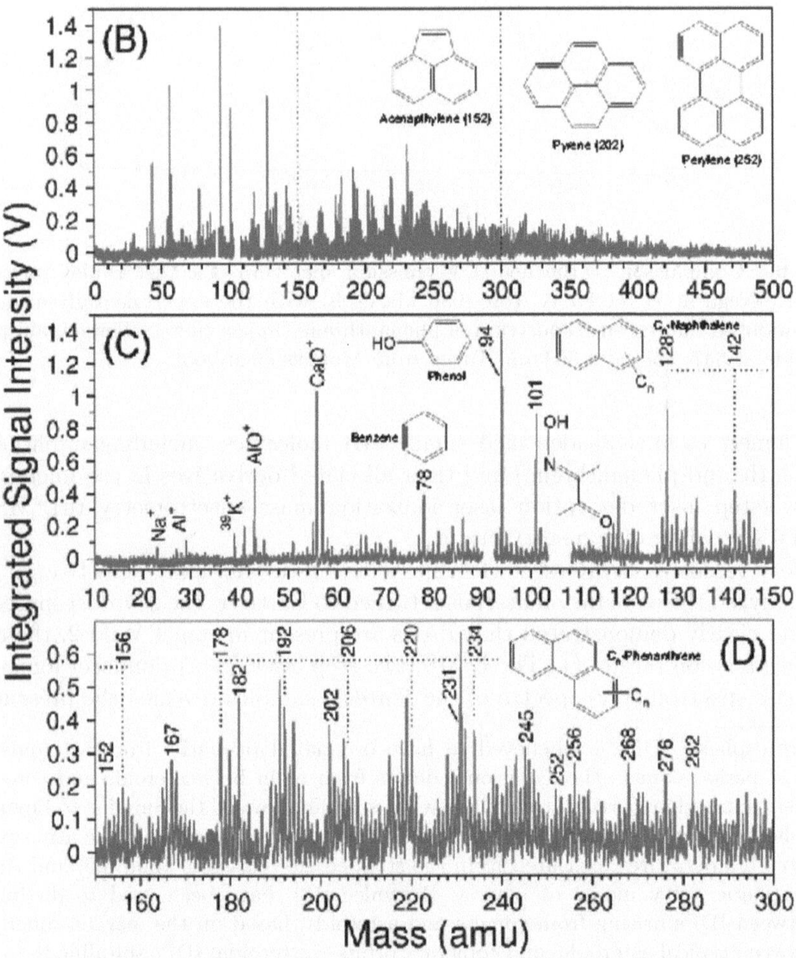

Fig. 7. PAH mass spectrum distribution for a *Stardust* sample obtained with the two-step laser mass spectrometry. The most commonly found PAH species are naphthalene ($C_{10}H_8$), phenanthrene ($C_{14}H_{10}$), pyrene ($C_{16}H_{10}$), perylene ($C_{20}H_{12}$), and their alkylated homologs. Interspersed within these species is a rich suite of auxiliary peaks which appears to represent the presence of O and N substitution, where the heterofunctionality being external to aromatic structure. Taken from Clemett et al. [14].

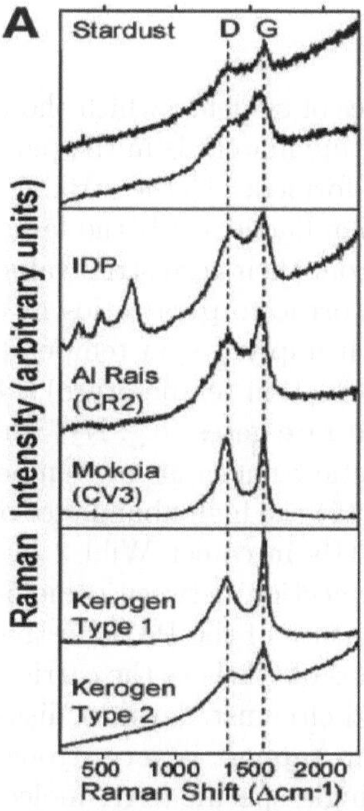

Fig. 8. Raman spectra of two *Stardust* particles (*top panel*) compared with the spectra of organics from extraterrestrial (*middle panel*) and terrestrial (*bottom panel*) carbonaceous materials. All exhibit "D" and "G" bands characteristic of disordered sp^2-bonded aromatic carbon. Taken from Sandford et al. (2006).

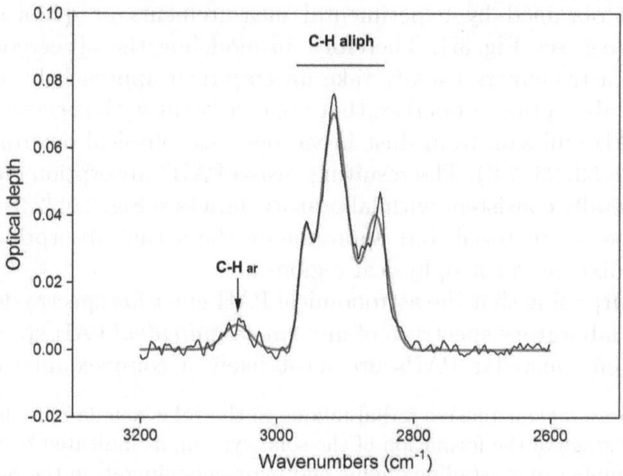

Fig. 9. Continuum-subtracted FTIR absorption spectrum of a *Stardust* sample. Both aromatic ($\sim 3050\,\mathrm{cm}^{-1}$) and aliphatic ($\sim 2967$, 2928, $2872\,\mathrm{cm}^{-1}$) C–H stretching bands are clearly seen. Taken from Keller et al. [49].

5 Discussion

There is a long series of pieces of evidence which show that comets have (at least partially) preserved the pristine materials in the parent interstellar cloud out of which the solar system has formed: (1) the striking similarities in the composition of cometary and interstellar ices; (2) the large deviations of the isotopic ratios for several elements from their terrestrial values (especially the high deuterium abundance); (3) the ortho-to-para ratios of cometary water, NH_3, and CH_4 in several comets which imply a spin temperature of ~ 30 K (which may be characteristics of their formation temperature); (4) the presence of volatiles, supervolatiles (e.g. CO), and rare gases (e.g. N_2) which may indicate that they were incorporated into cometary nuclei at low temperatures (e.g. 22 K for pure CO ice);[7] and probably also (5) the high abundance of cometary HNC (see [17]). The *definite* detection of PAHs in comet Wild 2 by *Stardust* provides another piece of evidence for the connection between comets and the ISM and has profound implications for the nature of the PAHs in the ISM and dust disks.

Admittedly, the hypothesis of PAHs as the carrier of the "UIR" bands widely seen in the ISM (Sect. 2) and circumstellar dust disks (Sect. 3) is still a hypothesis, although the evidence in support is very strong – so far there is no actual precise identification of a single specific PAH molecule in interstellar space or dust disks,[8] although the PAH model is quite successful in explaining the *general* pattern of band positions, relative intensities, and profiles observed in the "UIR" emission spectra, in terms of mixtures of highly vibrationally excited neutral and charged PAHs.

Details of the "UIR" spectra (precise band positions, bandwidths, and relative band intensities) remain hard to mimic exactly with the use of available PAH spectra obtained by experimental measurements or quantum chemical calculations (e.g. see Fig. 3f). Therefore, in modeling the observed PAH emission spectra, astronomers usually take an empirical approach by constructing "astro-PAH" absorption properties that are consistent with spectroscopic observations of PAH emission from dust in various astrophysical environments (e.g. see [22, 74, 52, 53, 23, 24]). The resulting "astro-PAH" absorption cross sections, although generally consistent with laboratory data (see Fig. 2 of [24]), do not represent any specific material, but approximate the actual absorption properties of the PAH mixture in astrophysical regions.

It is not surprising that the astronomical PAH *emission* spectra do not closely resemble the laboratory spectrum of any *single individual* PAH species since interstellar or circumstellar PAHs are most likely a complex mixture of many

[7] However, large-scale extensive radial mixing in the solar nebula must have occurred at the early stage of the formation of the solar system, as indicated by the detection of a large number of crystalline olivine and pyroxene minerals in the *Stardust* comet samples that, based on their solar isotopic compositions, appear to have formed in the inner regions of the solar nebula [9]. Also, the silicate dust in the diffuse ISM is predominantly amorphous [52].

[8] Cernicharo et al. [11] reported the detection of benzene (C_6H_6), the basic aromatic unit, in the protoplanetary nebula CRL 618.

individual molecules, radicals, and ions. As a matter of fact, Allamandola et al. [3] have demonstrated that the laboratory *absorption* spectra produced by co-adding different PAH spectra were able to provide a detailed match to the observed *emission* spectra.

One may still argue that the approach taken by Allamandola et al. [3] was not perfectly appropriate because they compared the astronomical *emission* spectra with the co-added laboratory *absorption* spectra, while the IR emission spectrum of a PAH molecule does not only depend on its IR absorption spectrum, but also depends on its absorption at shorter wavelengths, its heat capacity, and the intensity and spectral shape of the illuminating radiation field (e.g. see [23]). This, together with the detection of individual *specific* PAH molecules in the *Stardust* samples, suggests that it would be of great value to study the excitation, emission, and destruction of a large number of specific PAH molecules in the ISM, dust disks, and cometary comae, as well as the mechanism of releasing PAHs from the ice mantles of dust in comets and dust disks to the gas phase and their lifetime against photodestruction and photoionization (e.g. see [46, 55, 54]).

Acknowledgements I thank J. Crovisier, D.E. Harker, C.M. Lisse, M.J. Mumma, and D.H. Wooden for helpful discussions. I thank G.C. Clayton, V.C. Geers, M. Jura, K.H. Kim, S.L. Liang, B. Sargent, G.C. Sloan, and T.L. Smith for their help in preparing of this article. Partial support by Spitzer Theory Programs and a HST Theory Program is gratefully acknowledged.

References

1. Acke, B., & van den Ancker, M.E. 2004, A&A, 426, 151
2. Allamandola, L.J., Tielens, A.G.G.M., & Barker, J.R. 1985, ApJ, 290, L25
3. Allamandola, L.J., Hudgins, D.M., & Sandford, S.A. 1999, ApJ, 511, L115
4. Baas, F., Geballe, T.R., & Walther, D.M. 1986, ApJ, 311, L97
5. Beintema, D.A., et al. 1996, A&A, 315, L369
6. Beirão, P., et al. 2006, ApJ, 643, 1
7. Bockelée-Morvan, D., Brooke, T.Y., & Crovisier, J. 1995, Icarus, 116, 18
8. Brooke, T.Y., Tokunaga, A.T., & Strom, S.E. 1993, AJ, 106, 656
9. Brownlee, D.E., et al. 2006, Science, 314, 1711
10. Campins, H., & Swindle, T.D. 1998, Meteoritics & Planet. Sci., 33, 1201
11. Cernicharo, J., et al. 2001, ApJ, 546, L123
12. Clairemidi, J., Bréchignac, P., Moreels, G., & Pautet, D. 2004, Planet. Space Sci., 52, 761
13. Clemett, S.J., Maechling, C.R., Zare, R.N., Swan, P.D., & Walker, R.M. 1993, Science, 262, 721
14. Clemett, S.J., Nakamura-Messenger, K., McKay, D.S., & Sandford, S.A. 2007, Lunar & Planet. Inst. Conf. Abs., 38, 2091
15. Contursi, A., et al. 2000, A&A, 362, 310
16. Coulson, I.M., & Walther, D.M. 1995, MNRAS, 274, 977
17. Crovisier, J., 2006, Mol. Phys., 104, 2737

18. Crovisier, J., & Bockelée-Morvan, D. 2007, Molecules in Space and Laboratory, meeting held in Paris, France, May 14–18, 2007. Editors: J.L. Lemaire, F. Combes. Publisher: S. Diana, 19
19. Crovisier, J., et al. 2000, in ASP Conf. Ser. 196, Thermal Emission Spectroscopy and Analysis of Dust, Disks, and Regoliths, 109
20. Davies, J.K., Green, S.F., & Geballe, T.R. 1991, MNRAS, 251, 148
21. Derenne, S., Rouzaud, J.N., Clinard, C., & Robert, F. 2005, Geochimica et Cosmochimica Acta, 69, 3911
22. Désert, F.X., Boulanger, F., & Puget, J.L. 1990, A&A, 237, 215
23. Draine, B.T., & Li, A. 2001, ApJ, 551, 807
24. Draine, B.T., & Li, A. 2007, ApJ, 657, 810
25. Draine, B.T., et al. 2007, ApJ, 663, 866
26. Dwek, E. 2005, in AIP Conf. Proc. 761: SEDs of Gas-Rich Galaxies, 103
27. Elbaz, D., Le Floc'h, E., Dole, H., & Marcillac, D. 2005, A&A, 434, L1
28. Engelbracht, C.W., et al. 2005, ApJ, 628, L29
29. Engelbracht, C.W., et al. 2006, ApJ, 642, L127
30. Flynn, G.J. 1989, Icarus, 77, 287
31. Förster Schreiber, N. M., Roussel, H., Sauvage, M., & Charmandaris, V. 2004, A&A, 419, 501
32. Furlan, E., et al. 2006, ApJS, 165, 568
33. Geers, V.C., et al. 2005. Astrochemistry: Recent Successes and Current Challenges, Proceedings of the 231st Symposium of the International Astronomical Union held in Pacific Grove, California, USA, August 29- September 2. Poster sessions, 88
34. Geers et al. 2007, Astronomy and Astrophysics, 469, L35–L38
35. Gillett, F.C., Forrest, W.J., Merrill, K.M. 1973, ApJ, 183, 87
36. Greenberg, J.M. 1982, in Comets, ed. L.L. Wilkening (Tuscon: Univ. of Arizona Press), 131
37. Habart, E., Natta, A., Testi, L., & Carbillet, M. 2006, A&A, 449, 1067
38. Harker, D.E., Woodward, C.E., Wooden, D.H., Fisher, R.S., & Trujillo, C. 2007, Icarus, in press
39. Higdon, S.J., Higdon, J.L., & Marshall, J. 2006, ApJ, 640, 768
40. Hoban, S., Mumma, M.J., Reuter, D.C., Disanti, M., Joyce, R.R., & Storrs, A. 1991, Icarus, 93, 122
41. Houck, J.R. Roellig, T.L., & Van+Cleve, J. 2004, Optical, Infrared, and Millimeter Space Telescopes. Edited by Mather, John C. Proceedings of the SPIE, 5487, 62–76.
42. Hunt, L., Bianchi, S., & Maiolino, R. 2005, A&A, 434, 849
43. Irwin, J.A., & Madden, S.C. 2006, A&A, 445, 123
44. Jura, M., 2003, ApJ, 582, 1032
45. Jura, M., et al. 2006, ApJ, 637, L45
46. Joblin, C., Boissel, P., & de Parseval, P. 1997, Planet. Space Sci., 45, 1539
47. Joswiak, D.J., Brownlee, D.E., Pepin, R.O., & Schlutter, D.J. 2007, Workshop on Dust in Planetary Systems (ESA SP-643). September 26–30 2005, Kauai, Hawaii. Editors: Krueger, H. and Graps, A., 141–144
48. Kaneda, H., Onaka, T., & Sakon, I. 2005, ApJ, 632, L83
49. Keller, L.P., et al. 2006, Science, 314, 1728
50. Keller, L.D., et al. 2007, ApJ, in press
51. Léger, A., & Puget, J.L. 1984, A&A, 137, L5
52. Li, A., & Draine, B.T. 2001a, ApJ, 550, L213

53. Li, A., & Draine, B.T. 2001b, ApJ, 554, 778
54. Li, A., & Draine, B.T. 2007, in preparation
55. Li, A., & Lunine, J.I. 2003, ApJ, 594, 987
56. Lisse, C.M., et al. 2006, Science, 313, 635
57. Lisse, C.M., Kraemer, K.E., Nuth, J.A., Li, A., & Joswiak, D. 2007, Icarus, 187, 69
58. Love, S.G., & Brownlee, D.E. 1994, Meteoritics, 29, 69
59. Lutz, D., et al. 2005, ApJ, 625, L83
60. Lutz, D., et al. 2007, ApJ, 661, L25
61. Madden, S.C., Galliano, F., Jones, A.P., & Sauvage, M. 2006, A&A, 446, 877
62. Mathis, J.S., Mezger, P.G., & Panagia, N. 1983, A&A, 128, 212
63. Moutou, C., Léger, A., D'Hendecourt, L. 1996, A&A, 310, 297
64. Mumma, M.J., et al. 2001, ApJ, 546, 1183
65. Moreels, G., Clairemidi, J., Hermine, P., Brechignac, P., & Rousselott, P. 1994, A&A, 282, 643
66. Nier, A.O., & Schlutter, D.J. 1993, Meteoritics, 28, 675
67. O'Halloran, B., Satyapal, S., & Dudik, R.P. 2006, ApJ, 641, 795
68. Peeters, E., et al. 2002, A&A, 390, 1089
69. Peeters, E., et al. 2005, ApJ, 620, 774
70. Plante, S., & Sauvage, M. 2002, AJ, 124, 1995
71. Roche, P.F., Aitken, D., Smith, C., & Ward, M. 1991, MNRAS, 248, 606
72. Rosenberg, J.L., et al. 2006, ApJ, 636, 742
73. Sandford, S.A., et al. 2006, Science, 314, 1720
74. Schutte, W.A., Tielens, A.G.G.M., & Allamandola, L.J. 1993, ApJ, 415, 397
75. Sephton, M.A., et al. 2004, Geochimica et Cosmochimica Acta, 68, 1385
76. Siebenmorgen, R., Krügel, E., & Spoon, H.W.W. 2004, A&A, 414, 123
77. Sloan, G.C., et al. 2005, ApJ, 632, 956
78. Sloan, G.C., et al. 2007, ApJ, in press
79. Smith, T.L., Clayton, G.C., & Valencic, L. 2004, AJ, 128, 357
80. Smith, J.D.T., et al. 2004, ApJS, 154, 199
81. Smith, J.D.T., et al. 2007, ApJ, 656, 770
82. Sylvester, R. J., et al. 1997, MNRAS, 289, 831
83. Thuan, T.X., Sauvage, M., & Madden, S. 1999, ApJ, 516, 783
84. Tielens, A. G. G. M. 2005, The Physics and Chemistry of the Interstellar Medium (Cambridge: Cambridge Univ. Press)
85. van Boekel, R., et al. 2004, A&A, 418, 177
86. van Diedenhoven, B., et al. 2004, ApJ, 611, 928
87. van Kerckhoven, C., et al. 2000, A&A, 357, 1013
88. Verstraete, L., et al. 2001, A&A, 372, 981
89. Voit, G.M. 1991, ApJ, 379, 122
90. Voit, G.M. 1992, MNRAS, 258, 841
91. Watson, D.M., et al. 2007, in preparation
92. Werner, M.W., et al. 2004, ApJS, 154, 309
93. Wu, Y., et al. 2006, ApJ, 639, 157
94. Yan, L., et al. 2005, ApJ, 628, 604
95. Yan, L., et al. 2007, ApJ, 658, 778

53. Li, A., & Lunine, J.I. 2001b, ApJ, 554, 778 [illegible]
54. Li, A., & Draine, B.T. 2002, in preparation [illegible]
[illegible]
92. Wooden, D.H. [illegible]
[illegible]

Dynamical Modeling of the Deep Impact Dust Ejecta Cloud

T. Bonev[1] *, N. Ageorges[2], S. Bagnulo[2], L. Barrera[3], H. Böhnhardt[4], O. Hainaut[2], E. Jehin[2], H. U. Käufl[2], F. Kerber[2], G. LoCurto[2], J. Manfroid[5], O. Marco[2], E. Pantin[6], E. Pompei[2], I. Saviane[2], F. Selman[2], C. Sterken[7], H. Rauer[8], G. P. Tozzi[9], and M. Weiler[8]

[1] Institute of Astronomy, Bulgarian Academy of Sciences, Sofia, Bulgaria
tbonev@astro.bas.bg
[2] European Southern Observatory
[3] Universidad Metropolitana de Ciencias de la Educacion, Santiago de Chile, Chile
[4] Max-Planck-Institut für Sonnensystemforschung, Katlenburg-Lindau, Germany
[5] Université de Liège, Belgium
[6] Commissariat Energie Atomique, F-91191 Gif-sur-Yvette, France
[7] Vrije Universiteit Brussel, Belgium
[8] Deutsches Zentrum für Luft und Raumfahrt, Germany
[9] Istituto Nazionale di Astrofisica (INAF) Osservatorio di Arcetri, Italy

Abstract The collision of Deep Impact with comet 9P/Tempel 1 generated a bright cloud of dust which dissipated during several days after the impact. The brightness variations of this cloud and the changes of its position and shape are governed by the physical properties of the dust grains. We use a Monte Carlo model to describe the evolution of the post-impact dust plume. The results of our dynamical simulations are compared to the data obtained with FORS2[1] to derive the particle size distribution and the total amount of material contained in the dust ejecta cloud.

1 Introduction

Dynamical modeling of the dust coma is often used to constrain properties of the dust grains released from a cometary nucleus. An excellent summary of these efforts is given by [7]. In his review M. Fulle mentions that, in order to reach a better fit between model and observation, modelers are often pressed to make the assumption of time-independent particle size distribution (PSD). Observations of the Deep Impact (DI) dust ejecta cloud represent a special case in this respect. The time of the impact is exactly known (UT 5:52 on July 4, 2005) and the time interval in which particles excavated by the impact leave

* TB is indebted to the organizers for supporting his participation at this conference.
[1] FORS stands for FOcal Reducer and low dispersion Spectrograph for the Very Large Telescope (VLT) of the European Southern Observatory (ESO).

the circumnuclear region is relatively short (with some exceptions which are discussed below). Thus, time-dependence of the PSD could be excluded to the first approximation in model calculations (valid only if no further processing of the particles takes place later). The DI ejecta cloud was observed from the DI spacecraft [1] and from a great number of ground based observatories [11]. We present observations of comet 9P/Tempel 1 (hereafter 9P) obtained with FORS2 at the VLT of ESO in Paranal. We use a Monte Carlo model to describe the DI ejecta cloud observed during the 4 days after the impact. Inversion of our model allows to derive the PSD and, under appropriate assumptions, to make an estimation of the total mass of dust released by the impact and ejected with velocities higher than the escape velocity of 9P.

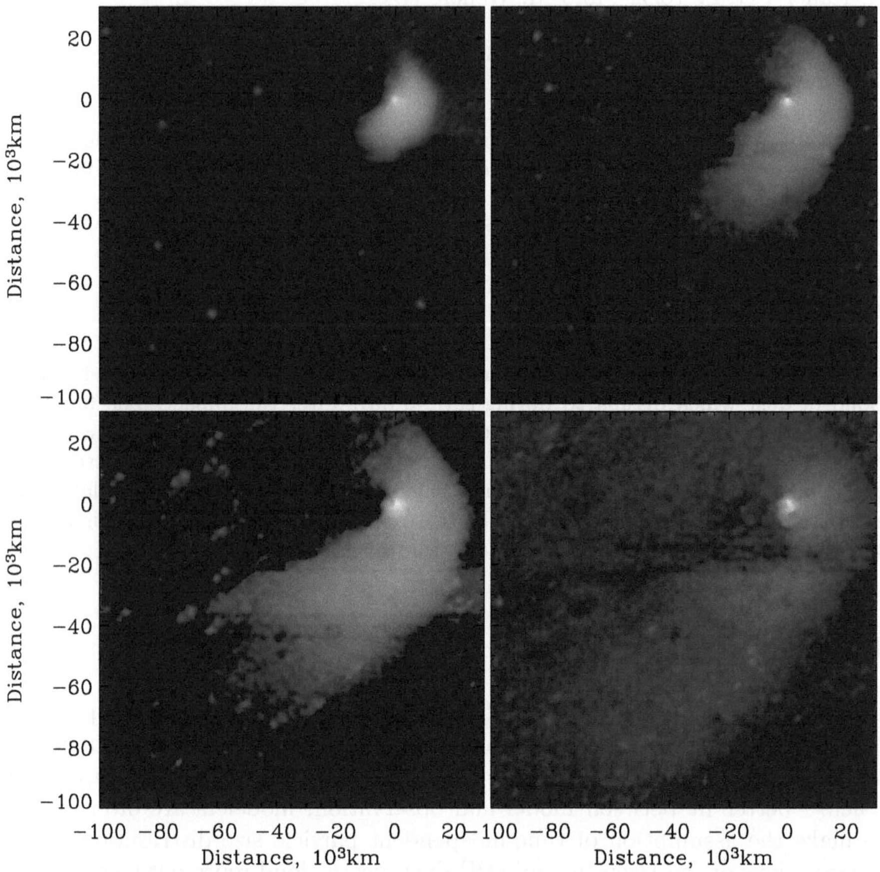

Fig. 1. Sequence of R-band images representing the dust cloud induced by the impact. From the left upper to the right lower corner the four panels show the changing dust distribution during the four post-impact days. Each panel contains the difference of the particular R-band image with a pre-impact image. North is up, East to the left. The projected direction to the Sun is at 291 degree, counted from North to East.

2 The Observations

The images used in this analysis were obtained with FORS2[2] [3], mounted at the VLT Antu. To secure a basis for comparison with the post-impact data several images were taken about 6 hours pre-impact, shortly before the time when the comet set at Paranal. During the next 4 nights images were obtained at 17.8, 42.4, 66.3, 89.4 hours after the impact. The heliocentric distance of 9P was 1.51 AU, almost constant during the observing period, as the comet was at perihelion on July 5.35. The geocentric distance increased from 0.89 to 0.91 AU, and the pixel scale changed correspondingly from 162 km/px to 167 km/px. The images were calibrated to fluxes and then transformed to Af, the Albedo-filling factor product [2]. Figure 1 shows the R-band post-impact images with subtracted contribution of the pre-impact coma. These images are used in the further analysis.

The quantity Af is convenient for comparison with theoretical models as it is directly related to the total cross-section of the scattering dust particles at any particular picture element. The total scattering cross-section × Albedo ($A \times S$) of all dust particles produced by the impact was obtained by integration over the whole area covered by the post-impact clouds shown in Fig. 1. The derived values are presented in Table 1. Note the reduction of $A \times S$ with time.

Table 1. Values of the product Albedo × total scattering cross-section ($A \times S$) derived from the post-impact images with subtracted pre-impact coma.

Time after impact, hour	$A \times S$ km^2
17.8	7.6
42.4	5.8
66.3	5.1
89.4	3.2

3 The Model

3.1 Initial Conditions

In the numerical model described below we use particles of radii from 0.25 to 250 μm. To run the model an initial guess for the velocity dependence on particle size is needed. The shape of the ejecta cloud (Fig. 1) shows that after the impact the dust expansion is initially at position angle (P.A.) 240° (the projected direction to the Sun is at P.A. 291°). Later it is deflected by the solar radiation pressure in antisolar direction. Four days after the impact most of the dust is spread over a large area in antisolar direction. But even on the fourth day (90 hour after the impact) dust particles are still found in direction to the Sun. We suppose that these are larger particles which are ejected with lower velocities and which are less influenced by the radiation pressure in comparison to smaller particles.

[2] for details see `http://www.eso.org/instruments/fors1/`

Assuming that the well expressed boundary in direction to the Sun is the stagnation region of particles ejected with velocity v, and measuring the distance d from the comet to this boundary, we can write $v = bt$ and $v^2 = 2bd$, where b is the radiation pressure acceleration. Measured values of d, and derived values for the velocity and acceleration are given in Table 2. The acceleration is used to derive values for β, the ratio of gravitational force to the force of radiation pressure. Finally, from β, we derive the radii of the particles, a, using the dependence [5]: $\beta = 0.585 \times 10^{-4} Q_{pr}/(\rho a)$, where ρ is the density of the particles. The rough estimation of the particles' radii, listed in Table 2, was made with an assumed value for the radiation pressure efficiency, $Q_{pr} = 1.7$, with a density, $\rho = 1\ \mathrm{g\,cm^{-3}}$ Our approach yields the initial velocities of the particles. Therefore

Table 2. Apex distance, measured in the four images, and derived values for the velocity, acceleration, and particle size.

UT Day of July 2005	Time after impact hour	Distance 10^3 km	Terminal velocity km $\mathrm{s^{-1}}$	Acceleration km $\mathrm{s^{-2}}$	β values	Particle radius µm
4.972	17.8	16.5 ± 2	0.51 ± 0.06	8.03E-6	3.01	0.3
5.995	42.4	21.0 ± 3	0.28 ± 0.04	1.81E-6	0.68	1.5
6.993	66.3	25.0 ± 4	0.21 ± 0.03	8.77E-7	0.33	3.0
7.955	89.4	29.0 ± 5	0.18 ± 0.03	5.60E-7	0.21	4.8

the values in Table 2 are about 2 times larger compared to the mean velocities given by many other observers ([11, 12]). The derived velocity-size dependence follows a power law with power index close to -0.4. This dependence resembles the velocities obtained from theoretical models of the natural gas-dust interaction in the vicinity of a cometary nucleus ([8, 6]), as well as the velocities derived from models describing the dust coma and tail ([7] and references therein). Initial guess for the location of the impact was found by the constraints coming from (a)the observed projected expansion direction of the ejected cloud, (b)the rotation axis orientation ([4]), and (c)the latitude of the impact (M. A'Hearn, private communication).

3.2 Direct Monte Carlo Calculations

The velocity law and the impact location were determined more precisely in a process of trial and error. We calculated a series of models with values around the initial guess until we reached a satisfying morphological reproduction of the observed dust distribution for the four observations.

In the model used for description of the impact cloud we use 1 million dust particles which are emitted for a period of 20 minutes starting at the moment of the impact. These particles are distributed in 100 emission events along 200 emission directions randomly spaced in a cone with full opening angle of 180 degree. The particles are assumed compact with density of 1 g/cm^3 and radii distributed logarithmically in 51 bins, in the range from 0.25 to 250 micrometer. After ejection the particles move along Keplerian orbits under the influence of gravity and radiation pressure. Their positions are calculated for the times of observation, the contributions of the different sizes are weighted with an initial guess for the PSD and integrated along the line of sight. The modeled brightness, at position (x, y) is described by:

$$B(x,y) = \sum_{i=0}^{50} K_i \times S_i(x,y) \qquad (1)$$

where $S_i(x, y)$ is the scattering area produced by the particles of one particular size, i, and K_i are the coefficients to be found.

3.3 Inversion of the Model

The coefficients K_i in Eq. (1) are derived by comparing the modeled brightness $B(x, y)$ to the observed one and by minimizing the differences through linear regression. Figure 2 shows the result of the fit to the dust cloud observed 17.8 hours after the impact. The full line in the left panel represents the initial guess for the PSD, a power law with power index −3.0. Triangles show the solution of Eq. (1) converted to the number of particles of given size. Particles larger than about 20 micron scatter strongly around the mean distribution. These large grains are expelled with lower velocities and their motion is less influenced by the radiation pressure in comparison to the smaller particles. Therefore large particles of several size bins coexist in a relatively small region around the nucleus and compete in the process of linear regression. It seems that we should simply restrict the upper limit of particle radii to smaller values. Indeed, this makes the solution more stable, but at the same time it removes the contribution to the brightness close to the nucleus. In order to reproduce the enhanced brightness of the ejecta plume observed in this region, particles of radii as high as 250 μm are needed. Extrapolation of our initial velocity law shows that these large particles have velocites still above the escape velocity of 9P, 1.7 $m\,s^{-1}$ [1]. The right panel in Fig. 2 shows the cumulative mass distribution of the dust. We derive a total mass of the dust in the ejecta cloud of 3.7 kiloton.

The inversion of the model succeeded only in the case of the ejecta cloud observed 17.8 hours after the impact, which is characterized by the highest signal-to-noise ratio. The dust distribution in the next images was reproduced with the parameters found from the fit to the first post-impact observation. The four modeled ejecta clouds corresponding to the four observations are presented in Fig. 3.

4 Results and Discussion

The PSD derived from the fit to the ejecta cloud observed 17.8 hours after the impact follows a power law with mean power index −3.0. In the detailed model

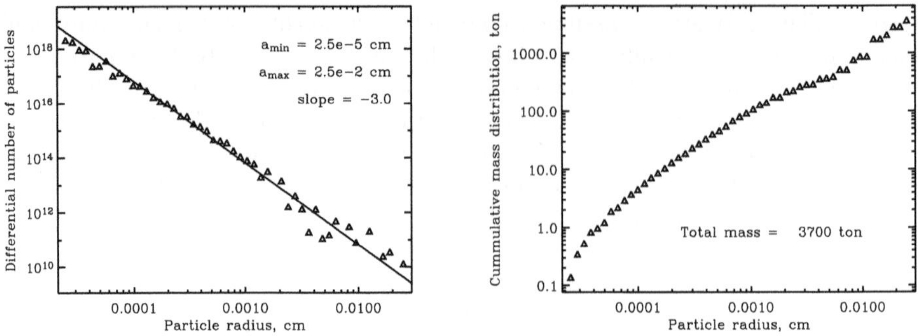

Fig. 2. The differential particle size distribution (*left*) and the cumulative mass distribution (*right*) derived from the fit of the model to the observations.

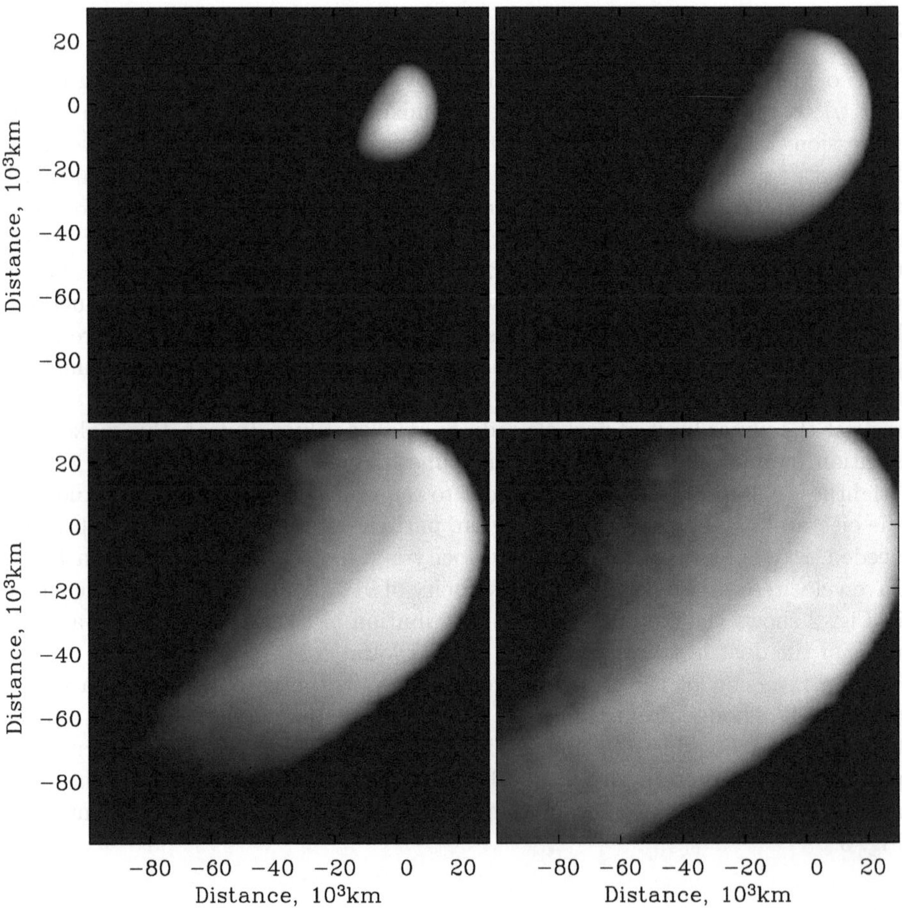

Fig. 3. Models of the DI dust ejecta cloud corresponding to the four observations.

developed by [9] particles of radii < 20 μm are used and a slope of the differential dust size distribution -3.2 is found. Light curves of the impact plume were obtained from space [10] and from the ground (Pittichova et al. (presentation at ACM'2005)). These light curves show similar behavior, first a sharp increase in a time interval dependent on the aperture used, and second, a gradual decrease of the brightness. The decreasing wing of these lightcurves is well described by subtracting the contribution of small particles leaving the detector diaphragm with the velocities used in our model and distributed in accordance with a PSD with slope -3.0.

The velocity distribution of the impact ejecta is similar to velocity laws which describe the natural activity of a comet. We came to this conclusion empirically, without analyzing possible mechanisms of particle acceleration related to the impact itself. Our conclusion is based on the particles having velocities greater than the escape velocity of 9P. At the same time our images with removed pre-impact coma show a brightness around the nucleus of the comet. Schleicher et al. [12] point to the same feature in their morphological analysis of the ejecta plume. They explain the enhanced brightness close tot he nucleus with the existence of large, heavy particles that have been ejected with velocities below the escape velocity of the comet. A future combined analysis of the motion of particles ejected with velocities above and below the escape limit could be of interes for understanding the acceleration mechanism of particles produced by the impact.

In Sect. 3.2 we have shown that the product scattering area × Albedo decreases with increasing time from the impact. It is interesting to extrapolate this trend back in time and to make a comparison with data obtained during the first hours after the impact. The Rosetta team registered a peak of the light curve 1 hour after the impact [10]. From the difference between this value and the pre-impact level these authors derived a total A × S of the newly created dust particles of 33 km^2. This is much greater compared to the value expected from the back extrapolation of our measurements, even if we fit them with an exponential law. This is an indication that the decrease of A × S should have been faster during the first hours after the impact. Possible explanation could be the fragmentation of particles with sizes comparable to the wavelength of observations. Their smaller products will scatter effectively at shorter wavelengths and, under given conditions, could become invisible in the R-band. Although this mechanism appears possible further work is needed to support it with quantitative arguments.

5 Conclusions

We used a dynamical model to describe the dust ejecta created by Deep Impact. The applied velocity-size dependence was derived empirically. Particle with radii in the range 0.25–250 μm were considered. We found a best fit to the dust cloud 17.8 hours after the impact with differential particle size distribution following a power law with index -3.0. The total mass of the dust dust ejecta having velocities greater than the escape velocity is 3700 ton.

References

1. M. A'Hearn, M. Belton, W. Delamere et al.: Science, **310**, 258 (2005)
2. M. A'Hearn, D. Schleicher, R. Millis et al.: Astronomical Journal, **89**, 579 (1984)
3. I. Appenzeller, K. Fricke, W. Furtig et al.: The Messenger, **94**, 1 (1998)
4. M. Belton, P. Thomas, B. Carcich et al.: The Spin State of 9P/Tempel 1. In *Proceedings 37th Annual Lunar and Planetary Science Conference, 2006*, ed by S. Mackwell, E. Stansbery, 1487
5. J. Burns, P. Lamy, S. Soter: Icarus, **40**, 1 (1979)
6. J.-F. Crifo: Hydrodynamic models of the collisional coma. In *ASSL Vol. 167: IAU Colloq. 116: Comets in the post-Halley era, 1991*, ed by R. Newburn, M. Neugebauer, J. Rahe, pp 937–989
7. M. Fulle: Motion of cometary dust, in *Comets II*, ed by M. Festou, H.U. Keller, H.A. Weaver, 565 (2004)
8. T. Gombosi: A heuristic model of the Comet Halley dust size distribution, In *ESA SP-250: ESLAB Symposium on the Exploration of Halley's Comet. Volume 2: Dust and Nucleus, 1986*, ed by B. Battrick, E. Rolfe, R. Reinhard, pp 167–171
9. L. Jorda, P. Lamy, G. Faury et al.: Icarus, **187**, 208 (2007)
10. M. Küppers, I. Bertini, S. Fornasier et al.: Nature, **437**, 987 (2005)
11. K. Meech, N. Ageorges, M. A'Hearn et al.: Science, **437**, 265 (2005)
12. D. Schleicher, K. Barnes, N. Baugh: Astronomical Journal, **131**, 1130 (2006)

Serendipitous Occultation of U0975-07195164 by 9P/Tempel 1 Witnessed from LaSilla

H. U. Käufl[1], I. Saviane[2], V. Ivanov[2], T. Bonev[3], and H. Boehnhardt[4]

[1] European Southern Observatory, D-85748 Garching, Germany hukaufl@eso.org
[2] European Southern Observatory, Santiago de Chile
[3] Astronomical Institute of the Bulgarian Academy of Sciences, Sofia, Bulgaria
[4] MPI for Solar System Research, Max-Planck-Str. 2, D-37191 Katlenburg-Lindau, Germany

Abstract Comet 9P/Tempel 1 was systematically monitored by ESO after reappearance in January 2005. Routinely various narrow-band filters in the 10 μm window were used with TIMMI2 at ESO's 3.6 m telescope. Quasi-simultaneously, the comet was also observed with an R-band filter from ESO's 3.5 m NTT. In the February 2005 period, the comet happened to pass accidentially rather close to a reasonably bright field star, unfortuately in non-photometric conditions. The NTT observing set-up was immediately changed from optical imaging to low-resolution spectroscopy, so that the reddening of the stellar light of this star by the inner coma could be measured with high precision. In combination with the thermal-IR data this may help for unique constraints on the dust particle size distibution as well as for the optical constants of the dust.

1 DI Pre-Impact Observations at ESO

From January to April 2005 multi-wavelength imaging observations[1] of comet 9P/Temple 1, the target of the Deep Impact space craft were performed. The mission is described in various contributions to these proceedings or in [1, 2]. The observations reported here using the thermal infrared mulitmode instrument TIMMI2 at ESO's 3.6 m telescope and SuSI2 – the *Superb Seeing Imager* – at the 3.5 m NTT were performed in service mode during three epochs. The results of these observations, basically a comparison of nuclear and coma fluxes in the R-band to thermal infrared fluxes at $\lambda \approx 10$ μm, have been reported elsewhere [3]. This campain was part of the overall ground based support observations at ESO, summarized in [4], which in turn were part of the worldwide multi-wavelength network of observations [5].

[1] ESO observing program ID 074.C-0782

2 Comet Occultations of Field Stars

To the best of our knowledge the first report about such an occultation event dates back to 1962 [6] where the passage of Comet 1921 II Reid in front of BD 50°3227 (^{m}9.2) as close as 7" is reported. The observer using a small refractor with an eyepiece noted a strong and rapid scintillation of the star during passage. Thereafter other reports of occultations have been published (e.g. [7, 8]). Observation of stellar cometary occultations culminated in the systematic multi site campaign to observe comet Hale-Bopp (C/1995 O1) [9], which provided for one systematic scan over the coma with modern CCD & PMT-technology, albeit plagued by non-photometric conditions.

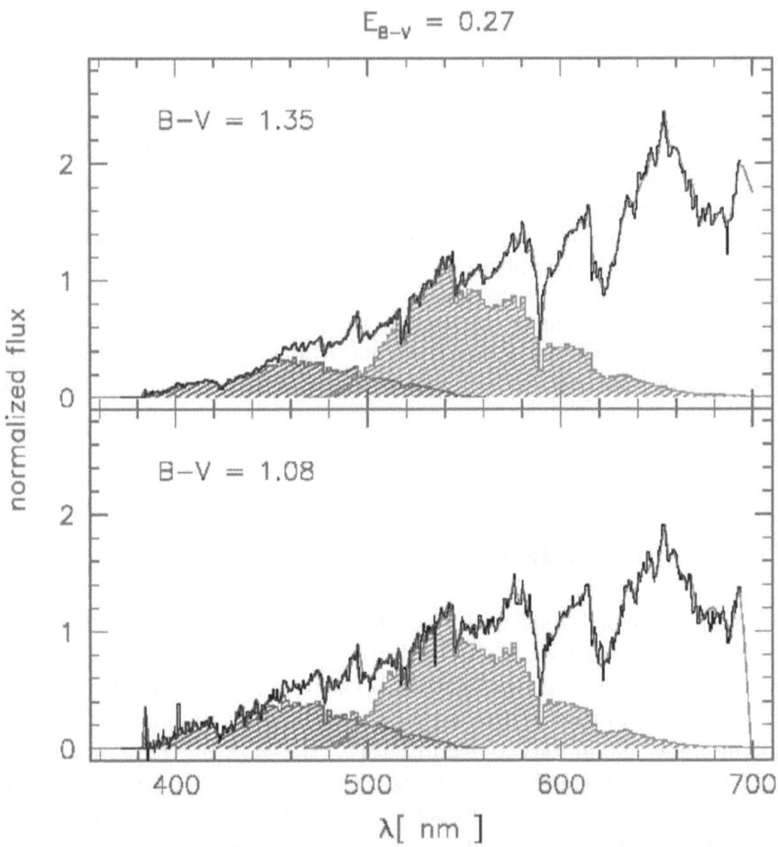

Fig. 1. EMMI-spectra of USNO U0975-07195164 at closest approach (15"), top, and the reference measurement, at ≈300" separation. The histograms hosw the convolution of the spectra with standard Bessel B and V filter curves. During closest apprach the cometary dust coma produces a reddening $E_{B_V} = 0.27$. Due to non-photometric conditions the spectra could not be normalized to also get the total extinction.

3 This Observation

During the the night 2005-02-22 (UT) at approximately 06h40 the comet appraoched the star USNO U0975-07195164 (aka GSC 12321332381 spectral type: unknown, best fitting template from Jacoby et al. M4.5 V mB = 16.8 / mR = 14.3) as close as ≈ 15" or 12500 km. The result of spectroscopy of the background star attenuated by the dust coma are shown in Fig. 1.

4 Discussion

Assuming that the clouds responsible for the non-photometric conditions are grey, then the reddening would be entirely due to the cometary dust. The reddening observed here, for the case of normal interstellar dust, would suggest a total extincition of 2.4 mag. In spite of all photometric problems this can clearly be excluded. The cometary dust produces for the same reddening noticably less extinction when compared to interstellar reddening. This is very consitent with the results of [9] who find for Hale-Bopp for a line of sight closer than 50 km (!) to the nucleus an extinction of 20%. In a next step the dust production rates of the two comets and thus the dust density in gm/cm^2 for the two geometries and epochs should be determined for a better intercomparison. Moreover, using the aquisition images a refined ephemeries will be constructed. It needs to be seen, if the three dust parameters measured, reddening, albedo, from scattered light and the upper limit for thermal emission can provide for new meaningfull constraints on cometary dust optical properties and the size-distribution.

References

1. A'Hearn, M.F., *Comets II*, ed by Festou, M.C., Keller, H.U., & Weaver, H.A. (2004) Univ. AZ Press, Tucson, AZ pp 17–22
2. A'Hearn, M.F., Belton, M.J.S., Delamere, W.A., et al. (2005), Science **310**, 258
3. Käufl, H.U., Bonev, T., Boehnhardt, H., Fernandez, Y.R., & Lisse, C. (2005), EMP **97**, 331
4. Käufl, H.U., Ageorges, N., Bagnulo, S., et al.(2005), The Messenger, **121**, 11
5. Meech, K., Ageorges, N., A'Hearn, M.K., et al. (2005), Science, **310**, 265
6. Danjon, A. 1962, Observatoire de Paris, Journal de Observateurs, **45**, 250
7. Combes, M., et al. (1982), IAUC 3751
8. Lecacheux, J., Thuillot, W., Encrenaz, T., et al. (1984), Icarus, **60**, 386
9. Fernàndez, Y.R., Wellnitz, D.D., Buie, M.W. et al. (1999), Icarus, **140**, 205

Part III

The Cometary Nucleus

The Size of the Artificial Explosive Crater on the Nucleus of Comet 9P/Tempel 1

K. I. Churyumov, V. G. Kruchynenko, and L. S. Chubko

Astronomical Observatory of Kyiv Shevchenko National University,
3, Observatorna str., Kyiv-053 Ukraine 04053 klim.churyumov@observ.univ.kiev.ua

Abstract The theoretical model of an explosive crater formation on the surface of the cosmic body has been considered. The model is verified for the Arizona crater. Calculations show that an explosion of a meteorite in a target occurs for the depth of submergence where it reaches maximal drag. It has been shown that as a result of the collision of the Deep Impact copper impactor with the comet 9P/Tempel 1 nucleus an artifical explosive crater with a diameter of 22–57 m and a depth of 4.8–5.7 m must have formed on the nuclear surface.

1 Introduction

The spacecraft Deep Impact has successfully started on January, 12th, 2005 from the American cosmodrome in Florida (Cape Canaveral) in a direction to short period comet Tempel 1 to which nucleus it has come nearer on July, 3rd, 2005. The main objective of mission was sending from a board of the orbital device of the special module – copper impactor by weight 372 kg to the 6 km ice nucleus of comet 9P/Tempel 1 with which impactor for the speed of 37014 km h^{-1} has collided on July, 4th, 2005.

2 Theoretical Model of Formation of the Explosive Crater

As result of falling meteorites on a surface of space bodies shock or explosive craters are formed. At not elastic collisions of bodies the condition of their destruction is defined by limiting value of hydrodynamical pressure on shock front which moves with speed V_1 in medium with density ρ . This value should not be less strength of substance on destruction, that is $\rho V_1^2/2 \geq \sigma_p$. If to accept, for example, strength of destruction d, which is equal to strength of compression for the Sikhote-Alin meteorite, that is $\sigma_p = \sigma_c = 4.1 \cdot 10^8$ N m^{-2} [1] at $\rho = 7.8$ g cm^{-2} necessary speed V_1 of collision for destruction of a crystal grate of substance should be not less than 0.32 km s^{-1}. At collisions in space this condition, as a rule, is always carried out.

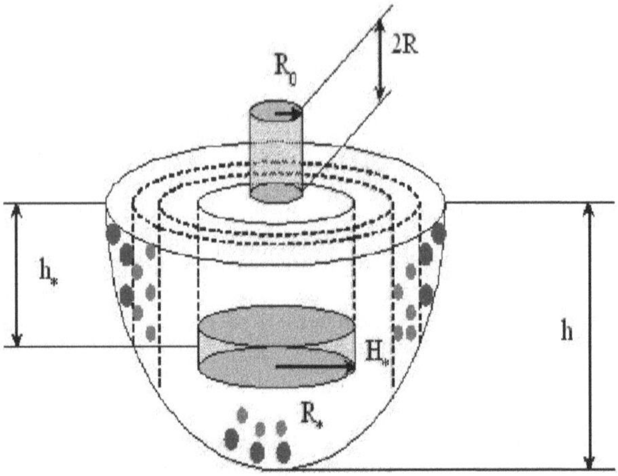

Fig. 1. Formation of the artificial crater on the surface of the comet 9P/Tempel.

According to [2], the basis of the theory of the crater formation is transfer by impactor to target of quantity of movement (impulse), instead of energy. The theory [2] is for today, in our opinion, the most exact, because is checked up and coordinated with experimental data (for speeds from 2 km s^{-1} up to 9 km s^{-1}) and according to lunar craters statistics is verified for speeds up to 40 km s^{-1} [3].

We accept, that impactor has the cylindrical form. The initial radius of the cylinder (and radius of a frontal surface) is equal R_0, its height $2R_0$. On the basis of this we accept, that formed both a liquid of impactor, and a liquid of a target – are incompressible, that is equality is right:

$$R^2 H = 2R_0^3 \tag{1}$$

where R and H – values of radius and heights of the cylinder during the subsequent moments of time. In dependence (1) we neglect change of weight of meteoroid, that, undoubtedly, rightly at the initial stage. Velocity V concerns to the center of weight of an impactor, V_1 – a velocity of movement of a frontal surface. Accepting, that the value of a velocity of different parts of an impactor linearly grows from a frontal surface up to back one, a velocity of last, obviously, will be equaled $2V - V_1$. Then a velocity of change (reduction) of height of the cylinder will be

$$\frac{dH}{dt} = -2(V - V_1)$$

Let $V_2 = \frac{dR}{dt}$ – a velocity of increase in radius of a cylindrical body. Change of quantity of movement for a time unit on unit of cross-section of an impactor we shall present in the form of:

$$P_1 = \frac{\rho V_1^2}{2} + \sigma_p \qquad (2)$$

where σ_p – strength of a target – small (in comparison with the first) an additional member which can play a role only at speeds smaller 1 km s^{-1}. Pressure upon a lateral surface of a cylindrical body

$$P_2 = \frac{\rho V_2^2}{2} + \sigma_p \qquad (3)$$

Accepting, that pressure upon a frontal surface of an impactor is equaled to the sum of internal and external pressure (on absolute value) on a lateral surface of an impactor, we shall write down a equality:

$$P_1 - P_1 = \frac{\delta V_2^2}{2} \qquad (4)$$

From formulas (2)–(4) we find dependence of speed V_2 on speed V_1,

$$V_2 = V_1 (\frac{\rho}{\rho + \delta})^{1/2} \qquad (5)$$

Dependence (5) shows, that a velocity of increase in radius of the cylinder – a value of such order, as a velocity of a shock surface. Lateral pressure will change the form an impactor: will be it flatten and with time can destroy it on separate fragments. Therefore the initial form of meteoroid has no basic value. As a result of flattening of an impactor its resistance that conducts to reduction of depth of penetration increases. Depth of penetration can approximately reach 4 diameters of an impactor instead of 10–20 which could be without flattening. The last can be received from extrapolation of data of ballistics.

From the Eq. (1) follows:

$$\frac{dH}{dt} = -2\frac{H}{R}\frac{dR}{dt}$$

or

$$V - V_1 = V_1 \frac{H}{R} (\frac{\rho}{\rho + \delta})^{1/2}$$

From here we receive the formula which connects a velocity of a frontal surface with a velocity of the center of mass and the characteristic sizes of an impactor,

$$V_1 = \frac{V}{1 + \frac{H}{R}(\frac{\rho}{\rho+\delta})^{1/2}} = \frac{V}{1 + 2(\frac{\rho}{\rho+\delta})^{1/2}(\frac{R_0}{R})^3} \qquad (6)$$

Obviously, the equation of braking of a meteoroid in the medium of a target can be presented in such kind:

$$m\frac{dV}{dt} = -\pi R^2 P_1 \qquad (7)$$

On the basis of formulas (1), (2), (6) and equality

$m = 2\pi {R_0}^3\delta$ the Eq. (7) we shall rewrite:

$$\frac{dV}{dt} = -\frac{R^2}{2R_0^3\delta}\langle\frac{\rho V^2}{2\delta\{1+2(\frac{R_0}{R})^3(\frac{\rho}{\rho+\delta})^{1/2}\}^2} + \sigma_p\rangle \tag{8}$$

In the right part of the Eq. (8) there are two functions of time: $R(t)$ and $V(t)$. Their communication follows from formulas (5) and (6),

$$V(t) = [(\frac{\rho+\delta}{\rho})^{1/2} + 2(\frac{R_0}{R})^3]\frac{dR}{dt}$$

At a derivation of the Eq. (8) any physical condition defining the moment of explosion was not considered, therefore its numerical decision is possible to continue until, when $V_1 = 0$. Developing the theory, Opik did not use any physical idea for definition of the moment (or depths inside of a target) explosion. At the coordination of the theory with various experiments (including with known nuclear explosions) it used approximations and approaches. From the analysis of different data it has received, what explosion of an impactor occurs on depth

$$h_*^{'} = 0.488h,$$

where h – depth of a crater from originally not destroyed surface. We consider, that explosion of a body occurs during the moment of achievement by it of the maximal braking in the given medium, that is depth of explosion of a meteorite follows from the same condition which we used at research of thermal explosions of greater meteoroids in an atmosphere of the Earth. Hence, on the basis of the Eq. (8) we shall equate the second derivative of speed of the center of mass of an impactor on time to zero. We shall designate: $K_1 = -0.25(\frac{\rho}{\delta})$, $K_2 = 2[\frac{\rho}{\rho+\delta}]^{1/2}$, and also neglected a small member σ_p also we shall receive dependence (condition) for depth of the maximal braking where, from our point of view, there should be an explosion:

$$V_*\left(\frac{dR}{dt}\right)_*\left(\frac{4K_2+R_*^3}{K_2+R_*^3}\right) + R_*\left(\frac{dV}{dt}\right)_* = 0$$

Instead of derivative of radius and velocity on time it is used received above the formulas. We shall receive the algebraic equation for radius R_* of a cylindrical body on depth of explosion,

$$2K_1R_*^6 + K_2R_*^3R_0^3 + 4K_2^2R_0^6 = 0 \tag{10}$$

Equation (10) is simple enough and also has one real positive root. The received decision we shall apply to the most investigated of the Arizona a crater-meteorite. According to estimations [4], mass of an iron meteorite ($\delta = 7.85$ g s^{-1}m^{-3}), which has formed crater Canyon Diablo ($\rho = 2.62$ g s^{-1}m^{-3}) depth $h = 170$ m and diameter $D = 1200$ m, make $2.0 \cdot 10^6$ tons at speed of falling of 16 km s^{-1} ($\sigma_p = 2 \cdot 10^8$ Nm^{-2}). For this case $K_1 = -0.0834$, $K_2 = 1.00$. From the Eq. (10) value $R_* = 2.06\ R_0 = 70.9$ m, because initial radius of meteorite

$R_0 = 34.4$ m. From resulted above the equations depth of the maximal braking $h_* = 2.36R_0 = 81$ m. The analysis of the numerical decision of the Eq. (8) for other cases testifies that the maximal braking is reached on depth, which it is equal (0.4–0.5) h, that is result it is close to conclusion of Opik. As a result of the detailed theoretical and empirical analysis of data on crater formation in the robot [2] settlement parities for the sizes of an explosive crater are deduced. The formula for depth of a crater from originally not destroyed surface depending on diameter d of an impactor can be written down in the form of:

$$\frac{h}{d} = 1.785(\frac{\delta}{\rho})^{1/2}(\frac{V^2}{\sigma_p})^{1/30} \cos Z \tag{11}$$

Where Z – angular zenith distance of a trajectory of an impactor. Dependence for definition of diameter D of a crater we shall present in such kind:

$$\frac{D}{d^{3/2}} = 1.20 \cdot (\frac{kV\delta}{h})^{1/2} \cdot (\rho\sigma_p)^{-1/4} \tag{12}$$

The dimensionless factor k transfers of a radial impulse is defined by the equation:

$$k \doteq C_1V^2(1 - 0.04k^2)^{1/2} + 2 \tag{13}$$

The minimal value k is accepted equal 2.0. Maximal – it is equal 5.0 and it is received on the basis of calculations of the maximal penetration of a shell in a target and it flattening. From here the ratio of the destroyed mass in "the central pipe" m_1 to mass of a meteorite m is certain: average value is equal 25. On the other hand, the ratio of kinetic energies of radial impact and a meteorite:

$$\frac{E_r}{E_m} = \frac{k^2m}{m_1}$$

Considering that this ratio cannot be more units, we receive: $k_{\max} \leq 5.0$. Therefore $2 \leq k \leq 5$. Constant $C_1 = 6.5 \cdot 10^{-13}$ s^{2}cm^{-2} at impact of iron on a stone, $4.2 \cdot 10^{-13}$ s^{2}cm^{-2} – at impact of a stone on a stone or iron on iron, $3.5 \cdot 10^{-13}$ s^{2}cm^{-2} – at impact of a stone on iron.

3 Conclusion

On the basis of the carried out calculations we [5] had been determined the possible sizes of a crater on the nucleus of comet 9P/Tempel 1 which should form artificial impactor. The velocity of collision was 10.3 km s^{-1}. We accept density of substance of a shell of 7.0–8.9 g cm^{-3}, density of substance of the cometary nucleus of 0.5–1.0 g cm^{-3}, strengths of substance of the cometary nucleus 10–100 kNm^{-2}. Equivalent diameter of an spherical impactor makes 43.0–46.6 cm, and the dimensionless factor of transfer of an impulse is equal 2.45. From dependence (11) have received, that depth of the formed crater will make 4.8–5.6 m, and the value of diameter of a crater which have calculated under the formula (12), 22–57 m. Volume of the destroyed substance (volume of a crater) will be equaled will make 810–7080 m^3. The corresponding mass of the destroyed ice in thousand times will exceed mass of an impactor. The average specific energy spent by an impactor on unit of the destroyed mass, will be equal 3.6–20 kJc^{-1}. This energy has not enough to fuse and sublimate all ice (energy is necessary for sublimation of an ice 2MJ kg^{-1}). Explosion will occur in the central cylindrical area and responsible for it sublimated mass will make less than 1 total erose mass: other mass will be thrown out in the form of pieces of ice. A conclusion which followed from our calculations: the sizes of a crater will be small – considerably smaller what at that time appeared in scientific editions.

Unfortunately our forecast [5] to compare there is nothing. Last value of mass received by [4] as the result of observation of meteors for strength so-called "dust spheres" – friable meteoroids, possibly, cometary origin which collapse in an atmosphere of the Earth under condition of when aerodynamic pressure is equal 1 k Nm^{-2}. At the same time, in recently published work [6] from the analysis of emission of very fine (micron) particles which make a surface of a comet as authors approve, by depth in tens meters, strength of such substance it is certain as extremely small – less than 65 Pa= 65 Nm^{-2}. If to accept such strength from the formula (15) follows, that diameter of a crater can be the order of 150 m.

References

1. Krinov E. *A meteoric rain.* (M: the Science, 1981.) pp. 192
2. Opik E. J. *Interplanetary encounters,* (New York: Elsevier scient. Publ. Comp. 1976.) pp. 155
3. Opik E. J.(1960): Mon. Notic. Roy. Astron. Soc. **120**, 404
4. Opik E. J. *Physics of meteor flight in the atmosphere,* (New York: Interscience Publ. (1958)) pp. 174
5. Kruchinenko V., Churjumov K., Dobrjanskij J., Walter A. (2005): Kinematics Phys. Celestial Bodies. Ph. **21**, 49
6. A'Hearn M. F., Belton M. J. S., Delamere W. A. et al. (2005): Science **310**, 310

Comparison of the Spectra of the Comets 9P/Tempel 1 and C/2004 Q2 (Machholz)

L. S. Chubko[1], K. I. Churyumov[1], V. L. Afanasiev[2], I. V. Lukyanyk[1], and V. V. Kleshchonok[1]

[1] Astronomical Observatory of Kyiv Shevchenko National University, 3, Observatorna str., Kyiv-053 Ukraine 04053 klim.churyumov@observ.univ.kiev.ua
[2] Special Astrophysical Observatory of RAS, Zelenchukskaya, Karachaevo-Cherkesia, Russia 369167 vafan@sao.ru

Abstract The results of spectral observations of comets 9P/Tempel 1 and C/2004 Q2 (Machholz) are presented. The real luminescent cometary continuum tied with the luminescence of the comet organic species which are in comet dust particles (CHON-particles) is detected.

1 Introduction

The comet 9P/Tempel 1 – the main target of the Deep Impact mission was observed with SCORPIO[1] installed in the Prime Focus of the 6-m telescope of the SAO of the RAS on March 15, 2005 (Mount Pastukhov, Nizhny Arkhyz, Northern Caucase, Russia). The observations were obtained 2.5 months before collision of the copper impactor of the Deep Impact spacecraft with the 6-m nucleus of short periodic comet 9P/Tempel 1. During the same nights we observed also spectra of new comet C/2004 Q2 (Machholz). We suppose that in the spectra of comets 9P/Tempel 1 and Machholz (C/2004 Q2) we detected the real luminescent cometary continuum tied with the luminescence of the comet organic species which are in comet dust particles (CHON-particles).

2 Observation and Processing of the Comet Spectra

The short period comet 9P/Tempel 1 ($r = 1.85$ AU, $\Delta = 0.91$ AU) was observed with SCORPIO installed in the Prime Focus of the 6-m telescope of the Special Astrophysical Observatory of the RAS on March 15, 2005 (Mount Pastukhov, Nizhny Arkhyz, Northern Caucase, Russia). During the same nights we observed also spectra of new comet C/2004 Q2 (Machholz) ($r = 1.43$ AU, $\Delta = 0.92$ AU).

[1] SCORPIO, Spectra Camera with Optical Reducer for Photometric and Interferometric Observations.

It is very interesting that comet C/2004 Q2 (Machholz) entered the field of view of the SOHO[2] on February 16, 2005 (a month before our observations).

Four spectra of comet C/2004 Q2 (Machholz) and two spectra of comet 9P/Tempel 1 were obtained on 15 March 2005 with the help of the 6m telescope of the Special Astrophysical Observatory of RAS. The focal reducer SCORPIO was used with long-slit (length of the slit is 2 arcmin, width of one is 1 arcsec).

3 Cometary Luminescence Continuum in the Spectra of C/2004 Q2 (Machholz) and 9P/Tempel 1

For the first time a luminescence continuum was detected in the comet Halley (1P) spectrum by G. Nazarchuk [1, 2]. The authors of this paper investigated spectra of comets 9P/Tempel 1 and Machholz (C/2004 Q2), with the aim to determine the real level of the non-solar-origin continuum in the spectral region 350–600 nm. Spectra of these comets were observed with the help of the 6-m BTA telescope and the spectrograph SCORPIO with the long slit at the Special Astrophysical Observatory of the Russian Academy of Sciences.

In the high spatial and spectral resolution spectra of several comets has been noticed, that the equivalent width of the basic Fraunhofer lines is less than in the solar spectrum [3, 4, 5, 6, 7].

Such phenomenon is connected with presence of an additional component of a continuous spectrum in cometary radiation. As a source of this additional radiation the luminescence organic cometary motes can serve. That fact testifies that spatial distribution of this source finds out very strong concentration to a cometary nucleus. Besides for organic materials the luminescent luminescence is characteristic at illumination by their ultra-violet radiation. Unfortunately, the luminescent luminescence in this case has a wide spectrum without significant features that complicates identification of a radiating material. Thus, a cometary spectrum I_{com} consists of three components:

$$\mathbf{I_{com}}(\lambda) = \mathbf{I_e}(\lambda) + \mathbf{I_s}(\lambda) + \mathbf{I_l}(\lambda) \quad (1)$$

where I_e is the emission cometary spectrum, I_s – the solar spectrum reflected by cometary dust, I_l – a cometary luminescent continuum.

Spectral regions with no strong emission lines are selected for determination of a level of a luminescent continuum $I_e \approx 0$. On these regions of a spectrum is accepted, that the level of a luminescent continuum does not vary. Thus, for the selected regions of a cometary spectrum it is possible to accept

$$\mathbf{I_{com}}(\lambda) = k \times \mathbf{I_f}(\lambda) + \mathbf{l} \quad (2)$$

Where $I_f(\lambda)$ – the known solar spectrum which is calculated taking into account the spectral resolution of a cometary spectrum and its discontinuity, k – factor which characterizes reflective ability of cometary dust, l – intensity of

[2] SOHO, SOlar & Heliospheric Observatory.

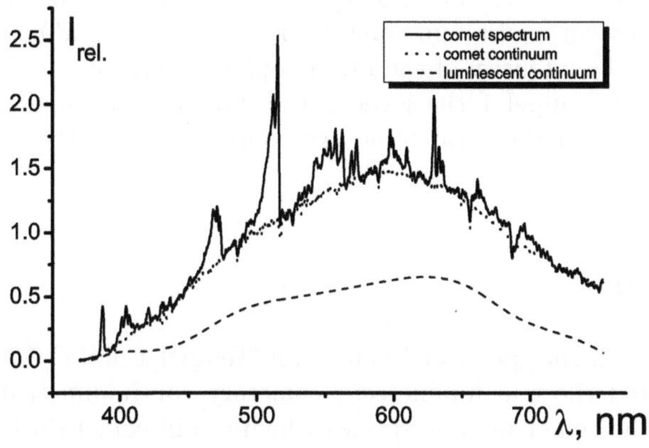

Fig. 1. A spectrum of comet C/2004 Q2 Machholz with dedicated cometary continuum and a luminescence of cometary dust.

a luminescent continuum. In practice parameters k and l are selected so that to receive the best accordance to a region of cometary continuum. Uniting data on a level of a luminescent continuum for different spectral regions, gives possibility to study behaviour of a luminescent continuum in all cometary spectrum.

The given technique was used for studying a luminescent continuum in spectra of comets C/2004 Q2 Machholz and 9P/Tempel 1 (Figs. 1 and 2).

The maximum of a luminescent continuum of comet C/2004 Q2 Machholz is close to 630 nm. In this region its intensity reaches 46% from a level of total cometary continuum. Maximum of intensity of the cometary luminescent con-

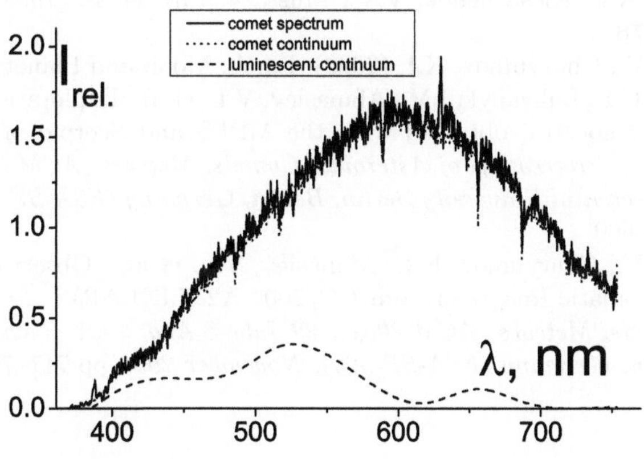

Fig. 2. A spectrum of comet 9P/Tempel 1 with a dedicated cometary continuum and luminescence of cometary dust.

tinuum is close to $\lambda \approx 525$ nm [4, 5, 7]. In this spectral region the intensity of cometary luminescent continuum equal to 30% from the total cometary continuum (42% from the intensity of the Fraunhopher spectrum).

For comet 9P/Tempel 1 the level of the luminescent cometary continuum is 30% of the level of the total cometary continuum with the maximum near $\lambda \approx 525$ nm.

4 Conclusion

We suppose that in the spectra of comets 9P/Tempel 1 and C/2004 Q2 (Machholz) we detected the real luminescent cometary continuum tied with the luminescence of the comet organic species which are in comet dust particles. For comet 9P/Tempel 1 the level of the luminescent cometary continuum is 30% of the level of the total cometary continuum with the maximum near $\lambda \approx 525$ nm and for comet C/2004 Q2 (Machholz) the level of the luminescent cometary continuum is 46% of the level of total cometary continuum with the maximum near $\lambda \approx 630$ nm. Comparison of spectra of two comets shows that new comet C/2002 Q2 in Oort's sense has the more high level of luminescent continuum and therefore more number of organic luminofors (CHON-particles) than old exhausted short period comet 9P/Tempel 1 has such luminescent motes.

References

1. Nazarchuk, H.K.(1987): Kometnyj Tsirkulyar **372**, 2
2. Nazarchuk, H.K. (1987): Kometnyj Tsirkulyar **377**, 2
3. Churyumov, K.I., Kleshchenok, V.V., Vlassyuk, V.V. (1994): Pisma v Astronomicheskij Journal **20**, 9
4. Churyumov, K.I., Kleshchenok, V.V., Mussaev, F.A. et al. (1999): Earth, Moon and Planets **78**, 1
5. Lukyanyk, I.V., Churyumov, K.I. (2002): Earth, Moon and Planets **90**, 1
6. Churyumov, K.I., Lukyanyk, I.V., Afanasiev, V.L. et al. :Exploration of the Comet 19/P Borrelly spectra, obtained with the MPFS and Scorpio of 6-m Telescope on Aug. 2001 : *Proceedings of Asteroids, Comets, Meteors (ACM 2002), 29 July-2 Aug. 2002, Technical University Berlin, Berlin, Germany (ESA-SP-500), November 2002* pp 657–660
7. Lukyanyk, I.V., Churyumov, K.I., Afanasiev, V.L. et al. : Observation of Spectra and Monochromatic Images of Comet C/2001 A2 (LINEAR) : *Proceedings of Asteroids, Comets, Meteors (ACM 2002), 29 July–2 Aug. 2002, Technical University Berlin, Berlin, Germany (ESA-SP-500), November 2002* pp 717–720

Impact Cratering on Volatile-rich Targets: Some Remarks Related to the Deep Impact Experiment

P. Claeys

Dept. of Geology, Vrije Universiteit Brussel, B-1050 Brussels, Belgium

Abstract The NASA Deep Impact experiment has important implications to better understand cratering processes on planetary bodies and the production and evolution of ejecta. This man-made impact of a solid Cu body on the nucleus of a comet fills the large gap existing between data derived from small-scale cratering experiments and large-scale field or remote sensing observations of craters. DI thus complements hydrocode modeling of cratering processes. The majority of cratering studies focus on solid silicate-rich targets rather than on porous, poorly consolidated and/or volatile-rich materials. However, volatile targets are common in the Solar System. The lessons learned from the DI collision with comet 9P/Tempel not only clarify the composition and physical properties of the cometary nucleus, but also can shed light on cratering mechanisms and evolution of plume and ejecta.

1 Introduction

On July 4 2005, the cylindrical shaped (1 m × 1 m) Deep Impact (DI) spacecraft weighing 364 kg (plus 7 kg of remaining hydrazine fuel) crashed on comet 9P/Tempel 1. The projectile was composed of 49% copper, its velocity was 10.3 km s^{-1} and the impact took place at an angle between 20° and 30° above the horizontal plane delivering 19 GJ of kinetic energy to the comet's surface. This collision excavated material from the surface and inner parts of the comet. The outcomes of this man-induced collision were observed by the nearby DI flyby spacecraft, by other missions currently cruising the solar system as well as by telescopes on the ground and in orbit around the Earth. The principal goals of the NASA DI mission were to investigate the inner composition of the nucleus of comet 9P/Tempel and the physical properties of its surface and interior. Exciting results have been published in several articles (see for example: [1, 22] Special volume ICARUS 187, 2007, these Proceedings).

2 Impact Cratering on Volatile-rich and/or Porous Surfaces

The DI mission also has implications to understand impact cratering on solar system bodies, including Earth. The DI mission represents one of the rare opportunities to fill the large observational gap between cratering experiments in the laboratory – most commonly at scale of a cm or smaller – and the direct study of ancient impact structures on Earth or their remote sensing observation on distant planetary surfaces at a scale of tens of meters to several hundred km. On Earth, impact-cratering has never been directly witnessed, and this is probably a good thing. DI offers the unique opportunity to observe "life" the production of ejecta, to closely monitor the evolution of a rising plume of dust and gas and to document the formation of a fresh crater, in this particular case in low-gravity conditions. In this sense, DI results are complementary to hydrocode modeling of impact events. The DI observational results could help refine and constrain existing cratering models. The nature of the cometary nucleus, probably a mixture of silicates and volatile-rich components such as ices [1, 36, 17] renders it a rather particular but not uncommon target. In most studies of impact cratering on Earth, and other planets silicate-rich lithologies are usually considered as the prime target rock rather than porous, poorly consolidated and/or volatile-rich materials.

However in the solar system, ices and poorly consolidated materials are abundant at the surface of planetary bodies. Ices, not necessarily water ice, cover the surface of several Moons around the giant gas planets (Europa, Ganymede, Enceladus, Triton etc.). In some cases, these icy surfaces can be "dirty" by the presence of silicate materials or other components. Thick layers of ice can also be found on planetary surfaces, like the Earth and Mars. On Earth, volatile materials such as ice and especially water oceans represent a large fraction of the surface that might be targeted by projectiles. Moreover, it was proposed that during extended period of its history the Earth might have been fully or in large part glaciated (so called Snowball Earth, some 600 million years ago, see Hoffman et al. [18]). It is thus evident that, through geological time, impact on terrestrial volatile-rich targets represented a significant proportion of the "cratering record".

On planet Earth, very few examples of impact events are known in the marine environment and so far none on ice. The oceans form $\sim 2/3$ of the planet, and ice covers a little less than 15% of its surface. These proportions have of course varied throughout geological history, with ice-free periods such as the Cretaceous alternating with intervals marked by extended glaciations (the Cenozoic since $\sim$ 35 million years ago). So far, the formation of impact craters, ejecta distribution and debris-plume production are essentially studied and understood in the case of solid and well consolidated silicate-rich targets. How ice-rich and porous targets react during an impact event has only started to be investigated in the last 10 years by groups of researchers, with special emphasis on the experimental approach.

3 Terrestrial Impacts on Volatile-rich Targets

The impact database (http://www.unb.ca/passc/ImpactDatabase/) lists ∼ 170 impact craters, of which only about 25 are identified as marine impacts, but all of them took place on the continental crust [12]. This proportion of marine impact is clearly too low for a planet composed at ∼ 70% by oceans. This limited representation is due the fact that because of plate tectonic activity, the oceans are mostly young features (< 200 million years) coupled with the lack of fine scale mapping of the seafloor's topography. Another reason is the limited knowledge of what a crater formed on the seafloor should look like. It must be considered that craters formed thin (∼ 6 km) oceanic crust do not reproduce identically the morphology and geophysical characteristics documented for craters on land. Marine craters are also expected to be quickly buried after formation, which is ideal for preservation, but also hides them from view. A rapid estimation, considering the roughly 2 to 1 proportion of ocean versus land indicates that a minimum of about ∼ 300 marine craters should have formed on Earth.

So far, the current knowledge of impact cratering in the marine realm is based essentially on computer modeling (see for example [32]). Seismic analyses and sedimentological studies reveal that near the crater rim, marine impact likely induces large-scale slope instability and collapse, fluidization of poorly consolidated sediments, slumping, and major debris flow or avalanches. This is clearly demonstrated at the periphery of the 200 km in diameter Chicxulub crater in Yucatan [33, 16] for example. Such large scale mass movements have also been observed at the smaller Montagnais crater [19], Mjølnir crater [37]; Lockne crater [35] and Chesapeake Bay crater [26]. The deposition of massive sand layers and chaotic debris flow can be attributed to the propagation of shock waves and earthquakes associated with impact cratering, coupled with the ease with which water-saturated sediments will become fluidized. Moreover, resurge effects and possible tsunami waves [38] could extend the active and rapid sedimentation in the area long after the end of the cratering process, and be important post-impact crater filling mechanisms.

There is so far only one known impact event in the deep marine setting, the Eltanin impact 2.5 million years ago in the remote Bellinghausen sea between Argentina and Antarctica [20, 14]. The size of the projectile is in the range of 500 m to 1 km in diameter. At the site, water depth was about 5 km and no trace of a crater was detected on the seafloor. Pistons cores drilled over the area encountered impact ejecta and evidence of major sediment disturbances, which are also clearly visible on seismic studies over an area of $\sim 80,000$ km^2. Different types of ejecta material produced by a pure oceanic impact are present; the unnmelted meteorite fragments, the vesicular impact melt, and the glassy spherules containing Ni-rich magnesioferrite spinels represent respectively the products of solid, liquid and vapor-phase ejecta [21]. The glass spherules are enriched in Na and Cl, indicating a contribution from the seawater target. Despite the lack of crater on the seafloor, this event produced a large volume of various types of ejecta and was capable to distribute them widely considering the size of the impactor.

Melosh [23] demonstrated that water vapor formed during an impact on a marine target would augment the amount of ejecta produced, and accelerate its distribution. Consequently, such an event will results in a much wider distribution of ejecta particles compared to dry silicate targets.

Cratering on sedimentary rocks, such as carbonates and evaporites with high porosity and the capability to release large volumes of volatiles (CO_2 and SOx) by shock vaporization, provide so far the closest analog to ice-rich targets. The 65 million years old Chicxulub impact excavated 3 km of shallow-water carbonate and evaporite platform sequence of Late Cretaceous age. Such lithology is estimated to have a range of porosity from 15 to 20%. Alvarez et al. [2] proposed that the release of large volumes of CO_2 during the impact event, lead to a two components plume rising above the crater. One component was formed by the vaporization of the projectile and the deep basement silicate lithologies. The other by the release of CO_2 (and SO_X) from the sedimentary upper layers of the target rock. The expansion of CO_2 within the plume contributed to lofting ejecta fragments, in the form of shock quartz grains, which originated from the deeper part of the target (silicate basement > 3 km), to great distance (> 5000 km) from the impact site. The degree of mixing between the two plumes is not well understood. More recently Artemieva and Morgan [3], using hydrocode modeling proposed that the majority of the ejecta material spread over 112 Cretaceous-Tertiary locations around the world [10] came principally from the sedimentary cover of the target.

4 Prediction of the DI Cratering Event

Before the DI experiment, there was a great deal of unknowns and uncertainties as to the outcome of the collision. Current knowledge of impact cratering on Earth, modeling and experimental results played a key role in the design of the experiment and of the impacting spacecraft [27]. Cratering scaling laws based on experiments offered the possibility to estimate the size and depth of the produced crater, as well as the duration of the event. Hydrocode modeling were used to document the excavation process and made it possible to approximately calculate the volume of material to be vaporized during the impact. Precise estimation of the excavation process, crater size and amount of ejected material depends to some extend on the composition and physical properties of the impacted target [25].

A detailed review of cratering mechanics is presented in Melosh [23]. Impact crater formation strongly resembles shallow depth explosions [15]. A crater forms by propagation of intense shock waves generated at the contact point between the projectile and the target lithologies. The passage of shock wave at high velocity vaporizes, melts, deforms, compresses and displaces a large volume of target material leading to the excavation of a cavity. The attenuation of the shock waves is rapid but depends on the composition and physical properties of the material encountered. Depending on the physical characteristics of the target rock, either strength of the excavated material or gravity stops the growth of the crater a few seconds or a few minutes after the collision. The extremely rapid, (a few seconds for a < 1 km to minutes for > 20 km size structures) and continuous cratering process can be divided in 3 phases (see [23, 13] for a more elaborate discussion):

1. contact and compression,
2. excavation, and
3. post-impact modifications.

A set of differential equations account for the conservation of energy, mass en momentum during impact and crater formation processes coupled with constitutive equations relating strain and shear stress in the impacted material. They are combined with a specific equation of state describing thermodynamics and pressure distributions in function of the density and internal properties of the target lithologies [23]. The latter is the least constrained and highly depends on the mineral composition, homogeneity and physical characteristics, such as amount of porosity for example, of the impacted material. The equation of state is still poorly known for heterogeneous materials such as rocks and minerals. Numerical modeling supported as much as possible by experimental studies carried out on material resembling the target lithologies provides the best illustration of the cratering process. Unfortunately, little experimental data exist for unconsolidated or porous material, or for materials rich in water, ice and/or volatile components, although the situation is improving quickly [5–7]. In particular, there is a lack of knowledge of the effects of high-velocity impact on the type of material forming the surface and inner parts of a comet.

The exact composition and density of comets is still controversial. Usually, they are considered to be rich in ice and/or volatile components and at least the upper lithologies are likely to be poorly consolidated and/or rather porous [4]. During an impact on the surface of a comet, the shock induced expansion, vaporization of the existing volatiles should contribute to greater extension of crater dimensions and to the formation of a larger debris, and vapor could as more material is removed from the crater during the excavation process.

Before the DI experiment, Richardson et al. [27] carried out numerical modeling of the collision using the SOVA hydrocode [31] and the ANEOS code for the equations of state [24]. They used serpentine, at different degrees of porosities (0, 50, 80%), to mimic the cometary surface. These experiments clearly demonstrated how the density of the target affects the vapor plume rising above the crater ([27], Fig. 3). In the high-porosity setting, the projectile penetrates deeper and the plume is projectile-free. In the case of a less porous target, the projectile appears to form an important component of the vapor and debris plume. At the comet surface, a distinct crater seems to form during the first phases of the excavation process. However in numerical modeling, it is difficult to account for porosity and material strength. The outcome of the cratering event appears not so well constrained, in particular in term of the final size of the crater [27].

To better constrain the DI crater size and appearance, Schultz et al. [30] carried out a detailed series of experiments using a wide range of target composition at the NASA Ames Vertical Gun Research Center facility. They concluded that a crater between 70 and 140 m in diameter could form if gravity-scaling relationships do apply. According to their experiment the DI mission should also produce a well-defined ejecta curtain. In case the crater growth is in part, strength limited a smaller (< 50 m) but deep crater would be expected. This would also be the case, if the inner part of the target has very high porosity (80%) and if compressibility inducing premature crater collapse. In such cases, little ejecta is expected and the resulting crater may not be clearly visible at the surface of the comet. On the other hand, very low surface density (<0.3 g cm^{-3}), a poorly consolidated surface and the presence of volatile-rich components at depth would result in a larger (> 180 m) crater being formed [30].

Richardson et al. [27] also modeled the evolution of the vapor and debris plume rising above the comet's surface. They presented (Fig. 10 in Richardson et al. [27]) the resulting plume in case of a gravity-dominated cratering event and in the case of a strength-dominated event. In the gravity-dominated scenario, the plume remains attached to the comet, while it separates from it in a strength-dominated event. Revising the modeling after the event, Richardson and Melosh [28] showed that the obtained plume could apparently be modeled as gravity-dominated impact on low-strength material.

5 The DI Cratering Observations

Before the DI impact, the nucleus of comet 9P/Tempel 1 was imaged at high resolution (<10 m pixel^{-1}), in particular in the area of the impact site [1]. On the surface of the nucleus, several dozen circular features ranging in size from 40 to 400 m stand out in term of topography. They are interpreted as ancient impact structures (see Fig. 1 in A'Hearn et al. [1]). Their general morphologies resemble those of the impact structures visible at the surface of asteroid such as Matilda and Vesta and according to A'Hearn et al. [1] their cumulative size-frequency is compatible with crater populations. It can then be expected that the DI experiment produces such an additional scar at the surface of 9P/Tempel. Documenting the size and morphology of the formed impact structure should provide key parameters to access the evolution of the cratering process but also will shed light on the composition and physical properties of the comet's surface and interior.

Unfortunately, the flyby spacecraft did not image the DI crater because the large cloud of fine dust ejecta obstructed its view [1]. Using image enhancement techniques, Busko et al. [8] attempted without success to detect the DI crater at the surface of comet 9P/Tempel. They concluded that no clear evidence for the presence of a crater can be found at the surface. The analyses of the images obtained after restorations techniques were applied are far from unambiguous. Busko et al. [8] advocated based on indirect arguments that a possible crater in the expected size range of 150–200 m existed on the comet surface. However, this indirect evidence seems to support a trough shaped crater. Such morphology is surprising considering the low obliquity the impact angle, and the existence of well-rounded craters at the surface of 9P/Tempel (Fig. 1 in A'Hearn et al. [1]).

Impact experiments do not support this elongated morphology [30]. Moreover, Busko et al. [8] stated that the expected impact site is located downstream and to the side of the would-be trough-shaped crater.

Two options can be considered here. First, and most probably, a crater formed and was hidden to observation by the ejecta material. This DI crater would be similar to the other impact structure observed at the surface. Second, it is possible to speculate that no stable or well defined crater was formed during the DI impact. The surface of the comet was excavated by the crater' s transient cavity but due to the overall low strength of the material (< 65 Pa), the low bulk density (~ 0.6 g cm^{-3}) of the nucleus, the abundance of poorly consolidated fine particles at the surface [1], and possibly the high porosity of the impact material, no stable impact structure formed. The high amount of volatile released by the impact let to the formation of a cavity much too large for the strength of the material present the comet's surface. The excavated structure simply collapsed on itself as the plume and ejecta were rising above the comet surface, and it was later covered with fallback material. Instead of a well-defined crater, a poorly marked depression formed at the surface. The result could form an extend low topography zone, which would be rather difficult to detect without detailed close up observations. The lack of a marked crater is difficult to reconcile with the large volume of ejecta produced, unless the comet surface was particularly volatile-rich as advocated by [1, 22]. Another problem lies in the existence of well marked impact craters at the surface of 9P/Tempel [1], which would then have to results from other types of collisions.

6 The DI Rising Plume

The evolution of composition of the dust and gas present in the hot plume and ejecta were determined directly after the impact. This composition can be compared with pre-impact and post-impact emissions from the comet. Continuous observations make it possible to follow the evolution of the plume and ejecta and to correlate them with the suspected internal stratigraphy of the comet. The impact ejecta material dissipated quickly and it took only five days to return to normal pre-impact emissions [22]. Directly after the collision, the plume composition included organic material, HCN, H_2O and CO_2, and fine size (μm) silicate dust [1, 17]. The dust to gas ratio liberated by the impact is larger than that observed before the collision, and the released material appears more crystalline [22]. Harker et al. [17] estimated that 10^4–10^6 kg of dust were released. The dust is composed of amorphous and crystalline olivine and pyroxene, and amorphous carbon grains [36]. These authors also advocate evaporation and expansion of volatiles to accelerate the dust grains in the inner part of the plume. The signal for H_2O and CO_2 seem to decrease faster in term of intensity and temperature, than the organic signal [1]. The quickly released volatile-rich early ejecta supports the presence of ice at or directly below the comet's surface, while the deeper material seem to be more silicate-rich and volatile-poor [1, 22]. The thermal difference are visible in the plume, with hotter zones located towards the center and the first ejecta appears much hotter than the material excavated at a latter phase [1].

7 Discussion and Perspectives

Such a direct observation of an evolving plume and ejecta is unique in term of impact cratering. It supplements hydrocode modeling for documenting the complex and heterogeneous plumes of gas and dust, which expands above a crater. The DI results clearly show temperature zones and clear short-term changes in the chemical composition of the plume and ejecta. In terms of impact cratering on Earth, these results have major implications to understand changes in compositional differences in the fallback material. In the Chicxulub case for example, the suevite (melt-rich breccias) landing back in the crater seem to be dominated by material coming from the upper part of the target rock [11, 34]. The lapilli particles accreted in the turbulent zones of this plume also seem to record changes in its chemical compositions and/or condensation conditions [9]. Another example is found in the marked chemical differences observed in the Ni-rich magnessioferite spinel crystals recovered at different sites of the world at the Cretaceous-Tertiary boundary [29]. The DI results make it possible to interpret the chemical variability as the result of the condensation of these crystals at different times in an heterogenous and chemically evolving plume and/or under different conditions of temperature and oxygen fugacity.

A large amount of data has already been extracted from the NASA DI experiment and will continue to be collected in the coming years. This information will significantly increase the current knowledge of cometary material. However, the implications in term of impact processes must not be neglected and could contribute to a better understanding of cratering mechanisms and especially in term of still-poorly known evolution of plume and ejecta.

Acknowledgements I am grateful to Chris Sterken for stimulating the writing of this manuscript and for his comments and encouragements. The "FWO Research Foundation – Flanders" financially supports impact-cratering research at the VUB through project FWOKN128: 1.5059.04.

References

1. A'Hearn, M.F., and 32 others., 2005, Deep Impact: Excavating comet Tempel 1: Science, v. 310, p. 258–264.
2. Alvarez, W., Claeys, P., and Kieffer, S.W., 1995, Emplacement of Cretaceous-Tertiary Boundary Shocked Quartz from Chicxulub Crater: Science, v. 269, p. 930–935.
3. Artemieva, N., and Morgan, J., 2007, Distal ejecta from the Chicxulub – Numerical model, Lunar and Planetary Science Conference, v. 38, Houston, Lunar and Planetary Institute, CD-ROM # 1543.

4. Brandt, J.C., and Chapman, C.R., 2004, Introduction to comets: Cambridge USA, Cambridge University Press, 249 p.
5. Burchell, M.J., Grey, I.D.S., and Shrine, N.R.G., 2001, Laboratory investigations of hypervelocity impact cratering in ice, Io, Europa, Titan and Cratering of Icy Surfaces, Volume 28: Advances in Space Research, p. 1521–1526.
6. Burchell, M.J., and Johnson, E., 2005a, Impact craters on small icy bodies such as icy satellites and comet nuclei doi:10.1111/j.1365-2966.2005.09122.x: Monthly Notices of the Royal Astronomical Society, v. 360, p. 769–781.
7. Burchell, M.J., Leliwa-Kopystynski, J., and Arakawa, M., 2005b, Cratering of icy targets by different impactors: Laboratory experiments and implications for cratering in the Solar System: Icarus, v. 179, p. 274–288.
8. Busko, I., Lindler, D., A'Hearn, M.F., and White, R.L., 2007, Searching for the Deep Impact crater on Comet 9P/Tempel 1 using image processing techniques: Icarus, v. 187, p. 56–68.
9. Claeys, P., 2006, Chicxulub, anatomy of a large impact structure: from impactite to ejecta distribution, 40th ESLAB-ESA First International Conference on Cratering in the Solar System, Volume SP-612, WPP-266: Noordwijk, NL, European Space Agency, p. 1–12.
10. Claeys, P., Kiessling, W., and Alvarez W., 2002, Distribution of Chicxulub ejecta at the Cretaceous-Tertiary Boundary, *in* Koeberl, C., and MacLeod, K.G., eds., Geological Society of America Special Paper v. 356, p. 55–69.
11. Claeys, P., Heuschkel, S., Louvejeva-Baturina, E., Sanchez-Rubio, G., and Stoeffler, D., 2003, The suevite in drill hole Yucatan 6 in the Chicxulub impact crater: Meteoritics & Planetary Science, v. 38, p. 1299–1317.
12. Dypvik, H., and Jansa, L.F., 2003, Sedimentary signatures and processes during marine bolide impacts: a review: Sedimentary Geology, v. 161, p. 309–337.
13. French, B.M., 1998, Traces of catastrophe: A handbook of shock-metamorphic effects in terrestrial meteorite impact structures: Houston, Lunar and Planetary Institute, 120 p.
14. Gersonde, R., Deutsch, A., Ivanov, B.A., and Kyte, F.T., 2002, Oceanic impacts - A growing field of fundamental geosciences: Deep-Sea Research II, v. 49, p. 951–957.
15. Gifford, A.C., 1924, The mountains of the Moon: New Zealand Journal of Science and Technology, v. 7, p. 129–142.
16. Grajales-Nishimura, J.M., Cedillo-Pardo, E., Rosales-Dominguez, C., Moran-Zenteno, D.J., Alvarez, W., Claeys, P., Ruiz-Morales, J., Garcia-Hernandez, J., Padilla-Avila, P., and Sanchez-Rios, A., 2000, Chicxulub impact: The origin of reservoir and seal facies in the southeastern Mexico oil fields: Geology, v. 28, p. 307–310.
17. Harker, D.E., Woodward, C.E., and Wooden, D.H., 2005, The dust grains from 9P/Tempel 1 before and after the encounter with deep impact: Science, v. 310, p. 278–280.
18. Hoffman, P.F., Kaufman, A.J., Halverson, G.P., and Schrag, D.P., 1998, A Neoproterozoic snowball Earth: Science, v. 281, p. 1342–1346.
19. Jansa, L., Pe-Piper, G., Robertson, P.B., and Friedenreich, O., 1989, Montagnais: a submarine impact structure on the Scotian Shelf, eastern Canada: Geological Society of America Bulletin, v. 101, p. 450–463.

20. Kyte, F.T., Zhou, Z., and Wasson, J., 1981, High noble metal concentrations in a late Pliocene sediment: Nature, v. 292, p. 417–420.
21. Margolis, S.V., Claeys, P., and Kyte, F.T., 1991, Microtektites, microkrystites, and spinels from a late Pliocene asteroid impact in the Southern ocean: Science, v. 251, p. 1594–1597.
22. Meech, K.J., and others 2005, Deep Impact: Observations from a worldwide Earth-based campaign: Science, v. 310, p. 265–269.
23. Melosh, H.J., 1989, Impact cratering: A geologic process: New-York, Oxford University Press, 245 p.
24. Melosh, H.J., 2000, A new and improved equation of state for impact computations, Lunar and Planetary Science Conference, v. 31: Houston, Lunar and Planetary Institute, p. CD-ROM # 1903.
25. Pierazzo, E., and Collins, G., 2003, A brief introduction to hydrocode modeling of impact cratering, *in* Dypvik, H., Burchell, M., Claeys, Ph., ed., Cratering in marine environments and on ice: Berlin, Springer, p. 323–340.
26. Poag, C.W., Plescia, J.B., and Molzer, P.C., 2002, Ancient impact structures on modern continental shelves: Chesapeake Bay, Montagnais, and Toms Canyon craters, Atlantic margin of North America: Deep-Sea Research, v. II49, p. 1081–1102.
27. Richardson, J.E., Melosh, H.J., Artemieva, N., and Pierazzo, E., 2005, Impact cratering theory and modeling for the Deep Impact Mission: from mission planning to data analysis: Space Science Review, v. 117, p. 241–267.
28. Richardson, J.E., and Melosh, H.J., 2006, Modeling he ballistic behavior of solid ejecta from the Deep Impact cratering event, Lunar and Planetary Science Conference, v. 37: Houston, Lunar and Planetary institute, p. CD-ROM # 1836.
29. Robin, E., Froget, L., Jehanno, C., and Rocchia, R., 1993, Evidence for a K/T impact event in the Pacific Ocean: Nature, v. 363, p. 615–617.
30. Schultz, P.H., Ernst, C.M., and Anderson, J.L.B., 2005, Expectations for crater size and photometric evolution from the deep impact collision: Space Science Review, v. 117, p. 207–239.
31. Shuvalov, V., 1999, 3 D hydrodynamic code SOVA for interfacial flows, application to the thermal layer effect: Shock Waves, v. 9, p. 381–390.
32. Shuvalov, V., Dypvik, H., and Tsikalas, F., 2002, Numerical simulations of the Mjolnir marine impact crater, Journal of Geophysical Research-Planets, v. 107, p. 1–12.
33. Smit, J., 1999, The global stratigraphy of the Cretaceous-Tertiary boundary impact ejecta: Annual Review of Earth and Planetary Sciences, v. 27, p. 75–113.
34. Stoffler, D., Artemieva, N.A., Ivanov, B.A., Hecht, L., Kenkmann, T., Schmitt, R.T., Tagle, R.A., and Wittmann, A., 2004, Origin and emplacement of the impact formations at Chicxulub, Mexico, as revealed by the ICDP deep drilling at Yaxcopoil-1 and by numerical modeling: Meteoritics & Planetary Science, v. 39, p. 1035–1067.
35. Sturkell, E.F.F., and Ormo, J., 1997, Impact-related clastic injections in the marine Ordovician Lockne impact structure, Central Sweden: Sedimentology, v. 44, p. 793–804.

36. Sugita, S., and 22 others, 2005, Subaru telescope observations of Deep Impact: Science, v. 310, p. 274–278.
37. Tsikalas, F., Gudlaugsson, S.T., Eldholm, O., and Faleide, J.I., 1998, Integrated geophysical analysis supporting the impact origin of the Mjølnir structure, Barents Sea: Tectonophysics, v. 289, p. 257–280.
38. Ward, S.N., and Asphaug, E., 2000, Asteroid impact tsunami: A probabilistic hazard assessment: Icarus, v. 145, p. 64–78.

Part IV

The Cometary Gas

Spectrophotometry of the Deep Impact Ejecta of Comet 9P/Tempel 1

K. W. Hodapp[1], G. Aldering[2], K. J. Meech[3], A. Cochran[4]

[1] Institute for Astronomy, University of Hawaii,
640 N. Aohoku Place, Hilo, HI 96720, hodapp@ifa.hawaii.edu
[2] Lawrence Berkeley Lab, Physics Div., MS-50/232,
One Cyclotron Rd., Berkeley, CA 94720, galdering@lbl.gov
[3] Institute for Astronomy, University of Hawaii, and NASA Astrobiology Institute,
2680 Woodlawn Drive, Honolulu, HI 96822, meech@ifa.hawaii.edu
[4] McDonald Observatory, University of Texas at Austin,
1 University Station C1402, Austin, TX 78712-0259, anita@astro.as.utexas.edu

Abstract We have obtained optical spectrophotometry of the evolution of comet 9P/Tempel 1 after the impact of the Deep Impact spacecraft [1], using the SNIFS Supernova Integral Field Spectrograph at the UH 2.2 m telescope. From the datacubes, we extracted both continuum flux distributions as well as emission line fluxes of the violet CN system and of [OI]. We found that the continuum brightness of the comet, i.e., scattered sunlight, started rising immediately after the impact, but that the ejecta were slightly bluer in color than the material normally released by the comet. The emission of [OI] at 630 nm, which is a tracer of water, rose similar to the scattered continuum light, but then remained nearly constant for several hours after impact. We found that CN emission at 388 nm centered on the nucleus was delayed compared to the rise of dust-scattered sunlight. This CN emission also expanded faster spatially than the cloud of scattering dust.

1 Introduction

As part of a coordinated simultaneous Earth-based observing campaign [2] of comet 9P/Tempel 1 before, during, and after the Deep Impact event, we used the Supernova Integral Field Spectrograph (SNIFS) [3] to obtain spectral datacubes covering the wavelength range from 350 to 1000 nm. This data acquisition method ensured that spatial, spectral, and temporal information was obtained over a wide wavelength range, without having to pre-select specific filter bandpasses or slit orientations before the event. After the impact event, it became clear that the most interesting aspects of our data were the spatial distribution and temporal evolution of the CN and [OI] emission and the evolution of the continuum scattered light color in the hours after impact.

2 Observations

Comet 9P/Tempel 1 was observed in the nights of July 2 – 9, 2005, at the UH 2.2 m telescope using the Supernova Integral Field Spectrograph (SNIFS) [3] that is permanently mounted at its bent Cassegrain focus. This instrument basically consists of a blue and a red spectrograph arm. After the two wavelength ranges are separated by a dichroic beamsplitter, the light in each arm is focused on a separate 15×15 lenslet array with 0.4″ lenslets, forming a contiguous 6″ × 6″ field of view. Each lenslet produces an image of the telescope pupil plane, which then becomes effectively the entrance point into a grism spectrograph. Each of the 15×15 pupil images is dispersed and produces a spectrum on the CCD detector. The lenslet array is rotated at such an angle around the optical axis that the 225 individual spectra do not overlap. SNIFS is therefore well suited to obtain spectrophotometry of the gas released from the comet nucleus against the background of the more extended coma.

Data were obtained on the nights of July 2 and 3 (UTC) prior to impact, on the night of July 4 (the impact night), when the first useful frame was exposed during the moment of impact, and on the nights of July 5, 7 and 8, after impact. The integration time used for spectra was 300s on July 2 and 3, 90 s on July 4 and 5, and 180 s on July 7 and 8. The nights of July 6 and 9 were lost due to poor seeing and cirrus clouds. From July 2 to July 5, non-sidereal open-loop tracking was used, while in the last two nights, non-sidereal guiding finally could be made to work. Intermixed with comet observations, we also observed a sky position 5′ from the comet and the solar analog standards P041C and P177D for photometric calibration, as well as internal flatfields and arc-lamp frames.

3 Data Reduction

Using flatfields, dark frames, and an arc-lamp spectrum (He/Hg/Cd for the blue arm and Ne/Ar/Xe for the red arm), the data were corrected, cosmic ray hits were removed, and the position of each spectrum on the CCD frame was determined. After wavelength and photometric calibration and standard Mauna Kea extinction correction, the extracted spectra were assembled into data cubes (x,y,λ) and stored as Euro-3D FITS tables. For the final analysis of the data we decided to bypass the photometric extraction functions of the SNIFS pipeline and instead extracted the photometric data from the Euro3D datacubes using IRAF scripts. Initially, the 15×15×λ data cubes were processed into 225×λ two-dimensional spectra. Figure 1 shows an example of these extracted, sky-subtracted spectra in the blue (top frame) and red channel (bottom two frames) in this two-dimensional format, which is essentially a set of 15 long (6″) slit spectra stacked in vertical direction. In this extracted, long-slit data format, the CN emission line is seen prominently. The C_3 emission band at 405 nm is also faintly indicated in the blue channel, but the signal-to-noise ratio is too low to analyze this emission in any more detail. The C_2 Swan bands were not clearly detected. The emission of [OI] at 630 nm is also clearly visible in the middle frame.

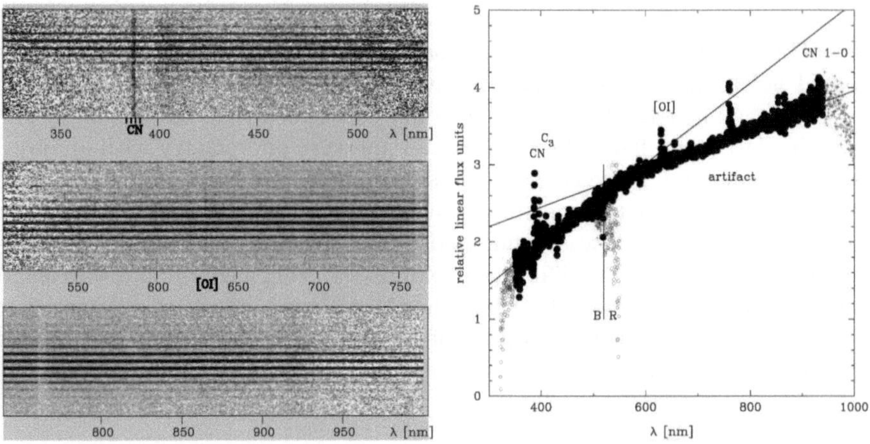

Fig. 1. 2D sky subtracted spectral frames (*left*) and extracted reflectivity spectrum.

An extracted sky-subtracted spectrum of the comet near the time of maximum continuum flux in a 2.4″ (6 pixel) diameter aperture (one half hour after impact) is shown in the right panel of Fig. 1. This spectrum is a reflectivity spectrum, obtained by dividing the comet spectrum by a spectrum of solar analog star P041C ([4]). The blue and red channel of SNIFS were joined at 520 nm, relying purely on the instrumental flux calibration without any further adjustments. The spectrum shows the emission of CN, the C_3 band, and the emission line of [OI]. The feature at 760 nm is an artifact from improper correction of the telluric O_2 absorption.

We also plot, as thin open symbols, the data points from both the blue and red channel at the limits of their respective wavelength ranges that we chose not to use. At their extreme wavelengths both SNIFS channels tend to give too small a signal due to calibration difficulties.

Spectrophotometric information on the violet band CN emission was extracted in a 10 pixel (2.9 nm) wide spectral window that included all the flux from the CN lines. The adjacent continuum was measured in two windows of the same size, at the adjacent shorter and longer wavelengths. Since the solar scattered light contains numerous absorption lines in this wavelength range, we normalized the 2-dimensional spectrum by the spectrum of solar analog star P041C ([4]). The underlying normalized continuum spectrum was estimated as the average of the two adjacent continuum wavelength intervals and subtracted from the spectral interval containing the CN emission. The resulting pure CN emission spectral frame was averaged in wavelength and rearranged into a spatial image of the CN flux distribution.

Information on the CN flux was extracted from this continuum-subtracted CN emission image in two complementary ways. First, the average flux over the full 6″ × 6″ field was computed using the IRAF task "imstat". This gives, in essence, the surface brightness of CN emission in the inner coma of comet

9P/Tempel 1, since the CN emission extends beyond the limits of our field of view. The varying centering of the comet coma in our images injects some additional noise even in these aperture averaged data, but for times before and long after the impact, when CN is much more extended than our $6'' \times 6''$ field of view, the data are reliable. Second, we did aperture photometry with an object aperture diameter of 7 pixels ($2.8''$) and sky annulus between $3.2''$ and $4.8''$ diameter. The CN photometry aperture was centered on the flux in the 350–400 nm integrated image, which was dominated by scattered continuum light and had enough signal-to-noise for reliable centroiding. By using a continuum wavelength range that included the wavelength of the CN emission, effects of differential atmospheric refraction on the position of the extraction aperture were avoided. More details on the observations and the data reduction is given in [5].

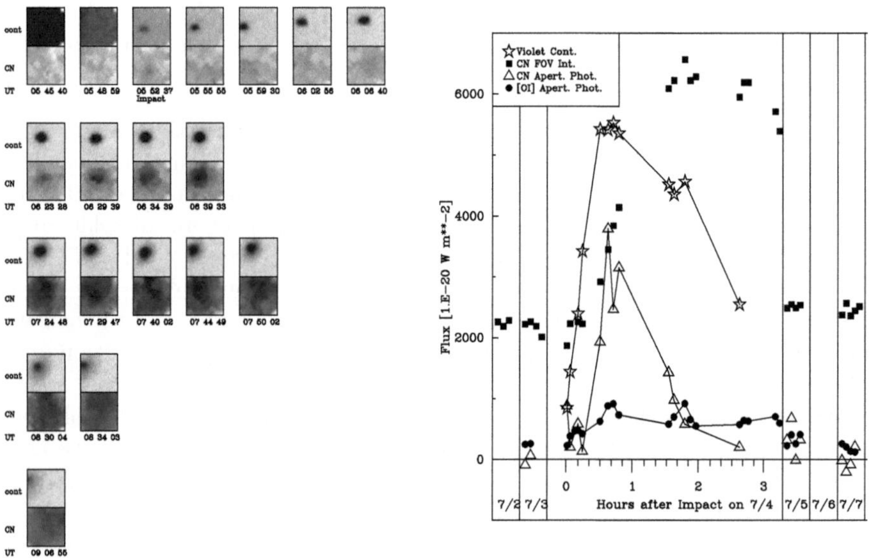

Fig. 2. *Left*: Blue continuum and CN emission images. *Right*: Lightcurves for CN aperture and field-integrated photometry, Blue aperture photometry, and [OI] aperture photometry.

Figure 2 (left panel) shows all the CN images and blue continuum images obtained on July 4. For each observation, the top panel is the 350–400 nm continuum image, and the lower panel is the continuum-subtracted CN flux image. The resulting CN lighturves, and also the [OI] and blue continuum light curves, are shown in the right panel of Fig. 2.

To study the broad color of the impact ejecta, the spectral datacubes were summed in 50 nm bins. We display the aperture photometry result in a logarithmic scale as magnitudes, to easily distinguish wavelength-independent brightness changes from changes in color. For each bandpass, the photometry aperture was individually centered to account for atmospheric dispersion shifts in the centroid

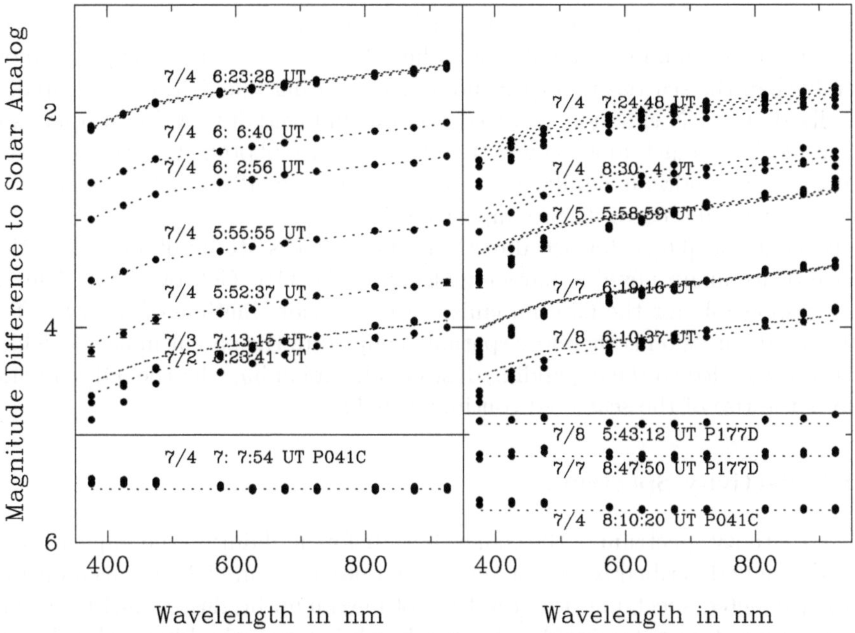

Fig. 3. Solar flux normalized relative magnitudes in 50 nm bins. Shown are the comet flux distributions in the nights prior to impact, and during the rise to maximum light on the left, and the decline in brighness and in the post impact night on the right.

position. In Fig. 3, on the left, we show the instrumental magnitudes of comet 9P/Tempel 1 in the 2 days prior to impact, and then from the moment of impact to maximum brightness. In the right panel, we show the instrumetal magnitudes during the decline in brightness and in the days following the impact. All magnitudes were normalized by subtracting the magnitude distribution of the solar analog star P041C, corrected for its deviations from the solar flux distribution [4]. The dotted line in both panels is the flux distribution at the time of maximum brightness, shifted to match the average brightness at any other time. It is clear that both before the impact and long after the maximum post impact brightness, the color of the comet was redder than that dotted line. We conclude that the material ejected by the impact scatters short wavelengths more strongly than the material released by the comet under steady state conditions.

4 Discussion

4.1 Gas Emission

It is not clear from what parent and/or grandparent molecules the CN radical is formed. However, the lifetime of those molecules are long compared to the few hours covered by our observations. This explains the delayed onset of CN emission concentrated on the comet's nucleus compared to the instantaneous rise in scattered continuum sunlight. Also, the CN emission spatially expands more rapidly than the scattering dust cloud, demonstrating that CN, once created, is long lived. The CN emission averaged over our full field of view (solid squares in Fig. 2) declined much more slowly than the aperture photometry (the open star symbols), and shows a significantly elevated level (by 12%) in the nights after the impact compared to its pre-impact level.

By contrast, [OI] is formed directly in its excited state by photodissociation of one of its many possible parent molecules (H_2, OH, CO, or CO_2). The ^1D state responsible for the 630 nm emission has a short lifetime of $\approx$ 110 s. The 630 nm [OI] emission is therefore spatially coupled to the parent molecule, which in turn is coupled to the expanding dust cloud, explaining why the [OI] emission follows the rise of the scattered continuum light.

4.2 Reflectivity Spectrum

The reflectivity spectrum of the comet, both before and during the impact event, can be rougly described by two slopes, with a break around 580 nm, as indicated in the extracted spectrum near the time of maximum brightness in Fig. 1. The slope change does not coincide with the break between the blue and red spectrograph channels. The reflectivity spectrum in the logarithmic instrumental magnitude representation of Fig. 3 also shows this steeper slope at short wavelengths. Further, Fig. 3 shows that the reflectivity spectrum was overall bluer while it was dominated by impact-generated ejecta. This can be explained if the material ejected during the impact contains either a larger fraction of very small particles, so that Raleigh scattering is more dominant, or contains very pure water ice crystalls, which tend to have blue color.

References

1. A'Hearn, M. F. and 31 co-authors (2005): Science **310**, 258
2. Meech, K. J., and 208 co-authors (2005): Science **310**, 265
3. Aldering, G. and 23 co-authors (2002): *Proc. of the SPIE*, Volume **4836**, 61
4. Bohlin, R. C., Dickinson, M. E., and Calzetti, D. (2001): AJ **122**, 2118
5. Hodapp, K. W. and 14 co-authors (2007): Icarus, **191**, 381

Submillimeter Wave Astronomy Satellite Observations of Comet 9P/Tempel 1

F. Bensch[1,2], G. J. Melnick[1] D. A. Neufeld[3], M. Harwit[4], R. L. Snell[5], B. M. Patten[1], V. Tolls[1]

[1] Harvard-Smithsonian Center for Astrophysics
60 Garden Street, Cambridge, MA 02138, U.S.A.
[2] Argelander-Institut für Astronomie, Universität Bonn, 53121 Bonn, Germany
fbensch@astro.uni-bonn.de
[3] Department of Physics and Astronomy, Johns Hopkins University
Baltimore, MD 21218, U.S.A.
[4] 511 H street, SW, Washington, DC 20024, U.S.A.; also Cornell University
[5] Department of Astronomy, University of Massachusetts
Amherst, MA 01003, U.S.A.

Abstract We present observations of the $1_{10} - 1_{01}$ ortho-water ground-state rotational transition in comet 9P/Tempel 1 made with the NASA Submillimeter Wave Astronomy Satellite (SWAS). The SWAS monitoring covered a 3-month period from June to August 2005 and included observations of the comet's activity during the collision of the Deep Impact spacecraft with the nucleus on 4 July 2005. No excess emission from the material ejected by the impactor was detected and we derive an upper limit of 1.8×10^7 kg on the vaporized water ice. However, the water production rate of the comet shows large natural variations of up to a factor of three throughout the 3-month monitoring period. The total amount of water vaporized between 5 June and 1 September 2005 is 4.5×10^{34} molecules ($\sim 1.3 \times 10^9$ kg) and we estimate that 9P/Tempel 1 has lost as much as $N \sim 10^{35}$ molecules during its 2005 apparition.

1 Introduction

Water is the most abundant species among the cometary volatiles and is responsible for much of a comet's activity in the inner Solar system ($\lesssim 3$ AU). Observations of water transitions are therefore well suited to monitor the cometary activity and for measuring the amount of volatiles released by the collision of the Deep Impact spacecraft with the nucleus of 9P/Tempel 1 [1, 2]. Most of the water vapor in the coma is rotationally cold [3]. Ground-state rotational transitions in the submillimetre regime are therefore particularly useful to measure the vaporization rate. Present-day heterodyne receivers provide the high velocity resolution ($< 1\,\mathrm{km\,s^{-1}}$) required to resolve cometary emission lines; however, the observations of water emission at submillimetre and far-infrared wavelengths have to be carried out in space because the Earth's atmosphere is opaque at these wavelengths.

The Submillimeter Wave Astronomy Satellite (SWAS) is a NASA small explorer mission in a low-Earth orbit and carries a heterodyne receiver for observations at submillimetre wavelengths [4]. Launched 5 December 1998, SWAS observes the $1_{10} - 1_{01}$ ground-state transition of ortho water ($H_2^{16}O$ and $H_2^{18}O$ at 556.936 GHz and 547.676 GHz, respectively) simultaneously with the transitions of three other astrophysically important species (neutral carbon, molecular oxygen, and the carbon monoxide isotopologue ^{13}CO). Constrained only by the comet's visibility from low-Earth orbit and a Moon-avoidance angle of 15°, SWAS monitored the ortho $H_2^{16}O$ line emission in 9P/Tempel 1. Observations of the $H_2^{18}O$ $1_{10} - 1_{01}$ transition were not made because the line emission of the rarer isotopologue was expected to be too faint in 9P/Tempel 1 to be easily detected with SWAS. The monitoring of the water vaporization rate is presented in Sect. 2, and the results from the observations of Deep Impact on 4 July 2005 are discussed in Sect. 3.

2 SWAS Monitoring of the Water Vaporization Rate

SWAS was brought out of an orbital hibernation on 1 June 2005, and after an initial check-out of the observatory a monitoring campaign was carried out for 9P/Tempel 1 from (UT) 5.29 June to 1.84 September 2005. Observations were made whenever the comet was 45° above the Earth limb, corresponding to segments of 20–38 min duration for each 97 min SWAS orbit. Typically, the spectra obtained over a period of 1–3 days have to be co-added to obtain a statistically significant detection of the water line. The exceptions are observations in July and August during phases when the overall outgassing of the comet was low. The weaker water vaporization rate together with the larger comet-Earth distance in these months (compared to June) resulted in a signal which was below the SWAS detection limit even for 5–10 day co-adds.

The SWAS-measured line-integrated intensities are converted to water vaporization rates using the model by Bensch & Bergin [3]. This model uses a spherically-symmetric density profile for the neutral water in the cometary coma, assuming a constant water vaporization rate and a constant expansion velocity (Haser model), and accounts for the finite life-time of the water molecules due to dissociation and ionization by solar UV photons. Water rotational-line excitation includes collisions of water with other water molecules, electron collisions, and the fluorescence of IR pumped ro-vibrational levels to rotationally excited levels of the vibrational ground state. The level population and the radiation transfer is calculated using `ratran`, an accelerated Monte Carlo code [5]. This code allows for a better treatment of the radiation transfer than models based on an escape-probability method as the optical depth of the water rotational transitions can be large.

The resulting water vaporization rate for 9P/Tempel 1 is shown in Fig. 1. (See [6] for a table of the SWAS observations and the corresponding Q_{H_2O}.)

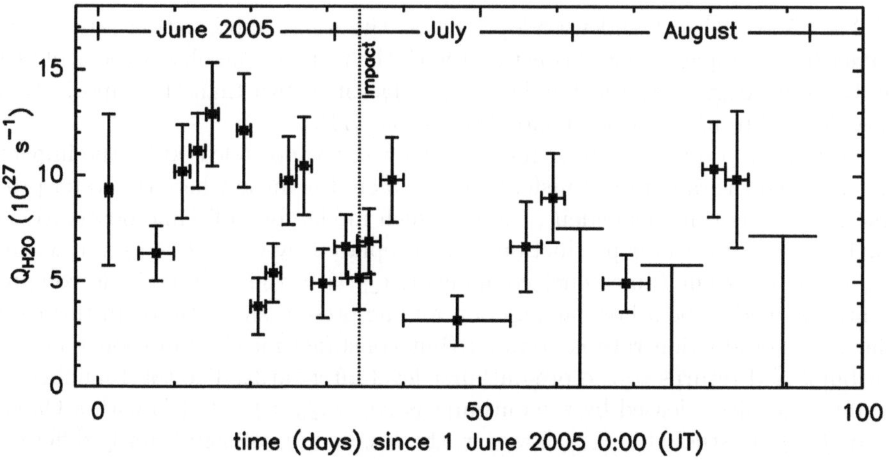

Fig. 1. Total water vaporization rate derived from the SWAS observations of the $1_{10} - 1_{01}$ ground-state rotational transition of ortho water. A table listing the SWAS observations and the corresponding Q_{H_2O} is available in [6].

The most striking feature is the rather large variation of the water vaporization rate of up to a factor of three on time-scales as short as 2–4 days. Generally, the water vaporization rate varies between periods with $Q_{H_2O} \gtrsim 10^{28}\,s^{-1}$ and periods where the outgassing is reduced to $Q_{H_2O} \lesssim 5 \times 10^{27}\,s^{-1}$. These fluctuations are much larger than any variation of Q_{H_2O} with the heliocentric distance over this 3-month period.

Comet 9P/Tempel 1 reached its perihelion on 5 July 2005 ($r_h = 1.506\,AU$), and we note that the ratio of the pre- to post-perihelion water production rate is 2.38 ± 0.13 (~1-month averages). We estimate that 9P/Tempel 1 vaporized a total of $N = 4.5 \times 10^{34}$ water molecules ($\sim 1.3 \times 10^9\,kg$) between 5 June and 1 September and possibly lost as much as $N \sim 10^{35}$ molecules during the 2005 apparition. Assuming a density of the nucleus between (0.35 and 0.6) $g\,cm^{-3}$ and a surface area of $113\,km^2$ [1, 2, 7], we calculate a net erosion per orbit of 4–8 cm (averaged over the total nucleus surface) and between 50 and 85 cm (assuming that 9% of the surface is covered by active areas [7]). This does not account for the dust, and the total erosion is larger by a factor $\gtrsim 2$ if the dust-to-gas ratio measured in the impact ejecta [8, 9] is an indication for the amount of dust dragged away by the gas vaporized during the 2005 apparition.

3 Water Vapor in the Impact Ejecta

No statistically significant increase in the water-line emission is detected by SWAS immediately following the impact. The average line-integrated intensity during the three days after the impact was $0.16 \pm 0.04\,K\,km\,s^{-1}$, virtually identical to the average intensity measured for the three days before the impact. The water vaporization rate for a 3-day interval centered on the impact date is

$(5.2 \pm 1.5) \times 10^{27}\,\mathrm{s}^{-1}$ and we adopt this as the water vaporization rate of the comet before impact. This is consistent with the results from observations of OH at UV wavelengths [8, 10], but lower by a factor of two than the vaporization rate derived from water hot-band observations [11].

In order to derive an upper limit on the water vapor released by the impact we extended the radiation transfer model to include comets where the water production rate is time-dependent, the first such model for radio-line observations [6, 12]. The total water production rate is replaced by $Q = Q_\mathrm{q} + Q_\mathrm{b}(t)$, where Q_q is the constant (quiescent) component. $Q_\mathrm{b}(t)$ is the time-variable component, assumed to be a box-car function for the present simulations. In this case the water production rate is elevated (but constant) for the duration τ of the outburst and returns to the pre-outburst level afterwards. The total number of water molecules released by the outburst is $N_\mathrm{b} = Q_\mathrm{b} \times \tau$. Models with a Gaussian $Q_\mathrm{b}(t)$ are studied in [12], however, the results are not significantly different in the present case.

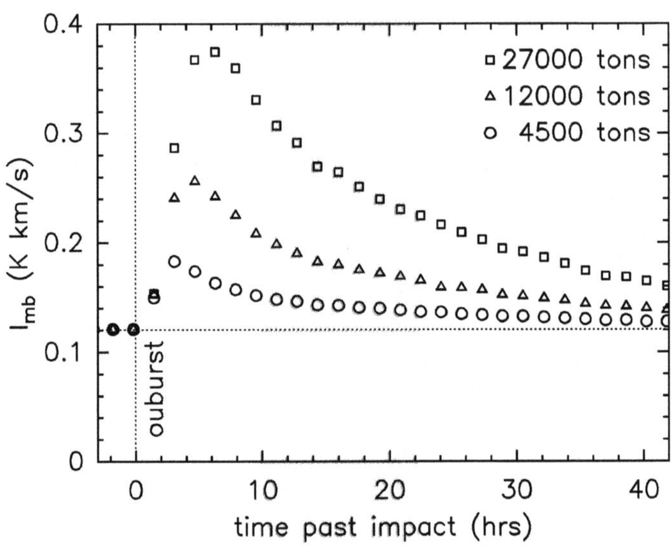

Fig. 2. Simulation of the line-integrated intensity for an outburst of $\tau = 30\,\mathrm{min}$ and three different N_b, indicated in the upper right of the panel.

Figure 2 shows the time evolution of the velocity-integrated intensity $I(t)$ for a simulated outburst of $\tau = 0.5\,\mathrm{h}$ and three different N_b. Within the first hour, the line-integrated intensity has only slightly increased and is nearly identical for the three models. This results from the large optical depth and the small beam-filling factor of the emission emitted by the water vapor in the impact ejecta. Thus, we expect that water radio-line observations made immediately following the impact ($\lesssim +1\,\mathrm{h}$) are not sensitive to the total amount of water released. The line intensity in the model peaks at $t_\mathrm{peak} \approx$3–7 h after the outburst. While the expansion of the ejecta increases the beam-filling factor the optical depth of the radiation decreases. In addition, the water molecules have reached the region in the coma where electron collisions are important, resulting in an additional boost for the line excitation. The actual time delay t_{peak} and peak intensity $I(t_{peak})$ depend on the total number of water molecules released (larger t_{peak} for larger N_b because the emission becomes optically thin at later times), the outburst duration (larger t_{peak} for larger τ), and the telescope beam size (smaller t_{peak} and larger $I(t_{peak})$ for telescopes with a larger primary reflector and thus a smaller main beam).

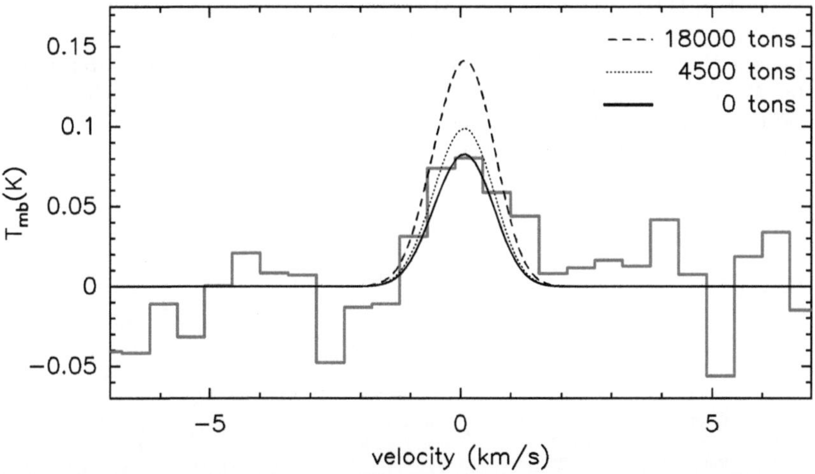

Fig. 3. Comparison of SWAS spectrum observed post-impact between July 4.3 and 6.0. The histogram gives a weighted average, where the individual spectra have been weighted by the square of the intensity predicted by the outburst model and the inverse square of the rms baseline noise [6]. The lines show the model spectra for $\tau = 30\,\mathrm{min}$ and two different N_b as well as for the pre-impact $Q_q = 5.2 \times 10^{27}\,\mathrm{s}^{-1}$ ($N_b = 0$).

No statistically significant increase in the water line emission was detected by SWAS immediately following the impact. Using the SWAS observations of up to 42 h post impact we derive a 3σ upper limit of 6×10^{32} molecules (1.8×10^{7} kg) for the water vapor in the impact ejecta, assuming an outburst duration of τ =0.5–2 h (Fig. 3; see [6] for details). A slightly better upper limit is obtained if we assume that the impact-generated outburst persisted for more than a few hours, with $N_b < 1.6\times10^{7}$ kg (1.2×10^{7} kg) for $\tau = 8$ h (16 h, respectively). These upper limits are consistent with the results presented by other authors in this volume which indicate that the impactor vaporized between 4.5×10^{6} kg and 9.0×10^{6} kg of water [8, 9, 13].

References

1. A'Hearn, M.F. et al. (2005): Science **310**, 258
2. A'Hearn, M.F. et al.: these proceedings
3. Bensch, F., Bergin, E.A. (2004): ApJ **615**, 465
4. Melnick, G.J. et al. (2000): ApJ **539**, L77
5. Hogerheijde, M.R., van der Tak, F.R.S. (2000): A&A **362**, 697
6. Bensch, F., Melnick, G.J., Neufeld, D.A. et al. (2006): Icarus **184**, 602.
7. Belton, M.J.S. et al. (2005): Space Sci. Rev. **117**, 137.
8. Küppers, M. et al. (2005): Nature **437**, 987
9. Bodewits, D. et al.: these proceedings
10. Schleicher, D.G. et al. (2006): AJ**313**, 1130
11. Mumma, M.J. et al. (2005): Science **310**, 270
12. Bensch, F., Bergin, E.A.: Rat4com: A radiation transfer model for water in comets. In: *Proceedings of the Dusty and Molecular Universe*, SP-577, ed by A. Wilson (ESO, Nordwijk 2005) pp 465–466
13. Biver, N. et al.: these proceedings

Gas Production by Deep Impact from Far-ultraviolet Observations

P. D. Feldman

Department of Physics and Astronomy, Johns Hopkins University, Baltimore, MD 21218, U.S.A.

Abstract We report on far-ultraviolet observations of comet 9P/Tempel 1 during the epoch of Deep Impact made from the *Hubble Space Telescope* and the ALICE instrument aboard the *Rosetta* spacecraft. The far-ultraviolet channel on *Galaxy Evolution Explorer* was not operational at the time of the observations but wide-field imaging spectroscopy at longer UV wavelengths (2200–3200 Å) was obtained.

1 Introduction

Jupiter-family comets, such as 9P/Tempel 1, observed at relatively large distance from the Sun and characterized by a low quiescent activity level, are faint in the far-ultraviolet region of the spectrum and consequently have rarely been observed from orbiting observatories. The spectral region below 2000 Å is rich in molecular and atomic emissions of the primary volatile components of the cometary coma. These include several band systems of CO and the strongest resonance transitions of the principal atomic species, C, H, O, N, and S, that have been observed in many comets, usually the most active to appear, over the past 30 years using orbiting observatories and sounding rockets [1]. These species were expected to show a large increase as a result of the excavation of volatile subsurface material by the impactor and so warranted a look by the available far-ultraviolet instrumentation in space. Time was awarded on two NASA orbiting observatories, the *Hubble Space Telescope* (*HST*) and the *Galaxy Evolution Explorer* (*GALEX*) to observe the comet before, during and after the impact. In addition, the ALICE far-ultraviolet spectrometer was one of the remote sensing instruments aboard the *Rosetta* spacecraft used in an extended observing campaign from June 28 to July 14, 2005. A summary of the results from each of these programs is presented below. More detailed reports may be found in the references cited.

2 Hubble Space Telescope

The *HST* campaign on Deep Impact observed periodic comet 9P/Tempel 1 both in the visible, at the highest spatial resolution (36 km) achievable from Earth, and in the far-ultraviolet to study the generation and evolution of gas, primarily CO, resulting from Deep Impact. The visible imaging results are discussed elsewhere in this Workshop and have been reported in detail [2]. For the latter, the Solar Blind Channel (SBC) of the Advanced Camera for Surveys (ACS) was used with the F140LP filter (bandpass roughly 1350–1700 Å) and the PR130L prism, both of which are sensitive primarily to emission from the CO Fourth Positive system. The SBC was used as a replacement for the Space Telescope Imaging Spectrograph (STIS) which failed in August 2004. However, because of the low observed count rate, the objective prism spectra are difficult to interpret and analysis to date has been limited to the F140LP images.

A weak signal was seen in all of the SBC filter images. Assuming that the spatially varying emission in the pre-impact images is due solely to CO fluorescence, we use a recent model [3] to derive a value of the quiescent CO production rate of 4–6 $\times 10^{26}$ molecules s^{-1}, or about 4–6% of the water production rate [4]. The first SBC observation was $\sim$2.5 h after impact and at the time we detected an increase in brightness of about a factor of two over the quiescent coma. The subsequent decay appeared exponential with a time constant of $\sim$5 h. If all of the excess emission is due to CO, this corresponds to 1.5×10^{31} molecules or a mass of about 650,000 kg, an amount that would normally be produced in 7 to 10 h of quiescent outgassing. Using 1.5×10^{32} molecules for the amount of water produced by the impact [5], this would give a ratio of CO to H_2O in the ejecta of 10%. Küppers et al. (this Workshop) have revised their estimate of the amount of water upward to $1.5 - 3.0 \times 10^{32}$ molecules, so the CO to H_2O ratio in the ejecta becomes <10%, and suggests that the volatile content of the material excavated by the impact did not differ significantly from the surface or near subsurface material responsible for the quiescent outgassing of the comet. A complete description of these results has been published [6].

3 ALICE/Rosetta

ALICE is a lightweight, low power spectrograph (700–2000 Å) on the *Rosetta* spacecraft designed for in situ imaging spectroscopy of the coma of comet 67P/Churyumov-Gerasimenko [7]. It has a spectral resolution of 8 or 12 Å depending on position along its 6° slit. Its large field-of-view enhances sensitivity to extended atomic emissions at the expense of gas emissions resulting from the excavation of the comet's surface. Continuous observations were made during the *Rosetta* Deep Impact campaign from June 28 to July 14, 2005.

At all epochs, only O I λ1304 emission from the comet is definitively detected. H I Lyman-β is seen as a weak enhancement above the interplanetary background. No variation is seen in the daily average of O I emission at the 2-σ

level following impact. There appears to be a slow secular decrease of this emission with time. A Haser model for O I fits the spatial distribution and gives a quiescent water production rate of 9×10^{27} molecules s^{-1}, in good agreement with other determinations [4]. The model is also consistent with the level of H I Lyman-β observed. The upper limits on C I and CO emissions are consistent with CO abundance determined from HST/ACS/SBC imaging. Full details of these results are given in [8].

4 GALEX

The *Galaxy Evolution Explorer* (*GALEX*) is a NASA Small Explorer mission designed to map the history of star formation using two modes: two-band photometry (FUV, 1350–1750 Å; NUV, 1750–3100 Å) and integrated field grism spectroscopy (10–20 Å spectral resolution). The 1.2° FOV diameter, 5″ spatial resolution, and high sensitivity to extended sources makes *GALEX* ideally suited to cometary coma studies. Its capability was demonstrated by March 2005 observations of comet C/2004 Q2 (Machholz) [9]. Unfortunately, the FUV channel was not operational at time of Deep Impact so that only NUV observations were obtained in both direct imaging and grism modes. These were made 2 days pre-impact, immediately following impact, and one day later.

Preliminary analysis was done with pipeline products produced shortly after the observations were made. Since then there have been significant improvements in the aspect solution of the spacecraft and new products have recently (October 2006) been delivered. These will enable us to take the comet motion out of the spectral images. Here we present some results from the preliminary analysis with the caveat that the numbers are likely to change over the next few months [10].

The extracted spectrum is typical of cometary spectra in this wavelength band, showing emission primarily from the CS (0,0) band at 2576 Å and the OH (1,0) and (0,0) bands at 2820 and 3085 Å, respectively, superimposed on a reflected solar continuum. There was a marked increase in all of these emissions following the impact. Fluxes were extracted from a 32,000 km square box, 5 h after impact. This box contains all of the material produced by the impact as an outflow velocity of 0.7 km s^{-1} carries molecules 12,600 km from nucleus. OH and CS abundances were determined using standard heliocentric velocity dependent fluorescence efficiencies. We find that the OH content after 5 h is $\sim 2.2 \times 10^{31}$ molecules. This leads to an estimate of the additional H_2O produced by the impact of $\sim 2.3 \times 10^{32}$ molecules (6.7×10^6 kg). After 25 h, water has moved 63,000 km from nucleus, outside our extraction box, and the OH emission returns to its quiescent level. The CS content is $\sim 4 \times 10^{29}$ molecules (2.8×10^4 kg), and is the same as its presumed parent, CS_2. The ratio of CS to H_2O (by number) is 1.7×10^{-3}, a typical value seen in many comets observed by the *International Ultraviolet Explorer* [11]. Our water estimate is consistent with the revised value of Küppers et al. (these Proceedings).

Acknowledgements The success of these campaigns owes much to the dedicated support of the mission operations teams at the Space Telescope Science Institute (*HST*), the ESA Rosetta Science Operations Center and Mission Operations Center (*Rosetta*), and the GALEX Science Operations Center at Caltech (*GALEX*). Special thanks go to my collaborators Stephan McCandliss, Hal Weaver, Roxana Lupu, Alan Stern, Joel Parker, and Andrew Steffl. This work was supported by grants from NASA.

References

1. Feldman, P. D., Cochran, A. L., Combi, M. R.: Spectroscopic investigations of fragment species in the coma. In: *Comets II*, ed. by M. C. Festou, H. A. Weaver, H. U. Keller (Univ. of Arizona, Tucson 2004), pp 425–447
2. Feldman, P. D., McCandliss, S. R., Route, M. et al.: Hubble Space Telescope observations of Comet 9P/Tempel 1 during the Deep Impact encounter, Icarus (2007), **191**, 276
3. Lupu, R. E., Feldman, P. D., Weaver, H. A., Tozzi, G.-P. (2004): BAAS **36**, 1433
4. Biver, N., Bockelée-Morvan, D., Boissier, J. et al. (2005): BAAS **37**, 710
5. Küppers, M., Bertini, I., Fornasier, S. et al. (2005): Nature **437**, 987
6. Feldman, P. D., Lupu, R. E., McCandliss, S. R. et al. (2006): Astrophys. J. **647**, L61
7. Slater, D. C., Stern, S. A., Booker, T. et al. (2001): SPIE **4498**, 239
8. Feldman, P. D., Stern, S. A., Steffl, A. J. et al., Ultraviolet spectroscopy of Comet 9P/Tempel 1 with Alice/Rosetta during the Deep Impact Encounter (2007) Icarus, **191**, 258
9. Morgenthaler, J. P., Harris, W. M., Combi, M. R. et al. (2005): BAAS **37**, 647
10. McCandliss, S. R., Feldman, P. D., Lisse, C. M. et al.: in preparation
11. Meier, R., A'Hearn, M. F. (1997): Icarus **125**, 164

Overview of Hubble Space Telescope Visible Imaging of 9P/Tempel 1 and Deep Impact

P. D. Feldman

Department of Physics and Astronomy, Johns Hopkins University, Baltimore, MD 21218, U.S.A. pdf@pha.jhu.edu

Abstract We present here an overview of the *Hubble Space Telescope* (*HST*) program to observe periodic comet 9P/Tempel 1 in conjunction with NASA's Deep Impact mission. Our objectives were to study the generation and evolution of the coma resulting from the impact and to obtain wide-band images of the visual outburst generated by the impact. All of the observations were made using the Advanced Camera for Surveys (ACS). For imaging, we used the ACS High Resolution Channel (HRC) with a broad V filter, F606W, which provided high sensitivity and a spatial resolution of 36 km (16 km pixel^{-1}) at the comet at the time of impact. Two observing campaigns utilizing a total of 17 *HST* orbits were carried out: the first occurred on 2005 June 13–14 and fortuitously recorded the appearance of a new, short-lived fan in the sunward direction on June 14. This outburst was also seen by the imagers on the Deep Impact spacecraft [1] and by ground-based observers [2]. The principal campaign began two days before impact and was followed by contiguous orbits through impact plus several hours and then snapshots one, seven, and twelve days later. Baseline images of the comet, made prior to impact, photometrically resolved the comet's nucleus. The derived diameter, 6.1 km, is in excellent agreement with the 6.0 ± 0.2 km diameter derived from the spacecraft imagers [1]. Following the impact, the HRC images illustrate the temporal and spatial evolution of the ejecta cloud and allow for a determination of its expansion velocity distribution. Velocities in the ejecta plume were found to be about twice those derived from the images of the natural outburst of June 14. One day after impact the ejecta cloud had passed out of the $\sim 30'' \times 30''$ field-of-view of the HRC.

Full details of the *HST* results are given in a paper currently in press in the *Icarus* special issue on Deep Impact [3].

Acknowledgements This work is based on observations with the NASA/ESA *Hubble Space Telescope* obtained at the Space Telescope Science Institute, which is operated by the Association of Universities for Research in Astronomy (AURA), Inc., under NASA contract NAS 5-26555. The success of the *HST* Deep Impact campaign owes much to the dedicated support of the mission operations team at the Space Telescope Science Institute. This work was supported by grant GO-10144.01-A from the Space Telescope Science Institute.

References

1. M. F. A'Hearn, M. J. S. Belton, W. A. Delamere et al. (2005): Science **310**, 258
2. L. M. Lara, H. Boehnhardt, R. Gredel et al. (2006): Astron. Astrophys. **445**, 1151
3. P. D. Feldman, S. R. McCandliss, M. Route et al. (2006): Hubble Space Telescope observations of Comet 9P/Tempel 1 during the Deep Impact encounter, Icarus, **187**, 113–122

Radio Monitoring of 9P/Tempel 1 Outgassing and Gas Released by the Impact

N. Biver[1], D. Bockelée-Morvan[1], J. Boissier[1], J. Crovisier[1], P. Colom[1], A. Lecacheux[1], R. Moreno[1], G. Paubert[2], D. C. Lis[3], M. Sumner[3], U. Frisk[4], Å. Hjalmarson[5], M. Olberg[5], A. Winnberg[5], H. Florén[6], A. Sandqvist[6], and S. Kwok[7,8]

[1] LESIA, CNRS UMR 8109, Observatoire de Paris, 5 pl. Jules Janssen, F-92190 Meudon, France `nicolas.biver@obspm.fr`
[2] IRAM, Avd. Divina Pastora, 7, 18012 Granada, Spain
[3] Caltech, MS 320-47, Pasadena , CA 91125, USA
[4] Swedish Space Corporation, PO Box 4207, SE-17104 Solna, Sweden
[5] Onsala Space Observatory, SE-43992 Onsala, Sweden
[6] Stockholm Observatory, SCFAB-AlbaNova, SE-10691 Stockholm, Sweden
[7] Dept. of Physics and Astronomy, University of Calgary, Calgary, AB T2N 1N4, Canada
[8] Inst. of Astron. & Astrophys., Academia Sinica, PO Box 23-141, Taipei 106, Taiwan

Abstract As part of the international observing campaign in support of the NASA Deep Impact mission to comet 9P/Tempel 1, we conducted a monitoring of its outgassing at radio wavelengths. OH observations conducted at the Nançay radio telescope provided a 4-month monitoring of the comet from March to July, followed by the observation of H_2O with the Odin satellite from June to August 2005. The peak of outgassing was found to be around 10^{28} molec. s^{-1} between May and July. Observations of HCN conducted with the IRAM 30 m radio telescope in May 2005 showed a variation of the production rate with a period of 1.73 ± 0.10 days, consistent with the rotation period of the nucleus.

The Odin satellite also monitored nearly continuously the H_2O line at 557 GHz during the 38 h following the impact on the 4th of July. Once possible periodic variations related to the nucleus rotation are removed, a small increase of outgassing related to the impact is present, which corresponds to the release of 5000 ± 2000 tons of water. Two other bursts of activity were seen on 23 June and 7 July.

1 Introduction

Observing comet 9P/Tempel 1 in support to the Deep Impact mission [13] was a major objective of ground-based radio observations and for the 2005 Odin space observatory program [16]. A full report of our radio observations of comet 9P/Tempel 1 appears in [7]. We show here the results of observations conducted with the Nançay radio telescope, the IRAM (Institut de radioastronomie millimétrique) 30 m telescope in May 2005, and the Odin satellite. Time variations of production rates are presented here and molecular abundances of the various molecules observed with the IRAM or CSO (Caltech submillimeter observatory) telescopes in a companion paper [9].

2 Data Analysis

Nançay observations were reduced and production rates calculated as has been regularly done for previous observations [8], assuming isotropic outgassing and taking into account collisional quenching of the OH maser. Intensities of millimetre and submillimetre lines are converted into production rates using the same modelling and codes as in previous papers (e.g. [4, 5, 6, 7]). A Haser model with symmetric outgassing and constant radial expansion velocity is used to describe the density. The variation of the photo-dissociation lifetimes of H_2O and HCN due to solar activity is taken into account. The expansion velocity (v_{exp}) is estimated from the line shapes of the spectra with highest signal-to-noise ratios. For May observations and after mid-July we use $v_{exp} = 0.65$ kms^{-1}. In June–July v_{exp} was estimated around 0.75 kms^{-1}. Our model of excitation of the rotational levels of the molecules takes into account collisions with neutrals at a constant gas temperature. Collisions with electrons also play a major role and are modelled according to [3] and [4] with an electron density scaling factor x_{ne} set to 0.3 for all data. We have weak constraints on the gas temperature: in May, from simultaneous observations of the two HCN J(3–2) and J(1–0) lines, we infer a rotational temperature of 12.2 ± 1.1 K. Given the assumed collision rate, this corresponds to kinetic temperature of 14^{+11}_{-8}K. For July data we do not expect a much higher temperature (given that the generally observed trend in previous comets is $T \propto r_h^{-1}$ to $r_h^{-1.5}$) and assume $T = 20$ K. The infered OH, H_2O and HCN (in May) "apparent" production rates (i.e. based on the stationary regime hypothesis) are given in Table 1.

3 Intrinsic Variations of the Outgassing and Molecular Abundances

3.1 Long-term Evolution of the Outgassing

Figure 1 shows the evolution of the water production rate from March to August 2005. The least squares fitting of a Gaussian to the water production rates, taking into account a r_h^{-2} heliocentric evolution due to increased solar irradiance as the comet gets close to the Sun, yields:

$$Q_{H_2O} = 10.0 \times 10^{27} \exp\left(-\left(\frac{t+23}{88}\right)^2\right) \times \left(\frac{1.506}{r_h}\right)^2, \qquad (1)$$

Table 1. Molecular "apparent" production rates of comet 9P/Tempel 1.

UT date [mm/dd.dd±d.dd]	$< r_h >$ [AU]	$< \Delta >$ [AU]	Q_{app}	$Q^1_{H_2O,ref}$ [$10^{27}s^{-1}$]
OH production rates (Nançay)			[$10^{27}s^-1$]	
03/11.6±7.5	1.89	0.98	< 6.0	5.0 **
03/28.8±8.0	1.78	0.82	4.2 ± 1.0	5.7 **
04/14.3±7.5	1.71	0.75	4.3 ± 0.9	6.2 **
04/27.3±4.5	1.65	0.72	4.9 ± 1.3	6.7 **
05/07.3±4.5	1.62	0.71	9.5 ± 1.4	6.9 **
05/17.8±5.0	1.58	0.72	6.2 ± 1.9	7.3 **
06/01.3±7.5	1.54	0.76	6.0 ± 1.5	7.7 **
07/06.3±4.5	1.51	0.90	< 37.0	8.0 **
H_2O production rates (Odin)			[$10^{27}s^-1$]	
06/18.25±0.18	1.516	0.818	11.5 ± 1.0	7.9 **
06/23.73±0.12	1.511	0.842	12.4 ± 1.3	8.0 **
07/03.93±0.28	1.506	0.892	9.1 ± 0.8	8.0 **
07/04.44±0.18	1.506	0.895	7.8 ± 0.9	7.4 ± 0.7 ***
07/04.84±0.18	1.506	0.897	9.6 ± 1.0	7.6 ± 1.0 ***
07/05.24±0.18	1.506	0.899	10.8 ± 1.0	9.0 ± 1.3 ***
07/05.64±0.19	1.506	0.901	8.7 ± 1.1	7.3 ± 0.9 ***
07/07.70±0.14	1.506	0.912	10.6 ± 1.5	8.0 **
07/09.34±0.12	1.507	0.921	7.8 ± 1.5	8.0 **
07/10.71±0.14	1.507	0.929	7.3 ± 1.3	8.0 **
07/12.07±0.12	1.508	0.937	9.0 ± 1.3	8.0 **
07/16.29±0.12	1.510	0.962	7.7 ± 1.4	8.0 **
07/25.35±0.14	1.519	1.020	6.8 ± 1.3	8.0 **
07/31.73±0.14	1.529	1.064	4.8 ± 1.3	7.8 **
08/07.68±0.13	1.543	1.116	8.4 ± 1.4	7.6 **
HCN production rates	(IRAM)		[$10^{24}s^{-1}$]	
05/04.90±0.14	1.625	0.711	6.8 ± 1.1	7.6 *
05/05.91±0.13	1.622	0.712	4.7 ± 0.6	5.0 *
05/06.92±0.12	1.618	0.712	10.1 ± 1.2	9.5 *
05/07.92±0.13	1.615	0.713	5.6 ± 0.8	4.5 *
05/08.92±0.13	1.611	0.713	10.1 ± 1.4	8.8 *

[1]$Q_{H_2O,ref}$ is from Eq. (4) (*), from $(8 \times 10^{27} \times (1.506/r_h)^2$ (**), or from measurements obtained for the same rotation phase between 7 July and 7 August (***).

where t =date$-T_{\rm perihelion}$ in days. The 23 June data point is not included in this analysis (cf. Section 5). The fit is marginally significant ($\chi^2 = 27$ versus 41 for no variation), and might be biased by the inhomogeneity of the data (Nançay versus Odin). A peak of activity before perihelion is suggested. This is in agreement with previous perihelion passages, though the peak in production was near 2 months before perihelion [12] rather than 3 weeks. This also suggests a mean production rate around the time of the impact of 9.3×10^{27} mols, in agreement with other measurements: 6×10^{27} from OH in the near UV [17], 7×10^{27} from H_2O observed with SWAS [2] to 10×10^{27} mols from H_2O observed in the IR [15].

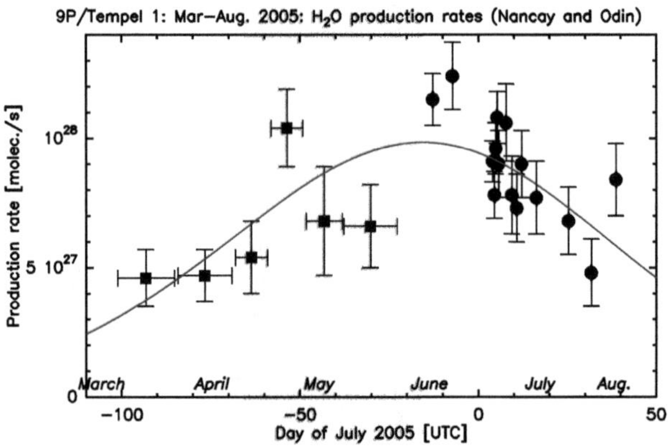

Fig. 1. Water production rates based either on Nançay observations of OH (*squares*) or Odin observations of H_2O (*circles*). The continuous curve corresponds to Eq. (1).

3.2 Periodic Variations of the HCN Production Rate

Figure 2 shows the HCN "apparent" (derived from a stationary regime assumption) production rate versus time observed in May. We split data into two time intervals per night and the production rates based on HCN(1–0) and HCN(3–2) are considereded separately. On the basis of simulations [7] we correct the observing time from the phase delay of the variations due to beam dilution ("beam delay" = 1.1–3.6 h). After these corrections, a 4-parameter (period T_p, phase origin T_0, mean production rate $Q_{m\rm HCN}$ and amplitude $\Delta\, Q_{\rm HCN}$) weighted least squares fitting over 14 points yields:

$$Q_{\rm HCN} = \left(7.5 + 2.9 \times \sin\left[2\pi \frac{t - 6.44 \text{ May } 2005}{1.73}\right]\right) \times 10^{24} \text{molec} s^{-1} \qquad (2)$$

$\chi^2 = 3.1$ and reduced $\chi^2_{\nu=10} = 0.31$. This curve is plotted in Fig. 2. The uncertainty on the period is 0.10 days. The simulations also suggest that HCN "apparent" production rate amplitude is 80% of the true one, so that these observed variation suggest a ±50% periodic variation in outgassing.

Periodic variations are also present in the HCN line shapes observed in May. The least squares sinusoidal fit adjustment to the evolution of the Doppler shift derived from daily averaged data yields:

$$\Delta\, v = -0.15 + 0.09 \sin\left[2\pi \frac{t - 7.18 \text{ May } 2005}{1.65}\right] \text{ km s}^{-1}, \quad (3)$$

with the date t in days corrected for "beam delay". The period of variation $T_p = 1.65^{+0.31}_{-0.14}$ days is in agreement with previous findings.

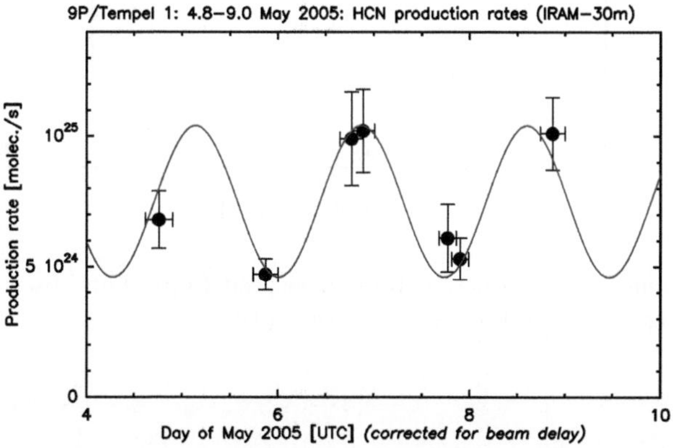

Fig. 2. HCN production rates based on IRAM 30 m observations in May 2005, based on isotropic outgassing. The sinusoidal fit to the data of Eq. (2) is plotted. All times were corrected for "beam delay" (cf. text).

3.3 Water Production in July

The evidence of periodic variation in outgassing seen in HCN May data or for several gaseous species in June–July [10] convinced us to look for variation in H_2O production rates measured with Odin. Due to larger beamsize, we expect the amplitude of the variation of the $H_2O(1_{10} - 1_{01})$ line observed by Odin to be 3.5 times lower than that for HCN observed at IRAM, i.e. $\approx 1 \times 10^{27}$ mols [7], similar to each individual measurement uncertainty.

Fitting a sine evolution (Fig. 3) with fixed phase, (T_0 = 3.90 July + 0.82 day delay and period (1.71 days), excluding the July 4.4–7.8 post-impact data, yields:

$$Q_{H_2O} = \left(8.1 + 0.7 \times \sin\left[2\pi \frac{t - 4.72}{1.71}\right]\right) \times \left(\frac{1.506}{r_h}\right)^2 \times 10^{27} \text{mols}^{-1} \quad (4)$$

Although weakly significant in comparison with a fit with no periodic variations, this results indicates that our data are compatible with strong periodic variations in outgassing still present in July.

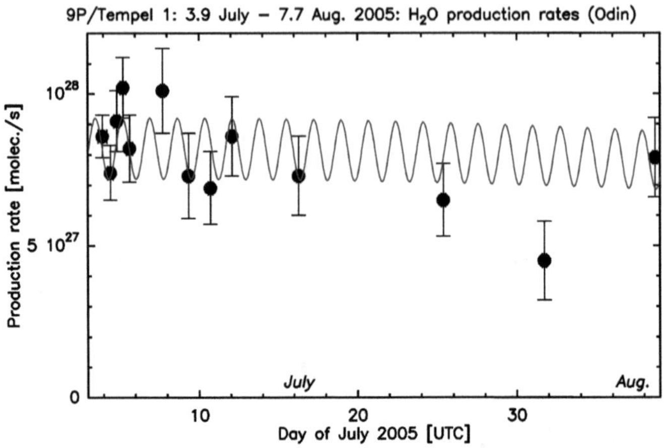

Fig. 3. Apparent H_2O production rates measured from Odin observations and the periodic least squares fit to the data from (Eq. (4)).

4 Amount of Gas Released by the Impact

For H_2O data, the simplest way (model (1), Table 2) to evaluate the outgassing excess due to the impact is to compute the difference between the observed apparent production rates and a reference production rate (Table 1). On the other hand, the main difference between the reference and impact spectra is a narrow spike around zero velocity which is superimposed on a broader component and is only seen on the 4.4–5.6 July high resolution spectra (Fig. 4). In this second approach we try to model the narrow spike component and estimate the amount of gas inside the ejecta (model (2)): we run our radiative transfer code with the excess molecules distributed inside a cone of opening angle of 90 degr, and expanding at 0.35 kms^{-1}. In parallel, we simulated outbursts in outgassing [7] that represent these observed variations, and both models converge towards a similar estimate of the total amount of water released by the impact of about $5 \pm 2 \times 10^6$kg ($N_t = 1.7 \times 10^{32}$ molecules) (Table 2). This value is similar to that found by [11] and the best match to our observations corresponds to a progressive release in 4–10 h.

Table 2. Impact residual H_2O production rate.

UT date [mm/dd.dd]	$\Delta\ Q_{app.H_2O}$ model (1) $[10^{27}s^{-1}]$	"Spike" area $\int T_b dv$ [K km s^{-1}]	$(\Delta\ Q_{H_2O})_{app}$ model (2) $[10^{27}s^{-1}]$
07/04.44	0.4 ± 1.1	0.02 ± 0.05	0.2 ± 0.5
07/04.84	2.0 ± 1.5	0.10 ± 0.06	0.9 ± 0.5
07/05.24	1.8 ± 1.6	0.09 ± 0.07	0.8 ± 0.6
07/05.65	1.4 ± 1.4	0.07 ± 0.06	0.7 ± 0.5
Average			
07/05.04	1.4 ± 0.7	0.08 ± 0.02	0.7 ± 0.2
Integrated mass of water [tons]:			
07/04.24–05.83	5700 ± 2800		5300 ± 1800[1]

[1] Integrated $\Delta\ Q_{H_2O}\ dt$ has been multiplied by 2 since only 30–70% of the molecules are seen in the hypothesis of a low speed, 1–10 h duration jet.

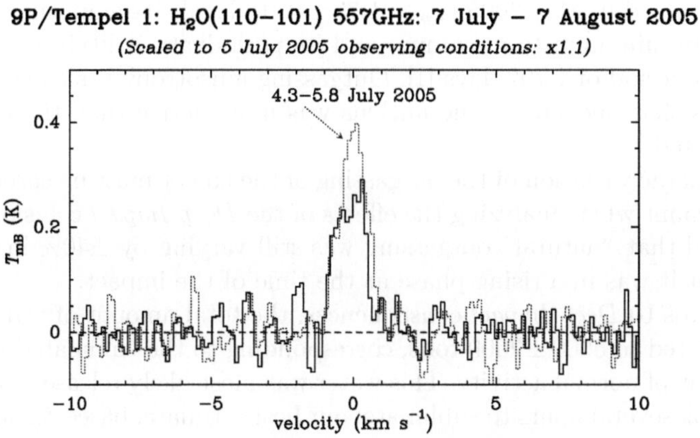

Fig. 4. The 557 GHz H_2O line observed by Odin after the *Deep Impact* collision. *Dotted line*: an average of observations obtained during the first rotation just following the impact. *Solid line*: an average of observations obtained between 7.7 July and 7.7 August, after dissipation of the impact ejecta cloud.

5 Other Outbursts of Activity

Several bursts of activity were reported by the cameras of the *Deep Impact* spacecraft before the encounter [1], or by other observations in visible to X-ray wavelengths (e.g. on 8 July [14]). Comparing the observed production rates of water with the expected natural variations, we found significant differences on two dates with $\Delta Q_{H_2O} = 4.0 \pm 1.2 \times 10^{27}$ mols on 23.7 June and $\Delta Q_{H_2O} = 2.9 \pm 1.4 \times 10^{27}$ mols on 7.7 July (Table 1). The stationary regime assumption is of course not relevant to analyze these outbursts, and the ΔQ_{H_2O} values given above are only approximate.

The *Deep Impact* team reported an outburst beginning on 22.38 June UT [1]. The excess in H_2O emission that we observed 32 h later is likely related to this outburst. In the spectrum, the extra signal due to this burst of molecules is broader than the impact-related spike (Fig. 4). From a simulation similar to those for the impact ejecta we infer a total release of $N_t \approx 10^{33}$ molecules corresponding to 1.4 day of normal activity. This natural outburst surpassed significantly the one created by *Deep Impact.*

6 Conclusion

Comet 9P/Tempel 1 is certainly one of the weakest comets extensively studied at radio wavelengths, with a peak outgassing rate around 10^{28} mols. This investigation campaign puts into evidence several characteristics of the comet:

- A strong regular variation of the outgassing rate is possibly present in HCN data obtained in May 2005 ($T_p = 1.73 \pm 0.10$ days). Its amplitude is a factor of 3 from minimum to maximum and its periodicity is likely related to the rotation period of 1.702 days [1]. Outgassing anisotropy is also evident. This suggests that one side of the nucleus was more active than the other when illuminated.
- The periodic variation of the outgassing of the comet must be carefully taken into account when analyzing the effects of the *Deep Impact* collision. It is not excluded that "natural" outgassing was still varying by ±40% in early July and that it was in a rising phase at the time of the impact.
- As regards to *Deep Impact* consequences, the total amount of water released is estimated to 5000 ± 2000 tons, corresponding to the cumulated production of 0.2 day of normal activity. This water was most likely released in icy grains that took several hours to sublimate: our best estimate, based on simulations, is around 4 h.
- The comet also underwent significant natural outbursts of outgassing, possibly of larger amplitude than the *Deep Impact* related burst. One took place on 22–23 June and may have released as much water as 1.4 day of normal activity.

Acknowledgements Odin is a Swedish-led satellite project funded jointly by the space agencies of Sweden (SNSB), Canada (CSA), Finland (Tekes) and France (CNES). IRAM is co-funded by France (CNRS), Germany (MPG) and Spain (IGN). The CSO is supported by NSF grant AST-0540882. The Nançay radio observatory is funded by the CNRS and the Région Centre in France.

References

1. M. F. A'Hearn, M.J.S. Belton, W.A. Delamere et al. (2005): Science, **310**, 258
2. F. Bensch, G.J. Melnick, D.A. Neufeld et al. (2006): Icarus, **184**, 602
3. N. Biver (1997): PhD Thesis, Université Paris VII
4. N. Biver, D. Bockelée-Morvan, J. Crovisier et al. (1999): Astron. J., **118**
5. N. Biver, D. Bockelée-Morvan, J. Crovisier et al. (2000): Astron. J., **120**, 1554
6. N. Biver, D. Bockelée-Morvan, J. Crovisier et al. (2006): Astron. Astrophys., **449**, 1255
7. N. Biver, D. Bockelée-Morvan, J. Boissier et al. (2006): Icarus, **191**, 494–512
8. J. Crovisier, P. Colom, E. Gérard et al. (2002): Astron. Astrophys., **393**, 1053
9. J. Crovisier, N. Biver, D. Bockelée-Morvan et al. (2006): *these proceedings*
10. E. Jehin, J. Manfroid, D. Hutsemékers et al. (2006): Astrophys. J. Letters, **641**, L145
11. M. Küppers, I. Bertini, S. Fornasier et al. (2005): Nature, **437**, 987
12. C.M. Lisse, M.F. A'Hearn, T.L. Farnham et al. (2205): Space Science Reviews, **117**, 161
13. K.J. Meech, M.F. A'Hearn, Y.R. Fernández et al. (2005): Space Science Reviews, **117**, 297
14. K.J. Meech, N. Ageorges, M.F. A'Hearn et al. (2005): Science, **310**, 265
15. M. J. Mumma, M. A. DiSanit, K. Magee-Sauer et al. (2005): Science, **310**, 270
16. H.L. Nordh, F. von Schéele, U. Frisk et al. (2003): Astron. Astrophys., **402**, L21
17. D. G. Schleicher, K.L. Barnes, and N. F. Baugh (2006): Astron. J., **131**, 1130

The Chemical Composition of 9P/Tempel 1 from Radio Observations

J. Crovisier[1], N. Biver[1], D. Bockelée-Morvan[1], J. Boissier[1], P. Colom[1], A. Lecacheux[1], R. Moreno[1], G. Paubert[2], D. C. Lis[3], M. Sumner[3], U. Frisk[4], Å. Hjalmarson[5], M. Olberg[5], A. Winnberg[5], H. Florén[6], A. Sandqvist[6], and S. Kwok[7,8]

[1] Observatoire de Paris, 5 place Jules Janssen, F-92190 Meudon, France
jacques.crovisier@obspm.fr
[2] IRAM, Avd. Divina Pastora, 7, E-18012 Granada, Spain
[3] Caltech, MS 320-47, Pasadena , CA 91125, USA
[4] Swedish Space Corporation, PO Box 4207, SE-17104 Solna, Sweden
[5] Onsala Space Observatory, SE-43992 Onsala, Sweden
[6] Stockholm Observatory, SCFAB-AlbaNova, SE-10691 Stockholm, Sweden
[7] Dept. of Physics and Astronomy, University of Calgary, Calgary, AB T2N 1N4, Canada
[8] Inst. of Astron. & Astrophys., Academia Sinica, PO Box 23-141, Taipei 106, Taiwan

Abstract In 2005, as part of a world-wide multi-wavelength investigation of comet 9P/Tempel 1 in support to the *Deep Impact* mission, we conducted radio spectroscopic observations with the Nançay radio telescope, the Odin satellite, the CSO 10-m and the IRAM 30-m telescopes. We report here our results concerning the chemical composition of the comet. The relative abundances of the detected species (H_2O, CH_3OH, H_2S, HCN) or their upper limits (CO, H_2CO, CS) are comparable to the mean values observed in other comets. No significant changes of the outgassing rates (except possibly for CH_3OH) or of the molecular abundances were observed following the impact.

1 Introduction

A world-wide multi-wavelength investigation of comet 9P/Tempel 1 was organized in support to the *Deep Impact* mission [1]. In the frame of this campaign, we conducted in 2005 radio spectroscopic observations with the Nançay radio telescope, the Odin satellite, the CSO (Caltech Submillimeter Observatory) 10-m telescope and the IRAM (Institut de Radioastronomie Millimétrique) 30-m telescope. We report here our results concerning the chemical composition of the comet. The analysis of the observations in terms of signal evolution and of a possible modification of the outgassing rate by the *Deep Impact* event is presented in a companion paper [2]. A full account of all these observations is published in the *Icarus* special issue on *Deep Impact* [3].

2 Observations and Results

Observations of the 18-cm lines of OH at Nançay were scheduled from March to July 2005. The comet could be detected since 20 March, but in July, the signal was beyond detection due to unfavourable excitation of the OH radical. Odin observations of the 557 GHz water line took place from June to August. Both Nançay and Odin observations provided a basis for the water production of the comet and its evolution with time [3].

Millimetric observations were conducted with the IRAM 30-m radio telescope around perigee on 4–9 May to assess the chemical composition of the comet before the impact. These observations resulted in detections of HCN, CH_3OH and H_2S with abundances relative to water (0.12, 2.6 and 0.5%, respectively) similar those observed in other comets (Fig. 1). Other species (CS, CO, H_2CO) were below detection limits.

The impact event and its aftermath were covered at IRAM by observations scheduled on 2–10 July. Only HCN and CH_3OH could be detected.

Additional observations were carried out at the CSO around the impact time (4 July, from 5 to 9 h UT) and the following day. HCN could not be detected, but CH_3OH was possibly detected following the impact (Fig. 2).

The post-impact observations at IRAM and CSO did not reveal a very significant increase or change of the outgassing rates and relative abundances. Only CH_3OH may have been more abundant by a factor 2–5 in the ejecta (from the CSO observation), while other variations may be linked to the natural variability of the comet, as discussed in [2, 3].

Table 1. Relative molecular abundances of comet 9P/Tempel 1.

Molecule	$Q_{\mathrm{molec.}}/Q_{\mathrm{H_2O}}$ May 2005	$Q_{\mathrm{molec.}}/Q_{\mathrm{H_2O}}$ 2–3 or 5–10 July 2005
HCN	$0.11 \pm 0.01\%$	$0.12 \pm 0.03\%$
CH_3OH	$2.8 \pm 0.9\%$	$2.7 \pm 0.4\%$
H_2S	$0.5 \pm 0.1\%$	
CS[1]	$< 0.13\%$	$< 0.25\%$
CO	$< 10\%$	$< 32\%$
H_2CO[2]	$< 1.5\%$	$< 2.3\%$

Adapted from [3].

[1] CS is assumed to come from CS_2 (600 km parent scalelength used here).

[2] H_2CO is assumed to come from a distributed source with a lifetime equals to 1.75 times the H_2CO lifetime [4].

The relative molecular abundances derived from our observations of 9P/Tempel 1 (or their upper limits) are summarized in Table 1.

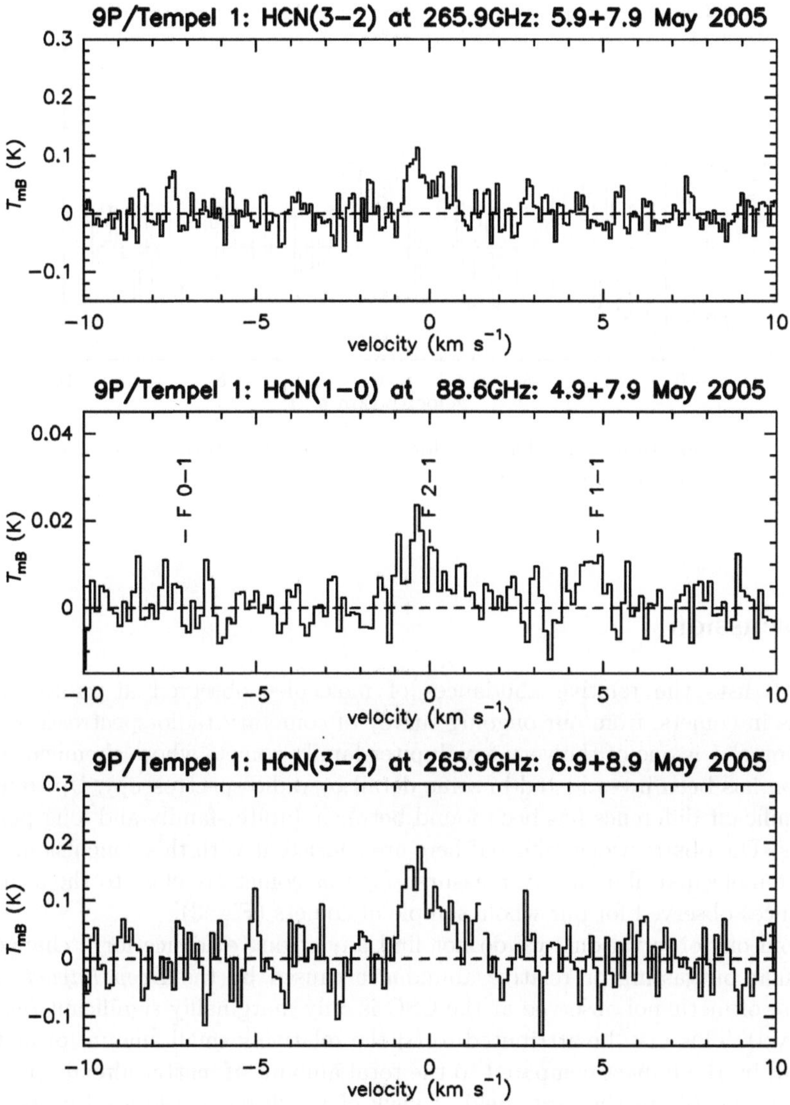

Fig. 1. Spectra of HCN observed in May at IRAM.

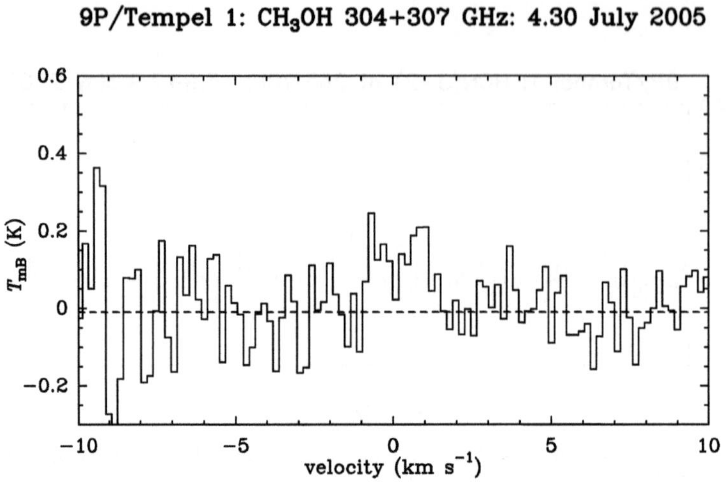

Fig. 2. CSO spectum of methanol following the impact (two methanol lines were co-added, integration time is 2.5 h).

3 Discussion

Table 2 lists the relative abudances of molecules observed at radio wavelengths in comets, from our ongoing survey of cometary radio spectroscopy [5]. 9P/Tempel 1 is one of the very few Jupiter-family comets whose chemical composition has been investigated in some detail by radio spectroscopy. Up to now, no significant difference has been found between Jupiter-family and long-period comets. The observations reported here are consistent with this conclusion. The relative molecular abundances measured for this comet are close to the average abudances observed for our whole sample of comets (Fig. 3).

From our observations, we do not find any strong evidence for a change of molecular outgassing or relative abundance caused by the *Deep Impact*. The increase of methanol observed at the CSO is only marginallly significant (at the 2–σ level). This may be attributed to (i) the relatively small amount of matter released by the impact compared to the total amount of matter already present in the coma; (ii) to the large field of view of Earth-based radio telescopes; or (iii) to the poor time resolution of the radio observations, which require large integration times due to their limited sensitivity.

4 Conclusion

- The relative molecular abundances observed in 9P/Tempel 1 are "classical", with (H_2O:CO:CH_3OH:H_2CO:H_2S:HCN:CS) = (100 : < 10 : 2.7 : <

Table 2. Volatile composition from radio observations. Comparison of Oort-cloud and Jupiter-family comets[a].

comet	CO	HCN	H_2S	CS	CH_3OH	H_2CO	Ref.
18 Oort cloud comets	< 1.7–23	0.08–0.25	0.12–1.5	0.05–0.17	< 0.9–6.2	0.4–1.3	[5]
22P/Kopff	< 6	0.13	0.58	< 0.15	2.5		[5, 6]
21P/Giacobini-Zinner	< 2.4	0.09	< 0.36	0.08	1.6	0.13	[5]
19P/Borrelly (1994)		0.11			1.7	0.4	[7]
19P/Borrelly (2001)	< 15	0.06	< 0.45	0.07		< 0.4	[7]
2P/Encke		0.09	< 0.4	< 0.06	4.1	< 1.4	[8]
9P/Tempel 1 before DI	< 10	0.11	0.5	< 0.13	2.8	< 1.5	[3]
9P/Tempel 1 after DI	< 31	0.13		< 0.27	2.7	< 2.3	[3]
73P/S.-W. 3	< 6	0.20	0.25	0.13	0.9	0.8	[9]

[a] Production rates are given relative to water = 100.

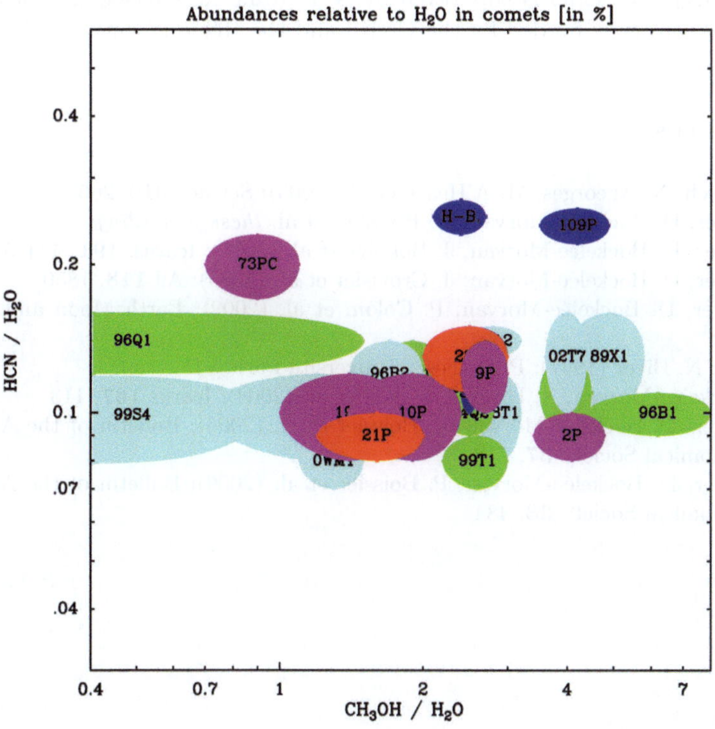

Fig. 3. The HCN and CH_3OH abundances relative to water from our spectroscopic database of comets. The ellipse sizes represent errors. Comet 9P/Tempel 1 stands right in the middle of the samples.

1.5 : 0.5 : 0.12 : <0.13). This is comparable to mean values observed in many comets.
- It was not possible to assess precisely the composition of the ejecta, except for a possible increase in the abundance of CH_3OH. No large excess of HCN was seen.
- Although the signal was rather weak to make a detailed compositional study of the impact ejecta, this observing campaign allowed us to observe in detail the behaviour of a Jupiter-family comet, and to put into evidence outbursts and periodic variations of outgassing poorly observed in any comet before.

Acknowledgements Odin is a Swedish-led satellite project funded jointly by the space agencies of Sweden (SNBS), Canada (CSA), Finland (Tekes) and France (CNES). IRAM is co-funded by France (CNRS), Germany (MPG) and Spain (IGN). The CSO is supported by NSF grant AST-0540882. The Nançay radio observatory is funded by the CNRS and the Région Centre in France.

References

1. K. Meech, N. Ageorges, M. A'Hearn et al. (2005): Science **310**, 265
2. N. Biver, D. Bockelée-Morvan, J . Boissier, et al. *these proceedings*
3. N. Biver, D. Bockelée-Morvan, J. Boissier et al. (2007): Icarus, **191**, 494–512
4. N. Biver, D. Bockelée-Morvan, J. Crovisier et al. (1999): AJ **118**, 1850
5. N. Biver, D. Bockelée-Morvan, P. Colom et al. (2002): Earth Moon and Planets **90**, 5
6. (1997): N. Biver (1997): PhD thesis, Université Paris VII
7. D. Bockelée-Morvan, N. Biver, P Colom et al. (2004): Icarus **167**, 113
8. N. Biver, D. Bockelée-Morvan, P. Boissier et al. (2005): Bulletin of the American Astronomical Society **37**, 710
9. N. Biver, D. Bockelée-Morvan, P. Boissier et al. (2006): Bulletin of the American Astronomical Society **38**, 484

Fluorescence Cascades of Water and Carbon Dioxide in the Emission Spectrum of Comet 9P/Tempel 1

J. Crovisier

Observatoire de Paris, 5 Place Jules Janssen, F-92195 Meudon, France
jacques.crovisier@obspm.fr

Abstract A simplified algorithm to compute the infrared fluorescence cascade of molecules in cometary conditions is presented. This method could be applied to observations of 9P/Tempel 1 by Deep Impact and to future cometary infrared observations such as those of comet 67P/Churyumov-Gerasimenko by Rosetta/VIRTIS. Results are presented for water and carbon dioxide.

1 Introduction

Cometary molecules have harmonic, combination and hot vibrational bands in the near infrared spectral domain. These bands are normally weaker than the fundamental vibrational bands. In order to assess the intensity of all these bands when excited in the cometary environment, the full fluorescence cascades of the molecules must be considered [1, 2, 3, 4]. Synthetic spectra can be simply constructed from available molecular databases, provided that the optical thickness of the lines are small, and that the molecules are concentrated on the ground vibrational state. Such synthetic spectra are presented here for water and carbon dioxide. This could be applied to observations of 9P/Tempel 1 by Deep Impact [5, 6] and to future cometary infrared observations such as those of comet 67P/Churyumov-Gerasimenko by Rosetta/VIRTIS [7].

2 Algorithm

The proposed algorithm proceeds according to the following steps:

1. Select a subset of transitions from the database (imposing an upper limit on rotational energies).
2. Sort the transitions by decreasing order of upper level energies.
3. Assume a rotational population distribution (e.g., thermal at a given temperature) for the ground vibrational state.

4. Read the transition list and compute excitation by solar radiation from the ground vibrational state. Compute upper level populations following this primary excitation.
5. Read again the transition list and compute fluorescence emission. Compute secondary population of lower levels during this process.

In this way, the inversion of a linear system of very large dimension is avoided. This algorithm is valid only:

- if the rotational population distribution is known (thermal);
- if only the ground vibrational state is significantly populated;
- if optical depth effects can be neglected.

Comprehensive molecular databases are required to cover the whole fluorescence cascade. The effect of optical thickness is expected to enhance the intensity of hot bands relative to bands originating from the ground vibrational state, especially fundamental bands of vibration.

3 Results

The method was applied to water using the database of Schwenke & Partrige [8]. An improvement can be expected by using the Barber et al. [9] database. It was applied to carbon dioxide using the GEISA [10] and HITRAN [11] databases. Examples of synthetic spectra can be found on the author's web page:
`http://www.lesia.obspm.fr/~crovisier/biblio/preprint/bru06_crovisier_poster.pdf`

References

1. J. Crovisier, T. Encrenaz: (1983): Astron. Astrophys. **126**, 170
2. J. Crovisier (1984): Astron. Astrophys. **130**, 361
3. D. Bockel'ee-Morvan, J. Crovisier (1987): Astron. Astrophys. **187**, 425
4. D. Bockelée-Morvan, J. Crovisier (1989): Astron. Astrophyis. **216**, 278
5. M. F. A'Hearn, M. J. S. Belton, W. A. Delamere, et al. (2005): Science **310**, 258
6. J. M. Sunshine, M. F. A'Hearn, O. Groussin, et al. (2005): Space Science Review **117**, 269
7. A. Coradini, F. Capaccioni, P. Drossart, et al. (1998): Planss **46**, 1291
8. D. W. Schwenke, H. Partridge (2000): J. Chem. Phys. **113**, 6592
9. R. J. Barber, J. Tennyson, G. J. Harris, R. N. Tolchenov (2006): MNRAS **368**, 1087
10. N. Jacquinet-Husson, N. A. Scott, A. Chedin, et al. (2005): Journal of Quantitative Spectrroscopy and Radiative Transfer **95**, 429
11. L. S. Rothman, D. Jacquemart, A. Barber et al. (2005): Journal of Quantitative Spectrroscopy and Radiative Transfer **96**, 139

Temporal Evolution of DI Ejecta Based on NIRSPEC Observations at Keck 2: Parent Volatiles and Dust

M. A. DiSanti[1], G. L. Villanueva[1,2], B. P. Bonev[1,3], K. Magee-Sauer[4], J. E. Lyke[5], and M. J. Mumma[6]

[1] Solar System Exploration Division, NASA Goddard Space Flight Center, code 693, Greenbelt, MD 20771, USA disanti@ssedmail.gsfc.nasa.gov
[2] NPP/NRC Resident Research Associate
[3] Dept. of Physics, Catholic University of America, Washington, DC 20064, USA
[4] Dept. of Physics and Astronomy, Rowan University, Glassboro, NJ 08028, USA
[5] W.M. Keck Observatory, Kamuela, HI 96743, USA
[6] Solar System Exploration Division, NASA Goddard Space Flight Center, code 693, Greenbelt, MD 20771, USA

Abstract The Deep Impact encounter with 9P/Tempel 1, a Jupiter Family Comet (JFC), on UT 2005 July 04 was observed at high spectral resolution ($\lambda/\Delta\lambda \approx 25,000$) using the cross-dispersed near-infrared echelle spectrometer (NIRSPEC) at Keck-2. This permitted a simultaneous measure of individual line intensities for several parent molecules in addition to dust continuum emission. Column abundances are presented for H_2O and C_2H_6 beginning 30 minutes prior to impact (T−30) and ending 50 minutes following impact (T+50), and for H_2O and HCN from T+50 until T+96, in time steps of approximately six minutes post-impact.

The ejecta composition was revealed by an abrupt increase in H_2O and C_2H_6 near T+25. This showed C_2H_6/H_2O to be higher than its pre-impact value by a factor 2.4±0.5, while HCN/H_2O was unchanged within the uncertainty of the measurements. The mixing ratios for C_2H_6 and HCN in the ejecta agree with those found in the majority of Oort cloud comets, perhaps indicating a common region of formation. This stands in contrast with the recently observed split JFC 73P/Schwassmann-Wachmann 3 and the disintegrated Oort cloud comet D/1999 S4 (LINEAR), both of which were depleted in most organic volatiles, but not HCN.

The expanding dust plume was tracked through the 3.5-μm spectral continuum and through 2-μm images acquired with the slit-viewing camera (SCAM). These showed a monotonic increase in continuum intensity following impact. The relatively sudden appearance of the volatile ejecta signature is attributed to heating of icy grains (perhaps to a threshold temperature) that were decreasingly shadowed by intervening (sunward) particles.

1 Introduction

Comets contain primitive icy material remaining from the epoch of solar system formation, but the extent to which they are modified from their initial state is uncertain. Measuring the natal composition of comets belonging to Jupiter's dynamical family (Jupiter-Family comets, or JFCs) is particularly problematic. JFCs have short orbital periods, so their surfaces could be highly processed compared with long-period (Oort cloud) comets. If so, the true initial composition of JFCs may be revealed only by measuring sub-surface material.

The NASA Discovery mission Deep Impact (DI) to the JFC 9P/Tempel 1 provided the first opportunity to examine material from the interior of a cometary nucleus resulting from a planned collision [1, 22]. A principal objective of the DI experiment was to compare the composition of dust and volatiles prior to and following the impact event.

Compositional differences were revealed both from ground-based and space-based data. Near-infrared spectra from the DI spacecraft showed a very pronounced cometary organics (CH-X) feature (centered near 3.4–3.5 μm) post-impact, in contrast to its much weaker pre-impact intensity [1]. Post-impact ground-based spectra (covering ~8–13 μm; [11]) and Spitzer Space Telescope spectra (covering 5–35 μm; [15]) indicated a large increase in grain emission within the first hour after impact, but declining in strength by 1.8 hours post-impact [11]. Ground-based near-IR spectra (~2.8–4.7 μm) revealed enhanced C_2H_6 but unchanged HCN and CH_3OH in the ejecta [17].

In this paper, we summarize our study of the evolution of parent volatiles (H_2O, C_2H_6, and HCN in time steps of ~5–6 minutes) and co-measured dust [9]. We also compare the ejecta chemistry with that seen for other comets in our database, including the recently observed split JFC 73P/Schwassmann-Wachmann 3.

2 Observations

Comet Tempel 1 was observed with the cross-dispersed echelle spectrometer (NIRSPEC) at Keck 2 [16], permitting simultaneous measure of H_2O and various other cometary parent volatiles (and the dust continuum) at sufficiently high spectral resolving power ($\lambda/\Delta\lambda \approx 25,000$) to conduct studies of individual spectral line intensities. Two echelle/cross-disperser configurations (denoted KL1 and KL2) were required to study abundances of H_2O, C_2H_6, and HCN. We observed the comet with the KL1 setting (sampling H_2O and C_2H_6) from approximately 30 minutes prior to impact (T–30) until 50 minutes after impact (T+50). We then switched to the KL2 setting (sampling H_2O and HCN) until T+96 (Table 1).

For all spectra, the telescope was nodded in an ABBA sequence, with A- and B-beams (each containing comet signal) separated by 12″ (one-half the slit length; e.g., see [18, 5]). Signal from both beams was combined, and spectra were extracted over 1.74″ centered on the nucleus. The spectral extracts contain continuum emission from dust and superposed emission from volatiles. Cometary line emission was isolated by subtracting a synthetic model of the observed continuum (Fig. 1).

Table 1. Log of comet 9P/Tempel-1 NIRSPEC observations (UT 4 July 2005)[a].

Setting Name	Frame IDs (#ABBA sets)[b]	Start-End (UT)[c]	Time on Source (min)	Molecules[d]
KL1 pre-impact	24–39(4)	05:22:12–05:52:13	16	H_2O, C_2H_6
KL1 post-impact	40–67(7)	05:53:04–06:41:06	28	H_2O, C_2H_6
KL2 post-impact	68–91(6)	06:43:16–07:24:50	24	H_2O, HCN

[a] During the observations, the heliocentric distance R_h=1.504 AU, and the geocentric distance Δ=0.894 AU.
[b] Frame designations and number of ABBA sequences for spectral observations, with each frame (each single A or B) representing 60 seconds of integration.
[c] Time of impact (including light-travel time to Earth) was UT 05:52:02 [1].
[d] Parent volatiles included in this study of temporal evolution. For additional molecules encompassed in these settings, see Mumma et al. [17].

3 Extraction of Quantitative Column Abundances

The impulsive nature of the impact event [1] caused the ejecta cloud (both gas and dust) to evolve rapidly and continuously. For this reason, we extracted abundances of H_2O, C_2H_6, and HCN from each post-impact ABBA sequence, for comparison with pre-impact abundances measured on UT 03 June and 04 July. Under such non-steady-state conditions, column abundances (defined as the number of molecules in the nucleus-centered beam) provide a more meaningful measure of the gas release rate than do production rates. The column abundance established by steady production (release) before impact is assumed to continue after (and to not be changed by) the impact. The expanding ejecta add to this, so a simple difference provides a measure of the column abundance in the ejecta cloud (if optically thin) and therefore represents an accurate assessment of the temporal evolution of parent volatiles.

For optically thin conditions, the column abundance is given by:

$$N = 4\pi\Delta^2 \frac{F_{\text{line}}}{g_{\text{line}}\, {R_{\text{h}}}^{-2}}, \tag{1}$$

where Δ(m) is the geocentric distance and R_{h}(AU) is the heliocentric distance. For each measured emission line, the parameters F_{line}, g_{line}, and GF are respectively the observed flux (W m^{-2}), the fluorescence efficiency g-factor (W molecule^{-1}) evaluated at R_{h} = 1 AU, and a factor that accounts for slit losses (see Table 2). We used a 3-pixel (0.432″) wide slit to observe Comet

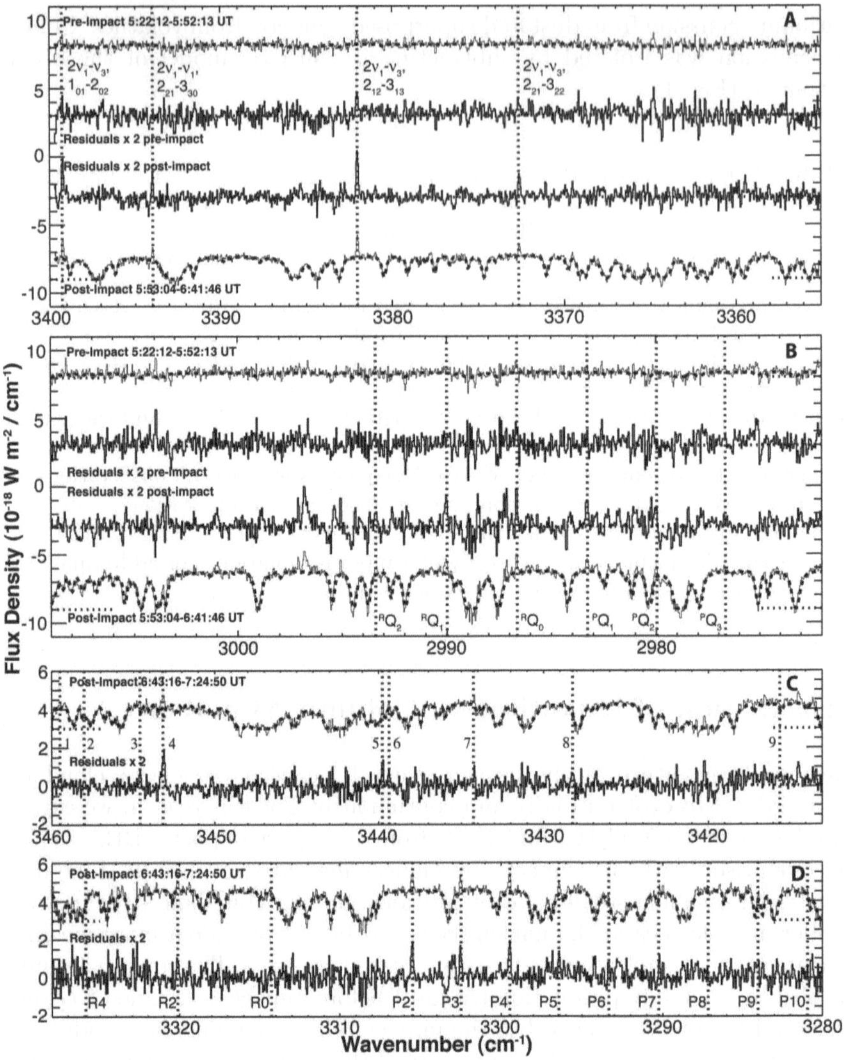

Fig. 1. Spectral extracts (with convolved synthetic atmospheric transmittance function, superimposed dashed traces) and residual emission in excess of the continuum for Comet 9P/Tempel 1 on UT 2005 July 04. Within each panel, the order of spectra (from top to bottom) is: Spectral extract, residuals, residuals,spectral extract (for A, B panels), and spectral extract, residuals (for C, D panels). Each tick mark on the ordinate axis corresponds (in units of 10^{-18} W m^{-2}/cm^{-1}) to 1.0 and 0.5 in panels (A, B) and (C, D), respectively, and zero offsets are indicated. A. Setting KL1 order 26 pre- and post-impact, containing H_2O emission. B. KL1 order 23 pre- and post-impact, containing C_2H_6 emission. C. KL2 order 26, post-impact H_2O. D. KL2 order 25, post-impact HCN. In all panels, the vertical dotted lines mark spectral lines used to measure molecular abundances. For the marked emissions, identifications (band ID, line ID) are indicated for the H_2O lines in panel A, as are line IDs in panels B (C_2H_6 ν_7 band) and D (HCN ν_3 band). In panel C, the H_2O line rotational ID and band ID (the

Table 2. Extracted column abundances on impact night.

Setting	Frame IDs	Mid UT	N	σN	N	σN
KL1pre	24–39	05:37:13	640	47	1.78	0.30
KL1post	40–43	05:56:00	557	124	1.58	0.78
	44–47	06:03:26	872	126	1.57	0.91
	48–51	06:10:57	781	68	1.48	0.57
	52–55	06:18:15	1301	129	4.36	0.76
	56–59	06:24:57	1046	76	4.80	0.65
	60–63	06:31:51	1448	258	5.71	0.71
	64–67	06:38:49	1102	147	4.48	0.57
KL2post	68–71	06:43:16	1177	337	3.39	0.42
	72–75	06:53:52	1599	318	3.63	0.31
	76–79	07:00:30	1084	315	2.36	0.41
	80–83	07:07:32	1341	193	2.22	0.54
	84–87	07:14:57	755	214	1.42	0.51
	88–91	07:21:55	914	139	2.29	0.38

[a] Values for the KL1 sequences pertain to C_2H_6; those for KL2 pertain to HCN. All entries include slit loss correction factors (Eq. (1)), $GF = 1.22$ (for KL1) and $GF = 1.35$ (for KL2). Uncertainties represent 1 σ, corresponding to the larger of stochastic and standard errors. Units of N, σN: 10^{28} molecules.

9P/Tempel 1, so the nucleus-centered beam includes signal within a rectangular aperture of dimension 280 x 1130 km at the comet on July 04.

The spectra encompass multiple lines of H_2O, C_2H_6, and HCN (Fig. 1) that sample a range of rotational energies. These lines were used to measure rotational temperatures ($T_{\rm rot}$) of ~40 K for the KL1 observations (both pre- and post-impact) and ~35 K for the KL2 observations [9]. Beam column abundances (N) are relatively insensitive to outflow assumptions, depending only on measured parameters ($F_{\rm line}, GF$), geometric factors ($R_{\rm h}$, Δ), and $g_{\rm line}$ (this depends only on $T_{\rm rot}$).

4 Evolution of Total Column Abundances

Table 2 lists column abundances of molecules in the nucleus-centered beam (corrected for slit loss) for each (ABBA) set. The observed total column abundance of C_2H_6 did not change (within error) during the first 20 or so minutes following impact; i.e., the values from the first three post-impact ABBA sets (frames

Fig. 1. (continued) roman numeral in square brackets) for numbered features are: (1) 111-202 [I], 211-212 [III]; (2) 000-111 [I]; (3) 211-220 [I]; (4) 202-321 [II], 110-221 [IV]; (5) 111-110 [III]; (6) 101-212 [I]; (7) 212-303 [I], 211-322 [IV]; (8) 212-211 [III]; (9) 303-422 [II]. Band legend: [I] = $\nu_1 + \nu_3 - \nu_3$, [II] = $\nu_1 + \nu_3 - \nu_1$, [III] = $2\nu_1 - \nu_3$, [IV] = $2\nu_1 - \nu_1$. Features having multiple entries denote blended lines, with the stronger component listed first; additional blended H_2O lines contributing <5% of the feature intensity are not listed (see Table 2 of Dello Russo et al. [6] for additional information).

40–51) were consistent with its mean pre-impact value (frames 24–39; see Table 2 and Fig. 2), and the total abundance of H_2O increased by only a very modest factor (1.18±0.12).

The spectral lines of H_2O and C_2H_6 brightened abruptly near $T+25$, as did the extracted column abundances (Fig. 2). This behavior is in stark contrast to the (simultaneously measured) spectral continuum, which increased monotonically until approximately $T+30$ (see Fig. 4B and related discussion).

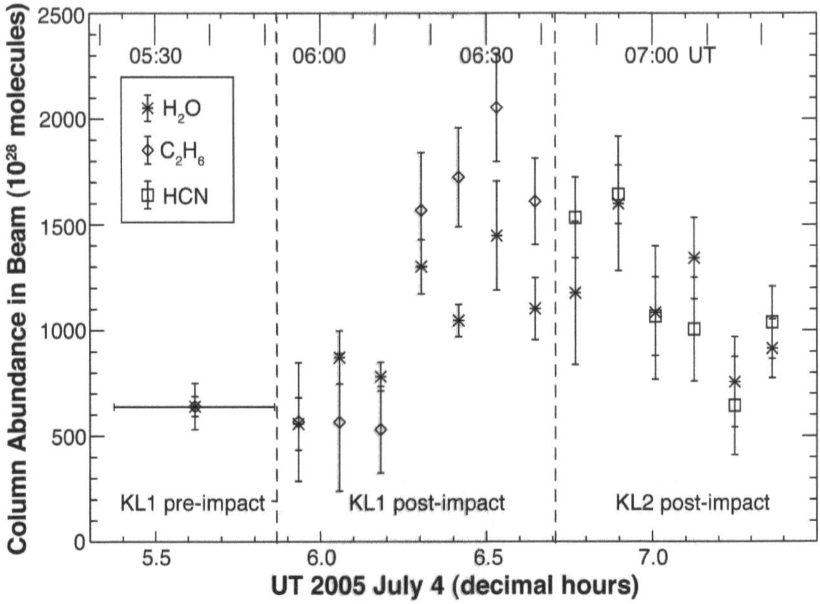

Fig. 2. Measured column abundances for H_2O, C_2H_6, and HCN. The pre-impact point represents 16 minutes on source, and each post-impact point represents one ABBA sequence (4 minutes on source). The points for C_2H_6 are scaled by 359 and those for HCN are scaled by 464, the reciprocals of their respective pre-impact mixing ratios relative to H_2O on July 04 (see Table 3).

5 Parent Volatile Composition of the Ejecta

Comparing observed column abundances (Table 2) provides a measure of the mixing ratios for C_2H_6 and HCN (Table 3, columns 3–6). These column abundances represent the *quiescent* (pre-impact) release of molecules plus (after impact) those ejected as a result of the DI event. The net number of molecules contributed by the ejecta is calculated by subtracting the measured pre-impact column abundances from the total column abundances.

The ratio of these differences for C_2H_6 and HCN to corresponding differences for H_2O provides a measure of mixing ratios in the ejecta (Table 3, columns 8–11). The mean value for C_2H_6 during the first three post-impact ABBA sets

Table 3. Abundances of C_2H_6 and HCN relative to H_2O^a.

Setting	Frames	X	σX	$\langle X\rangle$	$\sigma\langle X\rangle$	$N(H_2O)^b$	X	σX	$\langle X\rangle$	$\sigma\langle X\rangle$
KL1pre	24–39	0.278	0.052	0.278	0.052					
KL1post	40–43	0.283	0.154			–106.7	0.189	0.819		
	44–47	0.180	0.107			414.5	–0.050	0.231		
	48–51	0.189	0.075	0.199	0.057	320.5	–0.095	0.203	–0.067[c]	0.150[c]
	52–55	0.335	0.067			573.7	0.450	0.178		
	56–59	0.459	0.071			509.9	0.592	0.175		
	60–63	0.395	0.086			896.3	0.439	0.155		
	64–67	0.406	0.075	0.397	0.037	526.0	0.513	0.194	0.494[c]	0.087[c]
KL2post	68–71	0.288	0.090			537.4	0.419	0.287		
	72–75	0.227	0.049			959.5	0.260	0.101		
	76–79	0.218	0.074			444.5	0.275	0.234		
	80–83	0.166	0.047			701.4	0.155	0.105		
	84–87	0.188	0.087			115.6	0.245	0.728		
	88–91	0.251	0.056	0.215	0.025	274.6	0.420	0.300	0.236	0.066

[a] The values for KL1 pertain to C_2H_6; those for KL2 pertain to HCN. Columns 3–6 are total observed mixing ratios (abundances relative to $H_2O = 100$; Fig. 2); columns 7–11 refer to ejecta only (i.e., after subtraction of pre-impact column abundances; Fig. 3). The column abundances of ejected C_2H_6 and HCN are the product of these mixing ratios and the ejected water column. Mean abundances ($\langle X\rangle$) pertain to frames 40–51 and 52–67 (for C_2H_6), and to frames 68–91 (for HCN). For HCN, the pre-impact value for H_2O ($N = 640 \times 10^{28}$ molecules; Table 2) was multiplied by the HCN mixing ratio from June 03 (0.178±0.060%; see Mumma et al. [17]).
[b] Net nucleus-centered column of H_2O (10^{28} molecules) in the ejecta.
[c] These mean mixing ratios, from frames 40–51 (upper entries) and frames 52–67 (lower entries), are based on our measured pre-impact mixing ratio for C_2H_6 on July 04 (0.278%). Using the mean C_2H_6 mixing ratio from June 03 and July 04 pre-impact (0.227±0.041%) results in 0.013 ± 0.148% from frames 40–51 and 0.549±0.091% from frames 52–67. Using the mean C_2H_6 mixing ratio from June 03 only (0.194±0.041%) results in 0.064±0.149% from frames 40–51 and 0.583±0.094% from frames 52–67. See Fig. 3 and related text.

(frames 40–51) indicates no enhancement (within error) over its pre-impact value. This was followed by an abrupt increase to a mixing ratio consistent with those seen in most Oort cloud comets quantified to date (~0.5–0.6%, depending on the pre-impact mixing ratio used in the calculation; see Fig. 3A, B, C, and note *c* of Table 3). The corresponding mean mixing ratio for HCN in the ejecta throughout the KL2 observations (Fig. 3D) was consistent with its pre-impact value (Table 3).

The enriched mixing ratio of C_2H_6 (beginning near T ~20–25 minutes) remained constant (within statistical error) for the duration of the KL1 observations (~25 additional minutes). The mixing ratio for ejected HCN also remained constant (within error) over the measurement interval, however unlike C_2H_6 the ratio HCN/H_2O was unchanged from its pre-impact value (Fig. 3D). We conclude that the impact ejecta showed enhanced C_2H_6/H_2O but unchanged HCN/H_2O relative to the quiescent (pre-impact) source, and that the composition of the ejecta did not change with time.

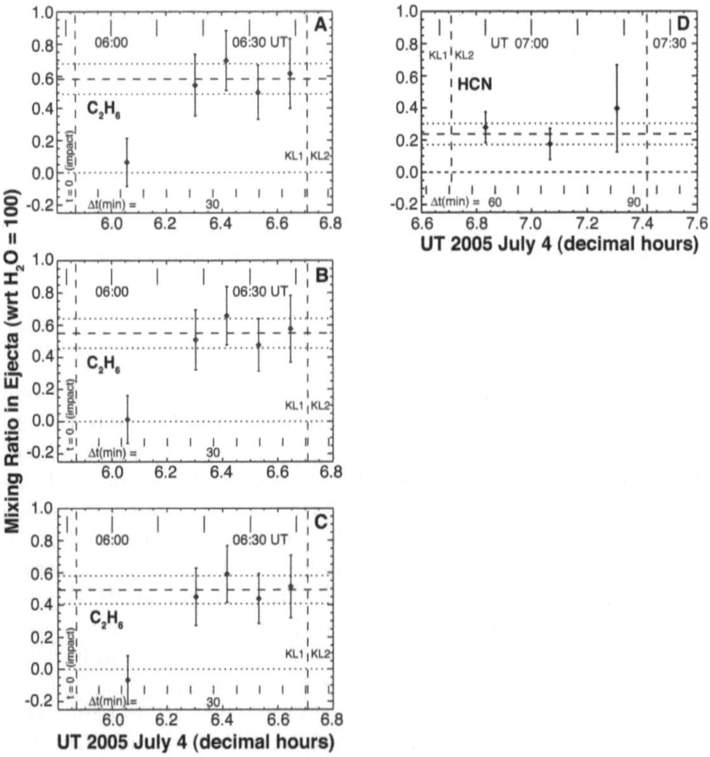

Fig. 3. Temporal evolution of column abundances for C_2H_6 (panels A–C) and HCN (panel D) relative to H_2O, after subtracting pre-impact column abundances. Assuming no change in *quiescent* (i.e., ambient) mixing ratios, these represent mixing ratios for C_2H_6 and HCN in the material ejected by the impact. In panels A–C, pre-impact mixing ratios for C_2H_6 are based on, respectively, the value measured from June 03 (0.194±0.041%; [17]), the weighted mean from June 03 and July 04 pre-impact (0.227±0.041%), and from July 04 pre-impact alone (0.278±0.052%; Table 3). The first point in panels A-C is the mean from frames 40–51 (the first three sets post-impact); these indicate no net change in column abundance for ethane (within error) over its pre-impact value. The subsequent points reveal an enhanced abundance of C_2H_6 in the ejecta. The mixing ratio for HCN in the ejecta was unchanged (within error) from that measured on June 03. See text and Table 3 for details.

6 Temporal Evolution of Dust

The spectra showed an increase in continuum intensity within the first minutes following impact, from essentially no detected continuum pre-impact (e.g., compare pre- and post-impact spectral extracts in Fig. 1A and B). After impact, the spectral continuum measured within the slit (280×1130 km footprint) increased strongly from 2.9–3.6 μm, indicative of grain thermal emission rather than reflected sunlight, and increased monotonically in time until peaking near $T + 30$ (see Fig. 4B). The 3.5 μm spectral continuum results require a dramatic increase in the number of μm-sized grains visible in the coma because they are most sensitive to warm particles of this size. The presence of pronounced post-impact emission from crystalline silicates also required small super-heated grains [11]. It is almost certain that these warm grains had lost whatever icy mantles they might have had when ejected.

The overall evolution of the impact event was also measured with the slit-viewing camera (SCAM) in the wavelength range 1.9–2.5 μm (scattered sunlight), providing a record of the temporal evolution of the coma morphology in steps of ∼40 seconds. The binned SCAM intensity (5x5″, or 3,240 km square) showed behavior similar to the 3.5-μm spectral continuum, but peaked later and decreased more slowly (see Fig. 4, and discussion below).

We developed a *straw-man* (Monte Carlo) model that adequately reproduces the observed evolution in continuum intensity, for both spectral and imaging (SCAM) observations; see DiSanti et al. [9] and Villanueva et al. (in preparation) for detailed discussions. We found this model useful for understanding the simultaneous physical phenomena affecting the ejected grains. The simulations qualitatively reproduce the temporal evolution of the dust coma observed in both spectral and SCAM modes (Fig. 4B). The emergent intensity from all particles follows the observed SCAM photometric measurements (*Simulated SCAM intensity* in Fig. 4B), while the emergent spectral intensity from particles within the slit follows the observed 3.5-μm continuum measurements (*Simulated spectral intensity*).

During the first few minutes after impact, the shadowing effect of the dust grains was severe, greatly reducing the observed continuum flux. As the dust coma expanded, the column density decreased and the fraction of particles *seen* in the respective apertures (slit or SCAM image) increased with time. The intensity maximum in the 3.5-μm spectral continuum (near $T + 30$) resulted from a balance between decreasing optical depth and decreasing number of particles remaining in the slit. If the dust coma were optically thin, the in-slit intensity would have remained constant until the slit no longer sampled the entire expanding cloud (*Total in-slit* in Fig. 4B). [The normalization in Fig. 4B is such that 1.0 corresponds to the total number of particles under optically thin conditions. The maximum of the *Total in-slit* curve (∼0.8) indicates that the slit would not *see* all the particles (even for optically thin conditions) because of seeing losses.] The 3.5-μm spectral continuum intensity dropped after $T + 30$ as the effect of grains exiting the slit increasingly dominated the decrease in dust optical depth.

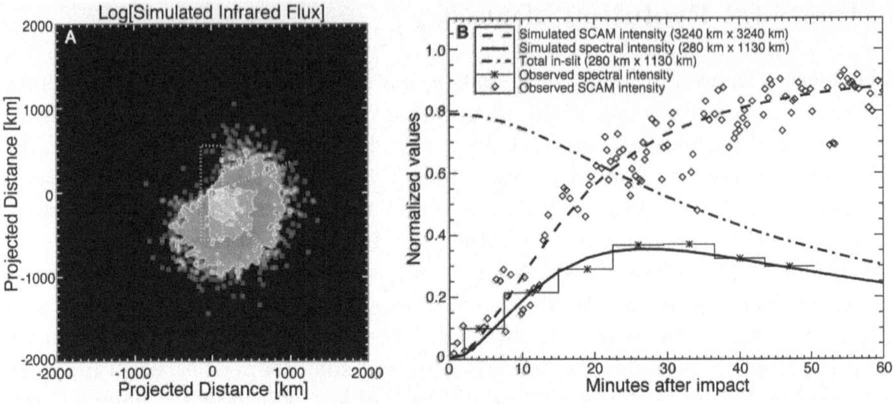

Fig. 4. Monte Carlo simulation of ejecta resulting from the DI event, based on observational parameters [21, 1]. **A**. Modeled plume evolution for T+1 hour (north is up and east is toward the left), approximately contemporaneous with Fig. 1b in Sugita et al. [21]. The spatial extent of our 9-row (1130 km) aperture is represented by the superimposed rectangle. The ejecta axis (at $PA = 225°$; [21]) is toward the lower right corner; the sunward direction is toward the right, 21° up from horizontal. **B**. Comparison between simulated and observed continuum intensities for spectral and SCAM observations (see text for detailed discussion).

7 Discussion

What factors could have caused the delay in enhanced observed molecular column abundances? The energy delivered by the impactor (1.9×10^{10} J; [1]) fell well short of that required to vaporize the estimated total amount of water ejected (4.4×10^6 kg; [13, 14]). We would not expect to see heating of ejected volatiles because H_2O cools rapidly though adiabatic expansion and rotational emission – our consistent T_{rot} values pre- and post-impact (~35–40 K) bear this out.

The maximum continuum intensity we observed in spectral mode (near T+30) coincided approximately with the time at which the column abundances of H_2O and C_2H_6 abruptly increased (Fig. 2). The volatile signature of the ejecta presumably signals the decrease in dust optical depth, coupled with the sublimation time required for icy grains having even a modest dust fraction (i.e., for "dirty grains;" [1]).

Our results suggest that the principal source of volatiles in the ejecta was sublimation from outflowing particles, and the rate of grain heating could be a primary controlling factor. Delayed heating of grains (to a threshold temperature) likely occurred, due to decreased shadowing with time (by intervening grains) as the ejecta plume expanded. It is also possible that a substantial increase in pumping photons seen by ejected volatiles may have given rise to the observed increase in column abundances, however the apparently abrupt nature of this increase would require a correspondingly rapid decrease in opacity to solar radiation.

The enhanced abundance of C_2H_6 in the DI ejecta (0.5–0.6% relative to H_2O) is consistent with its abundance measured in the majority of Oort cloud comets. The abundance of HCN in the ejecta also agrees (within error) with its abundance in other comets and (unlike C_2H_6) with its pre-impact value. The higher ethane abundance may more accurately reflect formation conditions for Comet 9P/Tempel 1, indicating a more nearly "interstellar" signature in that the C_2H_6 produced through H-atom addition to C_2H_2 (for example) was incorporated into the nucleus with little or no subsequent processing.

If the nucleus is internally homogenous, the lower pre-impact value for C_2H_6 (~0.2%) then suggests that surface (or near-surface) material lost during the *quiescent* (i.e., ambient) activity may have experienced considerable chemical processing compared to the ejecta. Alternatively, it is possible that the nucleus is chemically heterogeneous on large scales, and that the quiescent and impact material sampled cometesimals from different formation regions and of diverse chemistry.

Severely depleted C_2H_6 and nearly normal HCN were seen in the Oort cloud comet C/1999 S4 (LINEAR), however that comet was also depleted in virtually all other parent volatiles, suggesting its formation in the inner giant planets' region [19]. Likewise, highly depleted volatiles (excepting HCN) were observed in the recently split JFC 73P/Schwassmann-Wachmann 3 during its 2006 apparition [4, 23, 12, 8, 10, 7], also suggesting its formation closer to the Sun (or perhaps later in time) compared with 9P/Tempel 1 [23].

The depleted abundance of C_2H_6 and the *normal* abundances of HCN and CH_3OH observed pre-impact in Comet Tempel 1 [17] are similar to the C_2H_6 and HCN abundances seen in the JFC 21P/Giacobini-Zinner [20, 3]. If Tempel 1 returned to quiescent production on UT 5 July, the measured CO abundance (~4% at the 2–3σ confidence level, both on UT July 04 and 05; [17]) also falls in the normal category. This may indicate a similar processing history for the surface or near-surface material of these two JFCs (9P, 21P).

8 Summary

Using NIRSPEC on the Keck-2 telescope we measured, in time steps of approximately six minutes, the evolution of water, ethane, and hydrogen cyanide resulting from the Deep Impact collision with Comet 9P/Tempel 1. Relative to its pre-impact value, the post-impact column abundance of H_2O increased by only a modest factor (1.18±0.12) and that of C_2H_6 remained unchanged (within error) until $T+25$ minutes, after which they both abruptly increased. A plausible reason for this delay in enhanced column abundance may be the time required for outflowing (and possibly fragmenting) grains to reach critical (sublimation) temperatures.

The abundance of C_2H_6 (relative to H_2O) in the ejecta (0.55±0.09%) was a factor of 2.4±0.5 higher than its mean pre-impact value from observations on June 03 and July 04, and is consistent with abundances measured for the majority of Oort cloud comets. The abundance of HCN in the ejecta agreed with its pre-impact value and also with its abundance in the dominant group of Oort cloud comets. Provided the ejecta represent the internal composition of a homogeneous nucleus, this similarity suggests a common origin, perhaps in the outer giant-planets' region or the inner portion of the Kuiper belt.

Acknowledgements This research is supported by the NASA Planetary Astronomy Program, the NASA Astrobiology Program, the National Science Foundation, and the NPP Resident Research Associateship Program. We thank W. M. Keck Observatory Fred Chaffee for providing Director's Discretionary time on impact night, H. Kawakita and N. Dello Russo for their participation with the observations, and the staff of Keck Observatory for general support. The Keck Observatory is operated as a scientific partnership among the California Institute of Technology, the University of California, and the National Aeronautics and Space Administration. The Observatory was made possible by the generous financial support of the W. M. Keck Foundation. The authors recognize and acknowledge the very significant cultural role and reverence that the summit of Mauna Kea has always had within the indigenous Hawaiian community.

References

1. M.F. A'Hearn, and 32 colleagues (2005): Science **310**, 258
2. E.H. Beer, M. Podolak, D. Prialnik (2006): Icarus **180**, 473
3. N. Biver, D. Bockelée-Morvan, J. Crovisier, P. Colom, F. Henry, R. Moreno, G. Paubert, D. Despois, D.C. Lis (2002): EMP **90**, 323
4. G.A. Blake, C. Salyk, B.P. Bonev, G.L. Villanueva, M.A. DiSanti, M.J. Mumma, K. Magee-Sauer, E.L. Gibb (2006): IAU Circ. 8704
5. B.P. Bonev (2005): Towards a Chemical Taxonomy of Comets: Infrared Spectroscopic Methods for Quantitative Measurements of Cometary Water (With an Independent Chapter on Mars Polar Science). PhD Thesis, University of Toledo, Ohio
6. N. Dello Russo, R.J. Vervack, Jr., H.A. Weaver, N. Biver, D. Bockelée-Morvan, J. Crovisier, C.M. Lisse (2005): Bull. Am. Astronom.Soc. **38**, 503
7. N. Dello Russo, R.J. Vervack, Jr., H.A. Weaver, N. Biver, D. Bockelée-Morvan, J. Crovisier, C.M. Lisse (2007): American Astronomical Society, DPS meeting #38; #03.05; Bulletin of the American Astronomical Society, **38**, 485
8. N. Dello Russo, B.P. Bonev, M.A. DiSanti, M.J. Mumma, E.L. Gibb, K. Magee-Sauer, R.J. Barber, J. Tennyson (2006): Astrophys. J. **38**, 503
9. M.A. DiSanti, G.L. Villanueva, B.P. Bonev, K. Magee-Sauer, J.E. Lyke, M.J. Mumma (2007a): Icarus **187**, 240
10. M.A. DiSanti, W.M. Anderson, G.L. Villanueva, B.P. Bonev, K. Magee-Sauer, E.L. Gibb, M.J. Mumma (2007b): Astrophys. J. **661**, L101
11. D.E. Harker, C.E. Woodward, D.H. Wooden (2005): Science **310**, 278
12. H. Kawakita, H. Kobayashi, M.J. Mumma (2006): Bull. Am. Astronom. Soc. **38**, 503
13. H.U. Keller, and 11 colleagues (2005): Science **310**, 281
14. M. Kuppers, and 40 colleagues (2005): Nature **437**, 987
15. C.M. Lisse, and 16 colleagues (2006): Science **313**, 635

16. I.S. McLean, and 14 colleagues (2005): Proc. SPIE **3354**, 566
17. M.J. Mumma, and 13 colleagues (2005): Science **310**, 270
18. M.J. Mumma, and 18 colleagues (2001a): Astrophys. J. **546**, 1183
19. M.J. Mumma, N. Dello Russo, M.A. DiSanti, K. Magee-Sauer, R.E.Novak, S. Brittain, T.W. Rettig, I.S. McLean, D.C. Reuter, L.-H. Xu (2001b): Science **292**, 1334
20. M.J. Mumma, M.A. DiSanti, N. Dello Russo, K. Magee-Sauer, T.W.Rettig (2000): Astrophys. J. **531**, L155
21. S. Sugita, and 22 colleagues (2005): Science **310**, 274
22. J.M. Sunshine, and 22 colleagues(2006): Science **311**, 1453
23. G.L. Villanueva, B.P. Bonev, K. Magee-Sauer, M.A. DiSanti, C. Salyk, G.A. Blake, M.J. Mumma (2006): Astrophys. J. **650**, L87

Activity in Comet Tempel 1: Linking the Coma and the Nucleus' Surface

T. L. Farnham[1], D. D. Wellnitz[1], D. L. Hampton[2], J.-Y. Li[1], J. M. Sunshine[1], O. Groussin[1], L. A. McFadden[1], C. J. Crockett[1], M. F. A'Hearn[1], M. J. S. Belton[3], and C. M. Lisse[4]

[1] Department of Astronomy, University of Maryland, College Park, MD 20742
farnham@astro.umd.edu
[2] Ball Aerospace and Technology Corp., Boulder, CO 80301
[3] Belton Space Exploration Initiatives, Tucson, AZ 85716
[4] Applied Physics Laboratory, Johns Hopkins University, Laurel, MD 20723

Abstract We present an overview of the dust coma observations of comet Tempel 1 that were obtained in support of and during the Deep Impact mission and discuss how the data are being used to determine the source locations of the activity on the nucleus. Ground-based images covering the six-month period before impact show a long-lived southern asymmetry in the coma, though little detailed structure is seen. The wealth of Deep Impact images provides temporal coverage spanning a two-month period, with spatial resolution that continuously improves as the spacecraft closes on the nucleus. Three distinct jets are seen turning on and off and rotating with the nucleus' 1.7-day period. The brightest of these jets is likely to be a main contributor to the southern asymmetry. In the high resolution images obtained near close approach, intricate, primarily radial structures are seen in the coma. In the highest resolution images, small detailed jets can be seen emanating from the surface of the nucleus, and appear to be related, in some manner, to the patches of exposed water ice.

1 Introduction

Recent studies of comets suggest that the surfaces of most comet nuclei are essentially inactive, with active areas typically found to be a few percent of the surface [2, 7]. This information, combined with the observations of jets, arcs and other features that are seen in many comae [3, 9, 10, 11, 12], suggests that much of the activity from comets is generated from isolated active areas or vents that allow emission to percolate through otherwise mantled or inactive surfaces. Understanding the mechanisms by which these active areas form and develop can reveal important information about the evolutionary processes of comet nuclei and how they can change over time. Comet Tempel 1 was the target of the Deep Impact (DI) mission, and like many other comets, it exhibits features in its coma [1] (see Fig. 1). Because of the unprecedented amount of observational data, both ground-based and in situ, that was obtained as a part of this mission, comet Tempel 1 is an excellent candidate for the study of active areas.

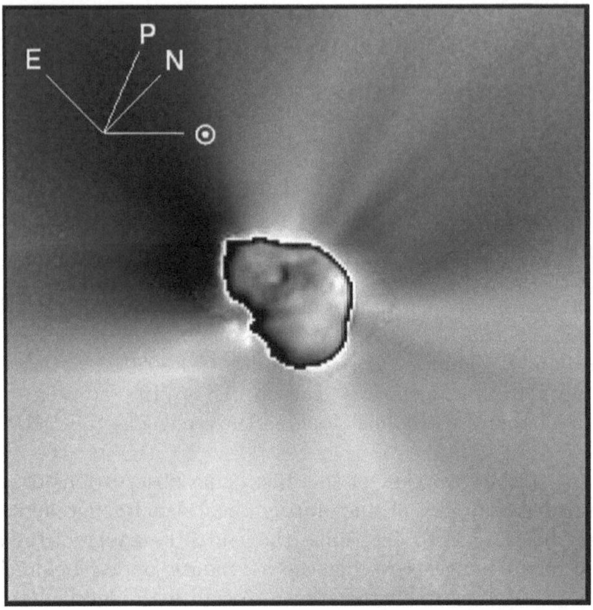

Fig. 1. Sample image of the coma of comet Tempel 1, enhanced by dividing out an azimuthally averaged radial profile. The nucleus is inset at a different contrast level to keep it from saturating the display. In the upper left, celestial north (N) and east (E) are noted, as well as the projected sunward direction (⊙) and the projected rotation pole (P).

In this presentation, we provide a brief overview of the observations of the ambient coma of comet Tempel 1 and review the preliminary findings regarging the images. We describe the morphology and how it changed throughout the approach and encounter with the nucleus and highlight some of the interesting features that were seen. We also discuss the analyses that are currently underway to investigate the origin of the features. A more extensive discussion of this work can be found in Farnham et al. 2007 [4]. In this review, we will focus on the features in the dust coma because the dust maintains a relatively predictable trajectory after decoupling from the gas and represents a long-term record of emission that can be followed to large distances. The gas species provide additional constraints on the activity, and these observations are discussed elsewhere [6, 8].

2 Observations

An extensive amount of data is available for this study. The spacecraft observations alone represent a data set that is unprecedented in its range of temporal and spatial scales. As the spacecraft approached the comet for its 4 July 2005 flyby and impact experiment, it obtained thousands of images of the coma and nucleus. Starting nearly two months before encounter, images were obtained every four hours to show the long term activity of the comet, with a dramatic increase in the observation rate in the days before encounter. This sequence is invaluable for revealing the temporal changes in the coma and illustrating how details become clearer and more obvious as the spatial resolution improves.

In support of the mission, a ground-based observing campaign was instituted to obtain images of the comet during the months before the encounter, to help in planning and to monitor the comet's normal activity. We obtained observations at roughly monthly intervals between January and July 2005, using the HB narrowband comet filters [5] to capture the morphology of the dust and gas in the coma. These images show the coma behavior on seasonal timescales, illustrating evolutionary variations as the sub-solar point and heliocentric distance change.

3 Coma Morphology

The images that capture the comet at the largest spatial scales, including both the ground-based and spacecraft images, reveal little evidence for detailed structure in the coma, though there is a strong asymmetry that exists from January through July (visible in Fig. 1). In this asymmetry, the surface brightness of a southern fan is up to twice as bright as the darkest regions to the northeast, with the highest contrast visible in the March/April timeframe.

A sequence of 221 High Resolution Instrument (HRI) frames from 1 May through 4 July 2005 was assembled to represent the appearance of the comet during the approach of the DI spacecraft[1] This approach sequence shows the development and temporal variations over a two-month period, including at least 10 outbursts of varying intensities [4]. In addition to the temporal information, the sequence shows apparent changes in the coma that result from the improving spatial resolution. As the spacecraft closed on the nucleus, individual features became distinguishable and the level of detail continued to increase until the spacecraft entered shield mode at the end of the sequence.

In late May, periodic variations produced by the rotation of the nucleus first became detectable. Three different jets were observed to turn on and off during the 1.7-day rotation of the nucleus (Fig. 2 with the jets labeled A, B and C). Given the orientation of the rotation axis as determined from the encounter images (R.A.=294°, Dec.=73° [14]), the main jet (A) appears to be a highly active extended source that arises from a latitude around −40°. This jet is likely

[1] available as a movie at
`http://pdssbn.astro.umd.edu/holdings/di-c-hriv_its_mri-5-movie-coll-v1.0`

to be a significant contributor to the southern fan that was observed throughout the first half of 2005. The two smaller jets (B and C) are much weaker and come from latitudes around +40°.

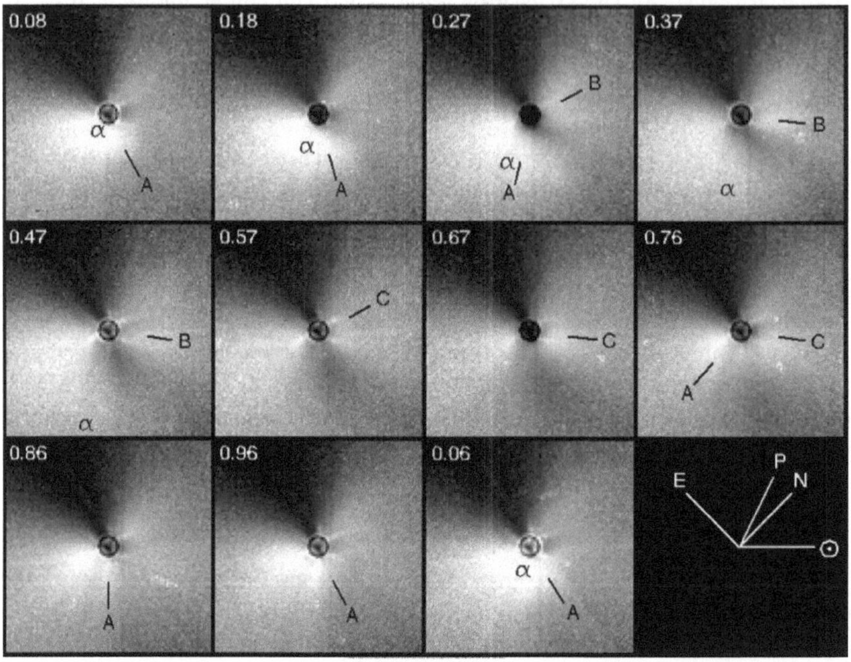

Fig. 2. Sequence of HRI images from 27 and 28 June showing the coma morphology changes throughout a full rotation of the nucleus (the rotational phase of each image is noted in the upper left corner, where 0 phase is defined as the time of impact). The images have been enhanced by removing an azimuthally averaged averaged radial profile. See the caption for Fig. 1 for more information.

During the encounter with the nucleus, highly detailed features were resolved in the coma, as shown in Fig. 1. Unfortunately, the short duration of the close approach precluded any detection of variability of these features. In the highest resolution images obtained as the spacecraft passed underneath the comet, a number of small well-defined jets became visible on the northern limb of the nucleus (see Fig. 3 and a movie posted at http://pdssbn.astro.umd.edu/holdings/di-c-hriv_its_mri-5-movie-coll-v1.0).

The locations of the bases of these features correspond well with the exposed patches of water ice observed on the surface of the nucleus [13, 4], suggesting that the jets are in some manner related to the ice patches. Further work will be done on this topic to investigate the relationship between the ice, the surface jets and the other radial features.

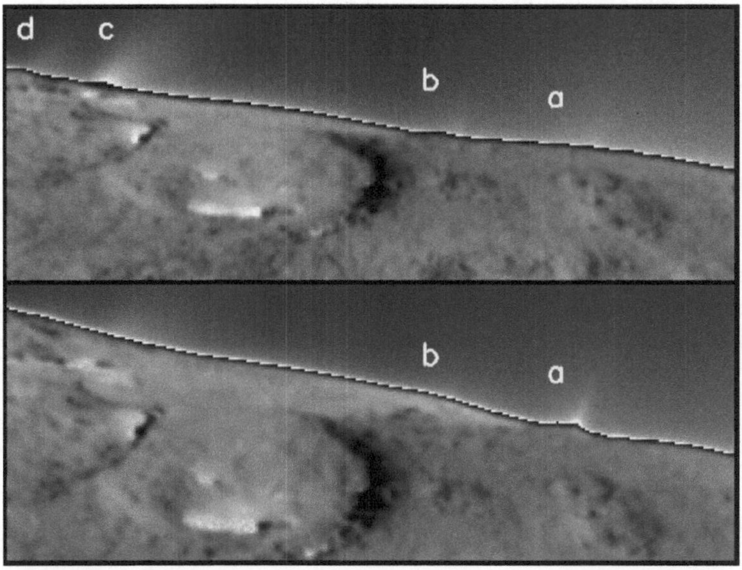

Fig. 3. Images of the limb of the nucleus showing at least four jets emanating from the surface. The apparent brightness of these small, well-defined features peaks when they are crossing over the horizon (as the spacecraft moves under the nucleus). The sources of these jets are somehow related to the patches of water ice observed on the surface. In this figure, the surface of the nucleus is inset at different display levels than the coma to avoid saturation.

4 Coma Analysis

The ultimate goal of this study is to use the features observed in the coma to locate the active regions on the surface and then investigate the conditions under which the activity is generated. The first part of this study is currently underway, using a variety of tools to determine the locations of the sources.

With the encounter images as a starting point, we note that the features are predominantly radial, and this fact can be used as a tool to constrain their sources. It is assumed that the origin of a linear jet lies somewhere along the extension of the feature backward across the nucleus. By itself, this provides a severe constraint on the location of the source, though it is not possible to derive a unique source from a single image. Fortunately, there are a number of additional types of information that can be incorporated to provide more precise constraints on the feature geometry.

The most powerful technique for this purpose involves stereogrammetry. Using the parallax produced by the spacecraft's motion (or from pairs of images from the impactor and flyby spacecraft) it is possible to uniquely determine where the jet intersects the surface of the nucleus [12]. However, this technique works only for features where parallax can be measured. In cases where parallax information is not available, other information must be incorporated to further constrain the jet sources. Temporal variations can be used to restrict the potential locations of active regions, under the assumption that a source will become active only when it crosses the terminator into sunlight (though crossing into darkness does not necessarily mean that the activity shuts off). The shape model of the nucleus also provides an effective tool for constraining the jet sources. Many potential locations can be rejected if the geometry of the surface cannot produce the observed features (e.g., does it face the right direction?).

Once the sources have been located, it will be possible to characterize the active regions, on the basis of the many types of data available from the DI mission (geomorphology, surface temperatures, color and albedo, infrared spectroscopy, etc.). The properties of the active regions can then be compared to the rest of the surface to look for similarities or differences.

References

1. M. F. A'Hearn, M. J. S. Belton, W. A. Delamere, et al. (2005): Science **310**, 258–264
2. M. F. A'Hearn, R. L. Millis, D. G. Schleicher, D. J. Osip, and P. V. Birch (1995): Icarus **118**, 223–270
3. D. E. Brownlee, J. D. Anderson, K. Atkins, et al. (2004): Lunar Planet. Sci. **35**, 1981
4. T. L. Farnham, D. D. Wellnitz, D. L. Hampton, et al. (2007): Icarus **187**, 26–40
5. T. L. Farnham, D. G. Schleicher, and M. F. A'Hearn (2000): Icarus **147**, 180–204,
6. L. M. Feaga, M. F. A'Hearn, J. M. Sunshine, O. Groussin, and T. L. Farnham (2007): Icarus, **191**, 134–145
7. H. U. Keller, D. Britt, B. J. Buratti, and N. Thomas: In situ observations of cometary nuclei. In *Comets II*, M. H. Festou, U. Keller, and H. Weaver, editors, (University of Arizona Press, Tucson, 2004) pp. 211–222
8. L. A. McFadden, C. J. Crockett, D. D. Wellnitz, et al. (2006): Bull. Am. Astron. Soc. **38**,1704
9. R. Z. Sagdeev, G. A. Avanesov, I. V.D Barinov, A. I. Debabov, and V. A. Kvasikov (1985): Adv. Space Res. **5**, 95–104
10. Z. Sekanina (1991): Astron. J. **102**, 350–388
11. Z. Sekanina and S. M. Larson (1986): Nature **321**, 357–361
12. L. A. Soderblom, T. L. Becker, G. Bennett, et al. (2002): Science **296**, 1087–1091
13. J. M. Sunshine, M. F. A'Hearn, O. Groussin, et al. (2006): Science **311**, 1453–1455
14. P. C. Thomas, J. Veverka, M. J. S. Belton, et al. (2007): Icarus **187**, 4–15

ESO Spectrophotometry of Comet 9P/Tempel 1

M. Weiler[1], H. Rauer[1,2], C. Sterken[3], J. Knollenberg[1], E. Jehin[4], E. Pompei[4], O. Hainaut[4], G. P. Tozzi[5], and J. Manfroid[6]

[1] Institut für Planetenforschung, DLR, Rutherfordstrasse 2, 12489 Berlin, Germany
[2] Zentrum für Astronomie und Astrophysik, Technische Universität Berlin, Germany
[3] Vrije Universiteit Brussel, Brussels, Belgium
[4] European Southern Observatory, Santiago de Chile
[5] Arcetri Observatory, Firenze, Italy
[6] Institut d'Astrophysique, Université de Liège, Liège, Belgium

Abstract The Deep Impact target comet 9P/Tempel 1 was observed by means of long-slit spectroscopy from two nights before impact up to eight nights after impact, using the ESO VLT UT1, UT2, and ESO NTT telescopes. Spectra covering the complete optical wavelength range were obtained, and information at different position angles in the coma was collected. The data were used to study the gas and dust activity of comet 9P/Tempel 1. Gas production rates before and after impact and the amount of material in the impact cloud were determined. The pre-impact $Af\rho$ parameter, the dust production rate and the dust-to-gas mass ratio were derived. A variation of the cometary gas activity with rotation of the nucleus was detected. A difference in the variation of the brightness of the CN gas emission band compared to the variation of the emissions by C_2, C_3, and NH_2 in the inner coma suggests compositional differences between different parts of the surface of comet 9P/Tempel 1's nucleus.

1 Introduction

As part of the ESO Deep Impact observing campaign, optical long-slit spectra of comet 9P/Tempel 1 were obtained from July 2 to July 12, 2005. For this purpose, the ESO VLT UT1 telescope was used from July 2 to July 8, and the VLT UT 2 from July 10 to July 12. Additional spectra were taken with the ESO NTT telescope on July 4. Observations were possible for about four hours in the beginning of each night. At the time of impact, the target comet had already set at the ESO observatories.

The instruments FORS2, FORS1, and EMMI were used, respectively. Two grisms were used to cover the whole optical wavelength range. The focus was on the use of a grism covering the wavelength range from 3700 to 6200Å, since the most prominent gas emission features in the cometary coma, originating from the radicals CN, C_2, C_3, and NH_2, fall into this region. With this grism, spectra at four different slit orientations were obtained. The center of the long-slit was placed on the optocenter of the coma, assuming this to be the position of the cometary nucleus. Therefore, information at the position angles with respect to the projected solar direction of 0°–180°, 90°–270°, 45°–225°, and 135°–315° were obtained in the coma. Furthermore, one spectrum per night was obtained with a second grism at one orientation of the long-slit (position angles 0°–180°). These spectra cover the wavelength range from 6100 to 9200Å.

The spectra obtained with the different grisms and instruments have a wavelength resolution between 570 and 780. A detailed observing log and a description of the reduction of the FORS data is provided by Rauer et al. [9].

The long-slit spectroscopic observations were performed to study the gas activity around the time of impact. Furthermore, the dust activity can be constrained from the continuum of scattered sunlight in the spectra.

2 Study of the Gas Activity

The results presented in this chapter represent a summary of Rauer et al. [9] and Weiler et al. [13].

The gas emissions of CN, C_3, C_2, and NH_2 could be detected in the spectra before and after the impact. In the four nights following the impact, an expanding gas cloud was observed. This cloud expanded all around the nucleus of the comet within about 18 hours after impact, when the first post-impact spectra were obtained. A projected velocity of about 1.2 km s^{-1} for the outermost dectectable parts of the CN and C_2 clouds was determined, while the center of the clouds moved with about 0.6–0.8 km s^{-1}.

No new gas emission bands could be observed after the impact. The Haser model [5] was used to determine the scale lengths and gas production rates in the two nights before impact and again from the fifth night after impact on. The scale lengths for the four detected radicals in the coma of 9P/Tempel 1 are typical compared to other comets. Gas production rates were about 11.3×10^{24} s^{-1} for CN, 15.9×10^{24} s^{-1} for C_2, 1.5×10^{24} s^{-1} for C_3, and 8.3×10^{24} s^{-1} for NH_2, averaged over for different position angles (0°, 90°, 180°, and 270°) and over two nights before impact and the fifth and sixth night after impact.

The amount of parent species in the impact cloud was determined by computing the total number of radicals in the impact cloud as a function of time and using a Haser-like two-step chemical model. The parent and daughter life times were computed using the determined maximum gas expansion velocities and the Haser scale lengths determined from observations of the undisturbed coma. This approach makes the assumption that all parent molecules are released at the same time, i.e. the time of impact, and that the formation chemistry is the same in the impact cloud and in the undisturbed coma. The averaged numbers of parent molecules released by the impact are 3.7×10^{29} for the CN parent, 4.7×10^{29} for the C_2 parent, and 0.8×10^{29} for the C_3 parent, respectively. Due to the low signal-to-noise ratio of the NH_2 emission, the number of parent molecules could not be determined for this species.

Since the used chemical model is the same for the undisturbed coma and the impact cloud, the ratios of the parent species of C_2/CN and C_3/CN before the impact, in the impact cloud, and after the impact could be determined. No significant differences in these ratios were observed, indicating no compositional differences of the impact cloud compared to the undisturbed coma as far as CN, C_2, and C_3 are concerned.

When plotting the flux of the emissions from CN, C_3, C_2, and NH_2 in the innermost coma versus the rotational phase of the nucleus, a variation of the gas activity with the nucleus rotation became obvious. The rotational light curve in the innermost coma shows two maxima, indicating the presence of at least two active regions on the nucleus surface. The light curves for the four different radicals show the same variation with rotation, except for CN. For the CN emission, one maximum is less pronounced compared to the lightcurves from the other species. This indicates a compositional difference of one active region compared to the rest of the nucleus surface.

3 Study of the Dust Activity

An expanding cloud was also observed in the continuum of sunlight scattered by dust particles in the cometary coma. The dust cloud is directed towards the west and south of the nucleus 18 hours after impact, and it moves with a projected velocity of about 200 m s^{-1}, showing a deceleration with time as a consequence of solar radiation pressure.

When assuming spherical symmetry of the dust coma, the dust activity parameter $Af\rho$ [1] can be determined from the spectra. This was done for the long-slit spectra of the night of July 3/4, 2005, resulting in a pre-impact $Af\rho$ value of 134.5 ± 54.6 cm. This value was obtained by averaging the values obtained from radial emission profiles at four different position angles in the coma (0°, 90°, 180°, 270°), and the large uncertainty results from coma asymmetries. The $Af\rho$ value was determined with $\rho = 5 \times 10^4$ km at a wavelength of 4422Å. This wavelength was chosen for comparison with other results, since it is the central wavelength of the ESO "blue" comet continuum filter. The $Af\rho$ value derived in this work is in agreement with results by previous publications (e.g. [10], and references therein).

From $Af\rho$, the dust mass production rate can be computed. This requires the knowledge of the size-dependent dust expansion velocity in the cometary coma and the maximum dust grain size in the coma [6]. In order to compute these quantities, first the gas flow from free sublimating water ice at a mean solar zenith angle of 60° was computed. In a second step, the equation of motion for test dust particles in this gas flow and under the influence of the nucleus gravity was solved. A summary of this approach is given by Weiler et al. [11]. A nucleus bulk density value of 800 kg m^{-3} and a value of the nuclear radius of 3 km [1] were used for the computation. For the assumed spherical dust grains a density according to Newburn & Spinrad [8] was assumed. After 100 nuclear radii, the dust has decoupled from the diluted gas flow and the nucleocentric dust velocities remain about constant. The resulting size-dependent final dust

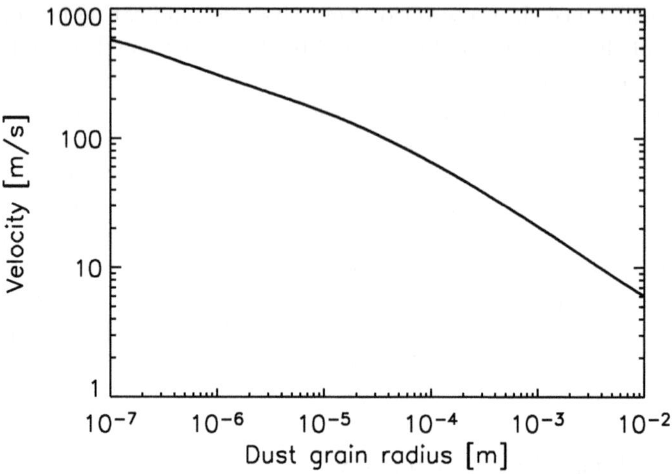

Fig. 1. Nucleocentric dust velocities as a function of dust grain radius in the coma of comet 9P/Tempel 1 before impact.

velocities are displayed in Fig. 1. This results in a maximum radius of 10.9 cm for dust particles that can be lifted off the surface by the gas flow.

For the computation of the dust production rate, a dust phase function according to Divine [3] is assumed. With a dust size frequency distribution according to Newburn & Spinrad [8] with a typical exponent of the size distribution of −3.5 and a maximum at a size of 0.54 μm, a dust mass production rate of 141.8 kg s^{-1} results, with the minimum and maximum values of 83.0 kg s^{-1} and 199.4 kg s^{-1} obtained with the lower and upper value of $Af\rho$. Assuming a water production rate of 3.4×10^{27} s^{-1} [7], this corresponds to a dust-to-water mass ratio of about 1.4. However, the uncertainty of this value remains large not only due to uncertainties in the model assumptions, but also due to the large uncertainty in $Af\rho$ and the water production rate. The water production rate might be as high as 10.4×10^{27} s^{-1} [4], meaning the dust-to-water mass ratio would be less than one.

4 Conclusions

No influence of the impact event upon the gas production rates of comet 9P/Tempel 1 on a timescale of several days was observed. No new active region of significant size on the nucleus surface was therefore created by the impact. Since no new gas emission features were detected in the spectra obtained after impact, no significant amount of species unusual for cometary comae were produced due to the impact. The ratio of C_2/CN and C_3/CN in the impact cloud is identical to the the corresponding values in the undisturbed coma, thus no indication for compsitional inhomogeneities of the cometary nucleus with depth was found. The large numbers of parent molecules in the impact cloud cannot be produced by sublimation due to the impact energy alone. The presence of sublimating grains in the coma after impact is therefore likely. Differences in the rotational lightcurve of the CN emission in the inner coma compared to the lightcurves of the C_2, C_3, and NH_2 emissions indicate compositional differences between different regions on the nucleus of comet 9P/Tempel 1.

The dust-to-gas mass ratio derived from the dust mass production rate is close to one, thus comet 9P/Tempel 1 is not a dusty comet (e.g. for comet 67P/Churyumov-Gerasimenko, the dust-to-water mass ratio was determined to about 4.8, see [12]).

A more detailed discussion of the results presented here is provided by Rauer et al. [9] and Weiler et al. [13].

References

1. M. F. A'Hearn, D. G. Schleicher, R. F. Millis, et al. (1984), ApJ, **89**, 579
2. M. F. A'Hearn, M. J. S. Belton, W. A. Delamere, et al. (2005), Science, **310**, 258
3. N. Divine: A simple radiation model of cometary dust for P/Halley. In: *ESA SP-174: The Comet Halley. Dust and Gas Environment*, ed by B. Battrick, E. Swallow, pp 47–53
4. P. D. Feldman, R. E. Lupu, S. R. McCandliss, et al. (2006), AJ, **647**, L61
5. L. Haser, Bulletin de la Scociete Royale des Sciences de Liege (1957) **43**, 740
6. L. Jorda (1995), Atmospheres cometaires: interpretation des observations dans le visible et comparaison avec les observations radio. PhD Thesis, Observatoire de Paris-Meudon, New York
7. M. Küppers, I. Bertini, S. Fornasier, et al. (2005), Nature, **437**, 987
8. R. L. Newburn & H. Spinrad (1985), AJ, **90**, 2591
9. H. Rauer, M. Weiler, C. Sterken, et al. (2006), A&A, **459**, 257
10. G.P. Tozzi, H. Boehnhardt, L. Kolokolova, et al. (2007), *Dust Observations of Comet 9P/Tempel 1 at the time of the Deep Impact*, A&A , submitted
11. M. Weiler, H. Rauer, J. Knollenberg, et al. (2003), A&A, **403**, 313
12. M. Weiler, H. Rauer, J. Helbert (2004), A&A, **414**, 749
13. M. Weiler, H. Rauer, J. Knollenberg, et al., Icarus, in press (2007), URL: `http://dx.doi.org/10.1016/j.icarus.2007.01.015`

Search for Ammonia Radio Emission in Comet 9P/Tempel 1 after the Deep Impact Event

G. P. Tozzi[1], F. Palagi[2], C. Codella[2], S. Poppi[3], and J. Crovisier[4]

[1] INAF–Osservatorio Astrofisico di Arcetri tozzi@arcetri.astro.it
[2] INAF–Istituto di Radioastronomia palagi@arcetri.astro.it, codella@arcetri.astro.it
[3] INAF–Osservatorio Astronomico di Cagliari spoppi@ca.astro.it
[4] Observatoire de Paris Jacques.Crovisier@obspm.fr

Abstract About 30 hours after the DI impact event, comet 9P/Tempel 1 has been observed with the 32 m dish of Medicina observatory (Bologna, Italy) to search for the NH_3 inversion transitions in the region around 24 GHz. The results show the presence of a line with S/N of about 6 and a FWHM equal to 1.35 km/s close to the NH_3(1,1) frequency. The cometary origin of the line seems sure, because the search for possible background sources of such a emission, performed a few days later along the same comet path, did not show any line at that frequency. Its identification is however dubious because of its too high outflow projected velocity and the too high ammonia production necessary to fit the line intensity. The frequency of the line is also close to a transition of methyl formate, a species already observed in comet Hale-Bopp and in hot cores. The outflow projected velocity would be smaller than in the case of the ammonia, but its abundance would be too high as well.

1 Introduction

The Deep Impact (DI) mission to the Jupiter family comet 9P/Tempel 1 (9P hereafter) aimed at the study of the primordial material present in cometary nuclei and the physics of cratering. On 4 July 2006, the impactor of the DI experiment had an artificial high-speed (about 10 $\mathrm{km\,s^{-1}}$) impact with the 9P nucleus, excavating a considerable amount of cometary material that was observed both in situ, by the DI fly-by spacecraft, and remotely, by Earth-based observations. First mission results are described in [1] and [2]. Due to the short time available for the spacecraft observations, the ground-based observations were integral part of the mission, not just a support to it, as for other space missions. Preliminary Earth-based and other space-based measurements of the event are published [3, 4, 5, 6, 7].

During the pre-mission phase a wide range of models were developed to describe the expected observable phenomena. Anything was expected, from the comet swallowing the impactor with virtually no effects to a complete disruption of the nucleus. However, the most likely models predicted an ejecta plume and a football stadium sized crater with high probability that pristine material from the inner, unevolved, layers would be released during the impact and the time thereafter, when the Sun will illuminate this newly formed active region on the nucleus. Since almost nothing was known about the composition of the material released by the impact, radio observations of NH_3 where programmed. Ammonia is a very difficult element to be observed in comets: it has been detected only in comets having either a close approach to the Earth or in bright comets (e.g. [8]). Without the effect of the impact, ammonia would have been for sure undetectable in a Jupiter family comet as 9P is. Unfortunately, observations scheduled on the day before and just after the impact (8–15 hours after) were not performed because of technical problems.

In this paper we describe the observations and the detection of a cometary line, probably unidentified.

2 Observations

Comet 9P was observed on July 25, 2005 from 14:43 UT to 21:17 UT (from 31:50 to 38:25 h. after the impact) with the Medicina 32m Radio Telescope in a spectral band covering the NH_3(1,1) and (2,2) lines. As already mentioned, observations prior and just after the event were also planned, but could not be performed because of technical problems. The observations lasted for 7 hours, giving an effective integration time on the source of 115 min. At that time the comet was at 1.51 and 0.90 AU from the Sun and Earth, respectively. The phase angle (Sun-Comet-Observer angle) was $\approx 41°$. The comet radial velocity with respect to the observer varied from 8.964, at the beginning of the run, to 9.508 $km\,s^{-1}$ at the end. Note that, during the observations time, the comet moved about 590 arcsec on the sky, i.e. more than 5 times the telescope HPBW. That means that any possible background source would appear only in a fraction of the whole run. Spectra were obtained with a dual polarization K-Band receiver feeding a dual section Autocorrelation Spectrometer.

The main characteristics of the used instrument are:

- Rest Frequency: 23694.496 MHz
- HPBW: 1.9 arcmin
- Spectral resolution: 0.253 $km\,s^{-1}$ (20 kHz)
- System Temperature: ≥ 100 K
- Spectral range: 8 GHz
- Antenna gain: ≈ 0.1 K/Jy

A position switching scheme (15 min ON + 15 min OFF) was adopted. Data were reduced using the CLASS (http://www.iram.fr/IRAMFR/GILDAS) package. First 2-3 consecutive ON and OFF scans were averaged and then the average scans denormalized to remove gain variations due to the instrument band-pass [9]. Any residual spectral baseline was then removed by a low degree polynomial interpolation.

3 Results

One of the two polarization was used to search for the $NH_3(2,2)$ line, were no emission was detected. The $NH_3(1,1)$ line was searched for in the other polarization. In this case, adding all the reduced spectra an intense (6σ) line appeared at 23693.78 MHz, as shown in Fig. 1. The spectrum is reported in units of main-beam brightness temperature (Tmb) versus Local Standard of Rest (LSR) velocity. The line parameters are:

- Integrated Area: 0.86 ± 0.10 K$\times$ km s^{-1}
- Tpeak: 0.60 ± 0.02 K (0.02 is the rms of the fitting)
- FWHM: 1.35 ± 0.16 km s^{-1}
- rms: 0.08 K

Assuming the rest frequency of $NH_3(1,1)$, the line is centred at –4.95 km s^{-1}, i.e. at +9.10 km s^{-1} from the comet LSR velocity at the epoch of the observation. In order to check if such line is due to a contamination of a background source, we derived three different spectra by considering three slots of 35, 45, and 35 minutes of integration time, at the beginning, middle and end of the run, respectively. The results are shown in Fig. 2: the same line appears in all the three spectra, thus excluding a background contamination. Moreover, the possible background source is also excluded by the repetition of observations of the same path covered by the comet, few days after the 9P observations. No lines were detectable out of 1σ.

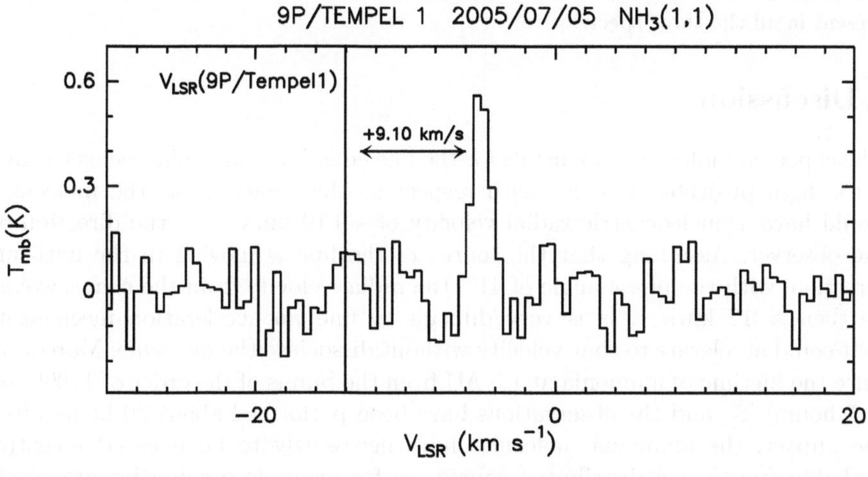

Fig. 1. Line emission observed towards 9P during the 2005 July 5th run with the 32-m Medicina radio telescope. The vertical line marks the LSR velocity (–14.05 km s^{-1}) of the comet. The spectrum is reported here in units of main-beam brightness temperature (Tmb). The emission line is centred at –4.95 km s^{-1}, i.e. it is at +9.10 km s^{-1} with respect to the comet.

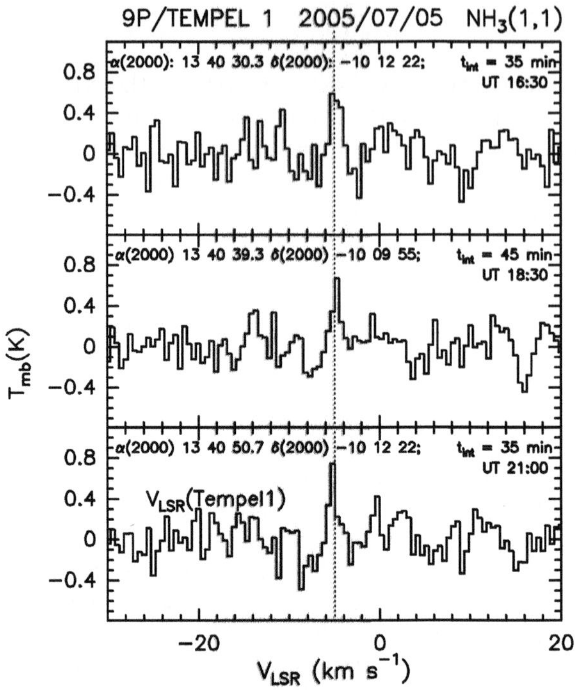

Fig. 2. The same of Fig. 1, but divided into three slots of 35, 45, and 35 minutes each. The three panels reports the obtained spectra as well as the corresponding UT and coordinates. It is possible to see that the emission at –4.95 $\mathrm{km\,s^{-1}}$ (*dotted line*) is present in all the three spectra.

4 Discussion

The expected molecule responsible for the line emission, ammonia, seems to have a too high projected velocity with respect to the comet, since the molecules would have a nucleocentric radial velocity of +9.10 $\mathrm{km\,s^{-1}}$, in the direction of the observer. Assuming that the source of the line is moving in the anti-sun direction, with the phase angle of 41°, the radial velocity from the comet would be then $\approx$ 12 $\mathrm{km\,s^{-1}}$. Is is very difficult to find an acceleration mechanism that could accelerate to that velocity without dissociate the molecule. Moreover, since the lifetime of ammonia at 1.5 AU from the Sun is of the order of 17000 sec ($\approx$5 hours) [8], and the observations have been performed about 30 hours after the impact, the ammonia molecules have necessarily to be released recently, probably from some distributed source, as for example organic/ice grains. A possible explanation for high outflow velocity in comets has been reported [10] for water in comet Hyakutake. The authors proposed that water was released by sublimation of charged, tiny ice grains accelerated by the solar wind. This mechanism could work also in the present case, but without any model it is difficult to check its validity. In addition, the narrow linewidth is not fully in agreement with such an acceleration mechanism. Even assuming that the force acting on the grains have similar strength, the acceleration, and then the final velocity, would depend on grain sizes, a, going as a^{-3}. A non-monochromatic size grain distribution would produce a much larger line FWHM. Only in the case of particles with about the same sizes a line so narrow could be justified. Another problem is the large amount of NH_3 necessary to have a such a line. Applying a LTE coma model with $T_{rot} = 30$ K [11, 12], the ammonia column density would be $\approx 3 \times 10^{13}$ cm^{-2}, which, for a normal comets, would correspond to an extremely high production rate (10^{29} s^{-1}).

Another candidate is the $HCOOCH_3$ (methyl formate) ($24_{4,20}-25_{5,20}$, E state) transition at 23694.090 MHz, corresponding to a cometocentric radial velocity in the observer direction of only $+3.97$ km s^{-1}. Note that [13] predicted that methyl formate can be present in comets as a sublimation product of organic grains (polyoxymethylene) ejected from the cometary nucleus. This molecule has been definitively observed in comet Hale-Bopp [14]), with an abundance of 0.08% that of water Methyl formate has been observed also in stellar hot cores, where it is possible to find an extremely rich organic chemistry with abundant amounts of complex O-bearing molecules (e.g. [15]). However, the identification of the cometary line with $HCOOCH_3$ can be ruled out because of the velocity mismatch and because the upper state of the transition has a too high energy (E_u = 202 K) to be significantly populated.

5 Conclusions

A cometary 6σ line emission at 23694.496 MHz was detected 31–37 hours after the impact. The identification of the line as due to NH_3(1,1) does not make sense, because of (i) the too high cometocentric radial velocity, (ii) the large production rate needed, and (iii) the short NH_3 lifetime with respect to the elapsed time from the impact. Methyl formate (previously observed in comet Hale-Bopp), which has a transition close to the line we detected, can be ruled out for similar reasons. Thus this cometary line remains unidentified.

References

1. A'Hearn M.F., Belton M.J.S., Delamere W.A. et al., 2005 Science **310**, 258
2. Sunshine J.M., A'Hearn M.F., Groussin O. et al., 2006 Science **311**, 1453
3. Meech K.J., Ageorges N., A'Hearn M.F. et al., 2005, Science **310**, 265
4. Sugita S., Ootsubo T., Kadono T. et al., 2005, Science **310**, 274
5. Harker D.E., Woodward C.E., Wooden D.H. 2005, Science **310**, 278
6. Lisse, C.M., VanCleve J., Adams A.C. et al., 2006, Science **313**, 635
7. Schleicher D.G., Barnes K. L., Baugh N.F., 2005, AJ **131**, 1130
8. Bird, M. K., Huchtmeier, W. K., Gensheimer, P., Wilson, T. L., Janardhan, P., Lemme, C., 1997, Astronomy and Astrophysics **325**, L5
9. Ables, J.G., Cooper, B.F.C., Hunt, A.J., Moorey, G.G., Brooks, J.W., 1975, Rev. Sci. Instrum. **46**, 284

10. Cosmovici, C., Montebugnoli, S., Orfei, A. Pogrebenko, S., Cortiglioni, S., 1998, P&SS **46**, 467
11. Crovisier J., Bockelée-Morvan D., Colom P. et al., 2004, Astronomy and Astrophysics **418**, 1141
12. Biver N., Bockelée-Morvan D., Boissier J. et al., 2007, Icarus **187**, 253
13. Cottin, H., Gazeau, M. C., Doussin, J. F., Raulin, F., 1999 Physics and Chemistry of the Earth **24**, 597
14. Bockelée-Morvan, D., Lis, D.C., Wink, J.E., Despois, D., etãl., 2000, Astronomy and Astrophysics **353**, 1101
15. Cazaux, S., Tielens, A. G. G. M., Ceccarelli, C., Castets, A., Wakelam, V., Caux, E., Parise, B., Teyssier, D., 2003 ApJ **593**, L55

Part V

Cometary Plasma, Cometary Space Missions, and the Future

The Strength of Cometary Surface Material: Relevance of Deep Impact Results for Philae Landing on a Comet

J. Biele[1], S. Ulamec[1], L. Richter[1], E. Kührt[2], J. Knollenberg[2], D. Möhlmann[2], and the Philae Team

[1] DLR, Institute for Space Simulation, Cologne, Germany jens.biele@dlr.de, stephan.ulamec@dlr.de, lutz.richter@dlr.de
[2] DLR Institute for Planetary Research, Berlin, Germany joerg.knollenberg@dlr.de, ekkehard.kuehrt@dlr.de, dirk.moehlmann@dlr.de

Abstract We discuss the Deep Impact estimates of strength of the surface material of comet Tempel 1. It appears doubtful that the tensile strength is as low as originally published. The method applied neglects the weakening by the shock wave, the acceleration by gas and the ongoing discussion whether gravity or strength craters are consistent with the observations. The various definitions of strength (tensile, shear, compressive, dynamic strength) and its size-dependency are discussed. Even for extremely low cohesion, a lower limit of $\approx$ 1 kPa for the quasi-static compressive strength at dm-scales is derived; far more likely is a compressive strength of the order of $\approx$ 10 kPa for very soft comet surfaces. The Rosetta lander, Philae, will touchdown on comet Churyumov-Gerasimenko in November 2014. A soft comet surface with a compressive strength $\geq$ 4 kPa indeed facilitates landing and leads to a penetration of the order of only 20 cm according to our model calculations.

1 Motivation

The mechanical strength of comets is a critical parameter for discrimination between models of their formation and evolution. Observations of, e.g., split comets have been analyzed for indications of a finite strength; recently, the strength of individual comets has also come into experimental focus (notably, for space missions Deep Impact and Rosetta). Any direct measurement of the strength of cometary material is thus highly welcome – the Deep Impact experiment seemed just to supply such a measurement. Since initial interpretations of Deep Impact (DI) mission results [1] have inferred an extremely low (< 65 Pa) (shear) strength of comet Tempel 1 surface material, questions were raised ('bad news'? [2]) whether a comet lander could land/dock and stay on such weak material. In the view of the ongoing Rosetta Mission, which was launched in March 2004 and will arrive at the target comet, 67P/Churyumov-Gerasimenko in 2014, when the

Lander, Philae, is going to be delivered, the results of the Deep Impact Mission (in particular regarding comet surface properties) have been acknowledged with highest interest.

It appears, however, necessary to discuss three principal issues in the interpretation of these data: By the impact shock itself, the material is stressed (fractured) and its tensile strength is modified. Thus, the pristine material properties can most likely not be determined with a method applied to the excavated plume material. Due to the impact a non-negligible amount of gas has been released from an extended source modifying the velocity distribution of the ejected dust particles. Thus, the detection of a minimum velocity of dust grains cannot be directly related to the material strength. Additionally, the simple formula used to derive strength appears only to be plausible for an order-of-magnitude estimation. The definition of strength in [1] needs to be defined more clearly in order to draw conclusions on, e.g., the penetration of a lander device with an impact speed of ≈ 1 ms^{-1}. Slow penetration ($\leq$ 50 ms^{-1}) into cometary material is depending primarily on the compressive strength, which is typically at least one order of magnitude higher than tensile or shear strengths. We will discuss the issues stated above and estimate the applicable compressive strength of cometary surface materials. Modeling the touchdown of Philae (the Rosetta Lander) results in a maximum depth of the order of 20 cm. Experimental studies are being conducted to investigate low velocity penetration of blunt bodies into dust-rich, fluffy comet analogue materials.

2 The Definition of Material Strength

Cohesion, tensile, shear and compressive strength: In (A'Hearn et. al, 2005 [1]), the DI team discusses shear strength $\cong$ cohesion = tensile strength. While for brittle materials, tensile strength is generally less than the shear strength (solid ice: shear strength $\approx$ 3 times the tensile strength), compressive strength is about one order of magnitude higher than tensile strength (see Sect. 5). In the case of soft landing ($v \approx 1$ m s^{-1}, typical area 500 cm^2, actually more a docking maneuver) compressive strength is the relevant parameter. In this paper, we will denote shear, tensile and compressive strength with σ_s, σ_t, σ_c respectively.

Dynamic and quasi-static strength: The strength values given above are derived from quasi-static experiments whereas, in principle, the dynamic strength for high strain rates is the relevant measure for impacts. During impacts, due to very high strain rates, the *dynamic* strength is typically higher than the (quasi-)static strength. It is known that the strength increases with strain-rate resulting in values about an order of magnitude higher (or even more) than the quasi-static strength for the same material. According to [3], generally the tensile strength σ_t is proportional to a power b of the strain rate $\dot{\varepsilon} = \varepsilon/\Delta t$, with a power law exponent typically around 1/4 to 1/3, depending on the material. For ice and strain rates between 1 s^{-1} and 2×10^4 s^{-1} (extrapolated to 1×10^6 s^{-1} and three terrestrial rocks in the high strain rate regime) Ai and Ahrens [3] find

$1/b = 3.97 \pm 0.05$. It is further interesting to note that this effect might be even stronger for porous materials [4].*Size dependence:*

Different theories indicate that the strength decreases with increasing size d according to $\propto d^{-q}$ where the exponent q is between 0.5 (fractal aggregate with fractal dimension $D = 2.5$ of ice [5]) and 0.6 (Weibull theory [6]). Thus, if extrapolated from typical lander (0.1 m), or impactor (1 m) to typical comet (1 to 10 km) scales, the size effect alone would produce a factor of 100 in the apparent strengths. This is in line with the observation that comets can often be described as essentially strength less bodies ('large cometesimal', 'rubble pile', 'swarm' models, cf. [7]) globally, while locally a significant material strength is to be expected.

3 Review of the DI Strength Estimation

The DI team supposedly calculated the (shear) strength σ_s by the formula

$$\sigma_s = \rho v^2$$

given ρ, the bulk density of the excavated material, and v, the minimum velocity of the dust ejecta. Here ρ may be 350 ± 120 kg m^3 (latest DI estimate, M. A'Hearn, priv. comm. 2006) or a 'worst case', sandlike 1500 kg m^3 [1]; the velocity is estimated from the fact that the ejecta cone remained attached to the surface throughout the encounter and even through the lookback imaging. Specifically one DI lookback image (no. 8, 45 min after impact, while the time of one hour after impact is employed in the calculations) is used to show that the ejecta curtain can be no higher than (500 ± 250) m above the surface. With again a 'worst-case' assumption on this, one obtains $v = 750/3600$ m s^{-1} and $\rho v^2 =$ 65 Pa. Note that with the nominal parameters, $\rho = 350 \pm 120$ kg m^{-3}, $v = (500 \pm 250)/2700$ m s^{-1}, one obtains $\sigma_s = 2 - -36$ Pa.

The derivation of the above formula is not given; apparently the strength has been regarded as an energy E per volume V and set equal to the kinetic energy of the same material volume:

$$\sigma = E/V = V\rho\frac{1}{2}v^2/V = \rho v^2/2$$

Note the factor 1/2 compared to the equation above. According to A'Hearn (priv. comm., 2006) this approach has been refined meanwhile (Richardson, J.E., Icarus, submitted 2006) leading to an improved shear strength estimate of (200 ± 100) Pa.

The ballisitc method used by the DI team to derive the shear strength at the surface of comet Tempel 1 neglects acceleration by gases and is, in our mind, therefore not conclusive (see below).

3.1 Discussion of DI Results

Due to the impact a non-negligible amount of gas (H_2O, CO_2, CO) has been released from an extended source [8] modifying the velocity distribution of the ejected dust particles. Note that while sunlit pure ice grains can survive in space for a long time, the dirty ice grains (albedo $\approx$ 0.1, [8]) in the ejecta cloud sublimate very quickly (order of minutes, [9, 10]). Gas dynamic (plus radiation pressure) dust/gas calculations [8, 11], constrained by OSIRIS measurements of ejecta velocities and gas mass show that the acceleration of dust by gas is a significant factor also during the first minutes after the impact; Keller et al. [8] derive a minimum dust velocity of 5 $\mathrm{m\,s^{-1}}$. Most of the ejecta do not follow ballistic trajectories, instead their trajectories are strongly influenced by the motion of the surrounding gas (reaching measured velocities of the order of 100–200 $\mathrm{m\,s^{-1}}$ after one hour; note that particles are also expected to be accelerated back to the surface). Nucleus density and strength could be constrained by the measurements of the DI spacecraft only if there is a dust component which is largely unaffected by gas drag.

Note that by the impact shock itself (before the material is actually excavated!) , the material is stressed (fractured) and its tensile strength is, thus, modified. Therefore, the pristine material properties can most likely not be determined with the applied method. Besides, it appears to be extremely model-dependent to infer quasi-static properties from supersonic impacts: Advocates of a possible strength-controlled cratering event during Deep Impact (Holsapple and Housen [12]) conclude that (while at least mentioning the role of gas pressure) any strength between 0 and 12 kPa is consistent with the DI observations. On the other hand, Keller et al. in following Housen et al. scaling laws [13, 14] and taking their derived minimum dust velocity arrive at a (unspecified) strength of about 5×10^4 Pa in order to have 10% of the ejecta reach 5 $\mathrm{m\,s^{-1}}$ or more.

Another issue is the vertical stratigraphy of a cometary surface, presently discussed, e.g., in [15]. Comets likely have surface crusts [16] that can withstand the gas pressure by their minimum strength given by van der Waals forces [17]. Different layers may have different strength. It is not clear of which layers the particles in the ejecta curtain seen by DI are representative in terms of velocity and density, but an excavation depth of a few meters is likely, while the relevant depth for landing a spacecraft is in the cm to dm-range (see below).

4 Other Methods to Estimate the Strength of Comets

Material strengths of lunar and Mars regolith has been measured several times by in situ and sample-return space missions (Lunokhod, Surveyor, Apollo, Viking, MER rovers); the soil cohesion is generally found in the 1 kPa and the compressive strength in the 10–100 kPa range (e.g., [18, 19, 20]).

The recent paper by Toth and Lisse [21] gives a good overview on strength estimates for cometary material. While most strength estimates rely on theoretical considerations (e.g., [22, 23]) and parameterizations, only very few observational results reliably constrain the strength of comets.

4.1 Breakup of Comets, Topography Observations

Tidal disruption of comets indicate low global tensile strengths in the order of 100–10,000 Pa. For example, the break-up of Shoemaker-Levy 9 during its perijove in 1992 set a rough upper limit of the tensile strength (on global/km scales!) of 100 Pa [7]. Note that this could correspond to a modified state of the comet caused by preshattering (cracking apart fragile bonds in a welded aggregate) SL9 by intercepted ring material as it crossed Jupiter's equatorial plane upon ingress. The tensile strength of sun-grazing comets has been estimated as 10 kPa or $> 100..1,000$ Pa [24], probably including some uncertainty due to thermal stresses. Images by Stardust from comet 81P/Wild-2 showed that the cometary surface must have a finite strength on short scales (< 100 m) to support the observed topographic features; because of the small gravity, some 10 Pa might suffice [25]. Otherwise, only lower bounds on the tensile strengths are available (see, e.g., Table 4 in [26]), in the order of 1..100 Pa.

4.2 Breakup of Meteoroids

Another source of information about possible strength values of cometary surfaces on mm-dm scales stems from the analysis of meteoroids associated with certain comets which enter the earth atmosphere at high speeds and finally break-up and create a light flash. Wetherill [27] gives values for tensile strengths of these fireballs ranging from 1 kPa to 1 MPa. More recently, Trigo-Rodríguez and Llorca [28] have studied a broad data base of meteor ablation light curves and arrive at tensile strengths between (400 ± 100) Pa and 40 kPa, clustering around 10 kPa for not too evolved and rather low density < 1 gcm^{-3} (if known) cometary meteoroids.[1]

4.3 Laboratory Measurements

The small scale (cm) shear and tensile strength of snow in the relevant density range of 300–500 kg m^{-3} is of the order of 10–100 kPa [6, 29]. The tensile strength of snow is nearly independent on temperature, while the compressive strength shows a remarkable increase with decreasing temperatures. Simulating possible cometary analogue material in the scope of the KOSI experiments, Jessberger and Kotthaus [17] conclude that the small-scale compressive strength of porous mixtures of crystalline ice and dust lies in the range between 30 kPa and 1 MPa with increasing strength for an increasing dust fraction. Bar-Nun et al.

[1] Note that the factor of 10 (1Pa$\equiv$ 10 dyn cm^{-2}) often leads to confusion in the units for strength.

[30] measure a limiting compressive strength of 20 kPa for their amorphous gas-laden 200 μm ice samples with a density of about 250 kg m^{-3} and derive a tensile strength of 2-4 kPa for that material.

Blum et al. [26] measured the tensile strength of highly porous (SiO_2 and diamond) dust aggregates produced by random ballistic deposition of individual μm-sized particles. They found values of 300 to 1100 Pa for very fluffy (packing density $\phi = 0.13..0.23$) and 2300 .. 2400 Pa for slightly compacted ($\phi = 0.4$) samples. With icy constituent grains, having higher inter-particle forces than SiO_2, they expect for cometesimals tensile strengths of $10^3...10^4$ Pa.

4.4 Limits Derived from Comet Size and Rotation

Stability against disruption due to rotation yields lower limits of a combination of bulk density and tensile strength [31]. Rotational periods and sizes for many comets are known, but bulk density is not well constrained, so until now this method does not give us a constraint on the strength of comets (e.g., fast rotating and big comet C/Hale-Bopp (1995 O1) could be a strengthless rubble pile with a bulk density as low as 100 kgm^{-3} – it probably is indeed 'damaged', according to [21]).

4.5 Theoretical Estimates, Conclusion

There are different approaches to describe the tensile strength of powders on the basis of van der Waals interactions, cf. Greenberg et al. [22] or Chokshi et al. [32], the latter model including the elastic deformation of contacting spherical grains. The theoretical tensile strength of fluffy aggregates depends on particle radii, contact areas, packing geometry and typically scales with the bulk density. Greenberg et al. estimate a tensile strength, for interstellar silicate dust/ice material with a density of 280 kgm^{-3}, of 270 Pa. Sirono and Greenberg [23] derive 300 Pa for the tensile and 6000 Pa for the compressive strength for a medium composed of ice grains linked into chains by intermolecular forces.

Kührt and Keller [16] derive a theoretical strength of 100 Pa and 100 kPa for grains of 1 mm and 1 μm, respectively. Note that 95% of the Deep Impact ejecta dust cross section is represented by particles $r < 1.4$ μm [8].

From the discussion above the conclusion can be drawn that the cometary surface on meter scales has a reasonable lower limit of the tensile strength of the order of 1 kPa whereas the probable upper limit can be taken as 100 kPa.

5 Penetration Mechanics

5.1 Compressive Strength in Soil Mechanics Theory

The theory of soil mechanics describes well the failure of geologic materials (soils – compressible or not compressible, with and without pore water) [33, 34].

A body penetrating a soil will always cause shear failure of the underlying and surrounding material. In soil mechanics, cohesion is the static part in shear strength; dynamically, there is a friction part and the total shear strength can be written as Mohr-Coulomb's law (equivalently, but more complex, by the Drucker-Prager criterion, see Holsapple and Michel [35]):

$$\sigma_s = c + F\tan(\phi)$$

with F: normal force, c: cohesion and $\tan(\phi)$: friction angle being material constants. These constants actually depend on the size of the bearing area and on the range of the normal force F. Now in bearing capacity theory, the following expression for the bearing strength (quasistatic compressive strength for e.g. circular or square loads) of a soil (if gravity can be neglected) is given as

$$\sigma_c = s_c c N_c \tag{1}$$

with the shape-factor s_c ($s_c = 1.3$ for circular discs, [33, 34]) and a bearing capacity factor $N_c = N_c(\phi)$. The latter factor is always > 5.7 and reaches values of about 30 for friction angles close to 40° (Fig. 1, [36]). Thus, the ratio between compressive and tensile strength is of the order of at least 10; this is also confirmed by the theory of fluffy aggregates for relative densities (1 – porosity) = $\psi \geq 0.34$ [23].

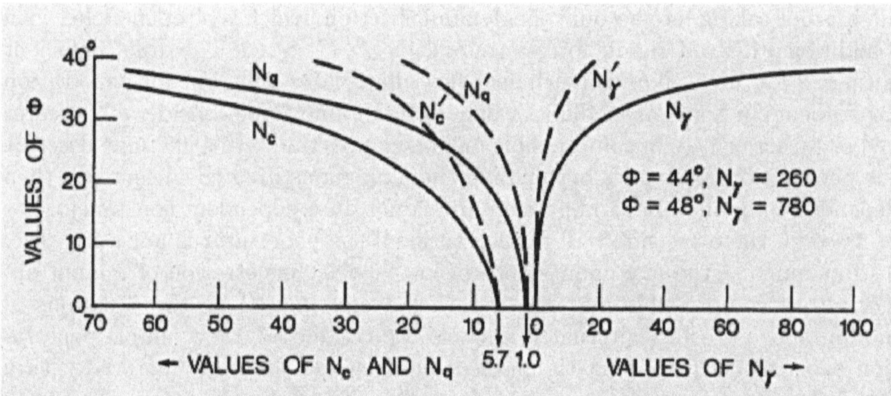

Fig. 1. Bearing strength coefficients after Wong [36]; for compressible soils, the dashed lines have to be used.

5.2 Provisional Equations of Motion

To estimate the landing penetration[2] of Philae, we use a provisional working model that includes breaking the compressive strength σ_c and accumulation

[2] $v \approx 1\,\mathrm{m\,s^{-1}} << c_{\mathrm{sound}} \approx 100\mathrm{m\,s^{-1}}$, [15].

of material in front of a flat penetrating surface (snow shovel effect). A simplification (negligible accumulation for not too low σ_c) is possible: final depth $y_e = v_0^2 m/2\sigma_c A$ (v_0: touchdown velocity; m_0: mass of lander; A: projected frontal area).

Otherwise, the penetration process is described by numerical integration of the equations of motion:

$$\begin{cases} \frac{dv(t)}{dt} = g - \tau\frac{A}{m(t)} - \frac{\rho A}{2m(t)}v(t)^2, \\ \frac{dm(t)}{dt} = \rho A v(t) \end{cases}$$

$$\begin{aligned} &v : \text{velocity } (v(0) = v_0) \\ &\sigma_c : \text{compressive strength} \\ &A : \text{projected frontal area} \\ &m = M + m' : \text{Mass of Lander } M \text{ plus accumulated mass } m' \\ &m'(0) = 0 \\ &\rho : \text{bulk density of comet surface material} \\ &g : \text{gravity, if not negligible} \end{aligned} \tag{2}$$

5.2.1 Improvement of Equations of Motion

While the above model may be a good starting point for low-velocity penetration into fluffy soil, it may well be underestimating the penetration resistance. First of all, it is not taking into account the dynamic friction, which is often modeled with a additional hydrodynamic force term $\propto C_d \times v^2$, $C_d >> 2$. The drag coefficient is $C_d = 2$ for 'ideal' liquids (turbulent flow, flat plate) only, but for penetration experiments in granular media its value is often found to be considerably greater [37] or rather $\propto 1/v$. In compressible media, compaction terms [38] may describe the penetration mechanics better than the ambiguous drag coefficient (C_d then depends on the material compaction and is not an independent constant).

Second, there are side wall friction terms if the penetrator is not a flat disc.

In summary, the maximum depth of low-velocity penetration of a blunt object into a compressible material is rather overestimated by the equations of motion above. To test alternative models, we conducted a few simple penetration experiments (dropping 10 cm-discs into regolith-analogue material with a very low cohesion).

5.3 Experimental Tests

In preliminary experiments we impacted three flat steel cylinders (mass = 0.6, 1.2, 1.8 kg; diameter 10 cm each) with a terminal velocity of 0.66–2.8 $\mathrm{m\,s^{-1}}$ (by varying the drop height) into 'Mars analog soil' (50% grounded olivine, 50% quartz sand, density 1400–1900 $\mathrm{kg\,m^{-3}}$). This material, in its mechanical properties, closely resembles the Mars regolith the MER rovers encountered and is normally used to test and improve the wheel design for Mars rovers. It is a very soft powder with a measured cohesion $c = 40$–400 Pa, $\phi \approx 25°$ from shear experiments on the surface; its predicted bearing capacity is $\sigma_c \approx 2000$ Pa for a few cm penetration. The depth of the material below the impact point was about 25 cm and the nearest wall of the 'sandbed' about 40 cm away. The material was stirred before each drop test. From the measured penetration depths and by inversion of the provisional equations of motion (Eq. (2)) an effective compressive strength in the 2–15 kPa range (higher than anticipated) was obtained, increasing with impact velocity v or penetration depth. This might be due to three effects:

- Influence of gravity and density (overburden terms) surpass cohesion term in compressive (bearing) strength
- Compaction of deeper levels provide much higher c than the measured surface layer
- Deviations of reality from provisional model (see Sect. 5.2)

Fig. 2. Test bodies and setup at DLR.

The preliminary experiments led to the following (not very surprising) conclusions: First of all, the reproducible preparation of a homogeneous, reproducible soil sample is very difficult. Second, the characterization of this soil is very difficult since c and ϕ are usually measured only at surface (mm-range); c can be vastly higher at a few cm depth due to unavoidable compaction in the Earth's gravity field ($1g$) and due to decreasing efficiency of mechanical decompaction (stirring) prior to the tests. Third, under $1g$, the compressive strength a few cm below the surface, according to soil mechanics theory, is significantly influenced by overburden terms in contrast to the simple relation Eq. (1); even without compaction, the effective compressive strength and penetration resistance then increases with depth (Fig. 1).

6 Landing Philae

In the following the Rosetta Lander, Philae, will be briefly explained, its landing strategy and anchoring philosophy. Rosetta is the third cornerstone mission of

ESAs science program. The Lander is provided by an international consortium [39, 40]. Philae has been designed to be particularly flexible in regard to various possible surface properties of the target comet; it was clear that reliable information on the nucleus of Churyumov Gerasimenko would only be available, when Rosetta reaches and observes the comet. (Actually, during the design phase it was still planned to go to comet Wirtanen.).

The nominal bulk density of 67P/Churyumov-Gerasimenko according to the analysis of Davidsson [41] is 100..370 $\mathrm{kg\,m^{-3}}$ (upper limit: 500..600 $\mathrm{kg/m^3}$). The local gravity on the surface is of the order of $3 \times 10^{-4}\mathrm{m\,s^{-2}}$.

6.1 The Mission, Technical Description

Philae is part of the International Rosetta Mission to comet Churyumov Gerasimenko, which was launched in March 2004 and will reach the target comet in spring 2014, after three swing-by manoeuvres at Earth and one at Mars. Rosetta will observe the comet from orbit with a set of remote sensing instruments, detect shape and mass of the nucleus and collect other data that will allow the selection of the preferred landing area, where Philae will be delivered at a heliocentric distance of about 3 AU, in November 2014.

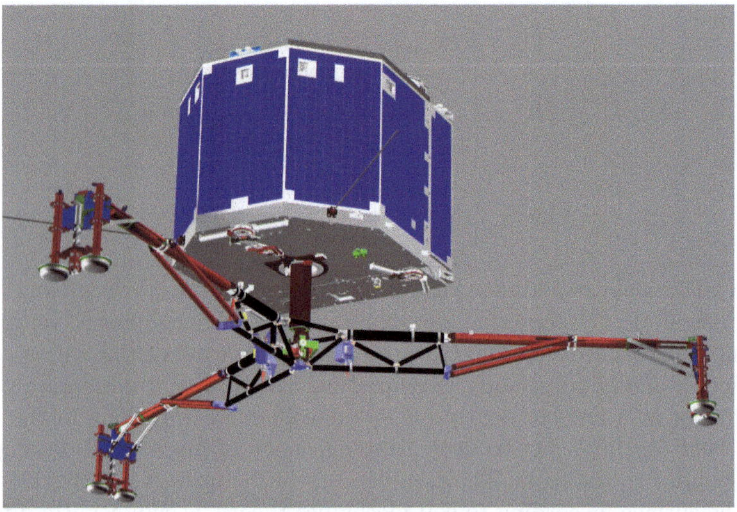

Fig. 3. Rosetta Lander, with extended Landing gear, in the configuration, prior to touch-down.

The Lander, which carries a payload of 10 scientific instruments, is powered with a primary battery (to guarantee a successful first scientific sequence) and a combination of a solar generator and rechargeable batteries for long term operations (during cruise it is connected via an umbilical with the Rosetta spacecraft). The Lander has an overall mass of 98 kg. It will communicate with Earth using the Rosetta Orbiter as a relay. Figure 3 shows a drawing of the Lander as seen from underneath, with the landing gear unfolded, as it could be seen from the surface of the comet during descent.

6.2 Delivery Strategy, Landing Scenario, Anchoring

Prior to the SDL sequence (separation–descend–landing), after a desirable landing site has been chosen, the Rosetta Orbiter needs to be brought into a dedicated delivery orbit. A decicated then will lead, after the separation, to a ballistic descent of Philae, with a touch-down vertical to the surface of the landing spot with the velocity vector parallel to the Lander z-axis (i.e. with the legs 'down'). The lander's vertical attitude during descend is maintained with a flywheel. Depending on the actual comet parameters (which can only be determined accurately after the arrival of Rosetta) a typical separation would take place in about 2 km altitude and the impact velocity at the surface would be somewhere between 0.5 and 1 $\mathrm{m\,s^{-1}}$. The separation velocity from the Orbiter can be pre-adjusted between 5 and 52 $\mathrm{cm\,s^{-1}}$.

The Lander will touch down on the landing gear, a tripod with a dedicated damping device in its central part to avoid re-bounce, and anchoring screws in its feet, which will avoid gliding on a potential slope as well as provide some anchoring force in the case of medium-hard surface material.

Immediately after touchdown a (redundant) anchoring harpoon is fired, which is connected to a tether (max. length 2.5 m), that will be tensioned (with pre-adjustable force, between 5 and 30 N) and secure fixation to the ground during all operations over the following weeks or months.

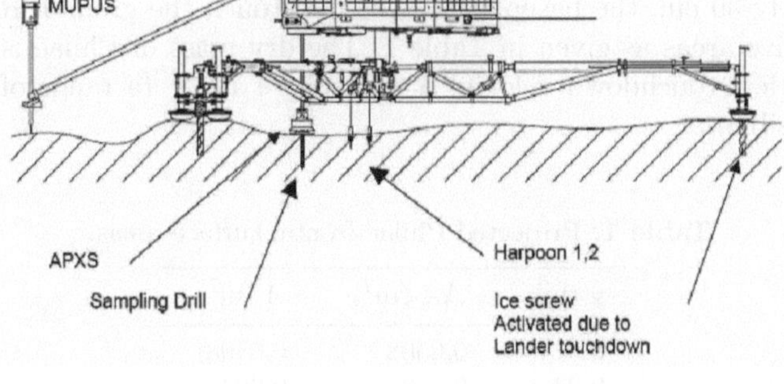

Fig. 4. Rosetta Lander side view after deployment of all units.

Figure 4 shows a drawing of the lander geometry from the side, after landing and deployment of all instruments.

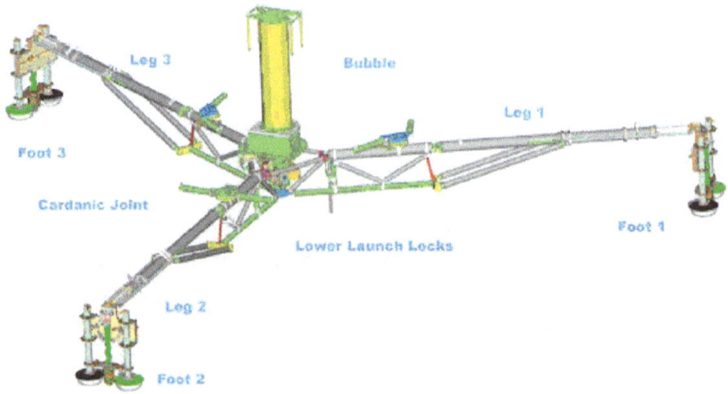

Fig. 5. Philae landing gear unfolded for comet surface phase.

6.3 Penetration Depth of Philae as a Function of Compressive Strength, σ_c

In order to evaluate the chance that Philae would penetrate unfavorably deep into the comet's surface material during touchdown, the provisional equations of motion (Eq. 2, see above), realistic estimates of the compressive strength of the comet's surface material and the geometry of Philae (Figs. 5 and 6, Table 1) have been considered.

Lander geometry, mass and velocity: Philae's projected frontal area A depends on the depth y: first the 2×3 feet plus the ice screws and their brackets penetrate (0.0508 m^2); after approximately 20 cm, the legs (0.153 m^2) and after approximately 30 cm, the baseplate (0.626 m^2) touch the ground. Relevant are the cumulative areas as given in Table 1. The dry mass of Philae is 97.588 kg, and the vertical touchdown velocity is typically 1 ms^{-1} (a range of 0.50 .. 1.2 ms^{-1} is possible).

Table 1. Projected Philae frontal surface areas.

y (m)	ΔA (m^2)	A (m^2)
0	0.0508	0.0508
0.23	0.153	0.204
0.30	0.626	0.830

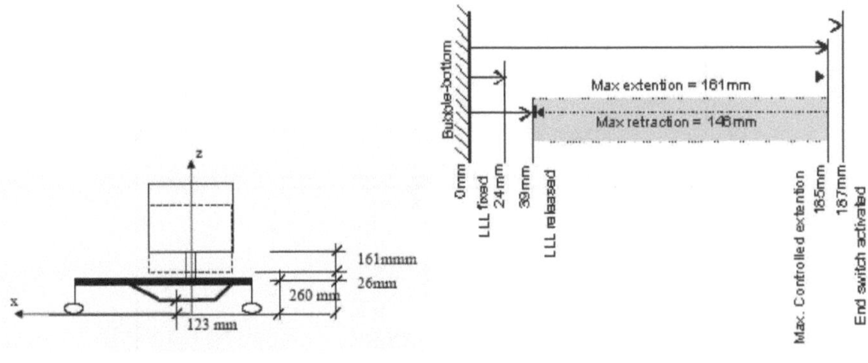

Fig. 6. Philae, schematic heights (partly adjustable).

Figure 7 shows how deep the Lander would penetrate as a function of the compressive strength σ_c of the surface material (touchdown velocity 1 m s^{-1}). Obviously, with σ_c greater than 4 kPa only the feet would penetrate; with σ_c smaller than 2 kPa the baseplate would touch and penetrate the surface. In no realistic case, the complete Lander would sink more than 30 cm into the ground. Note the bends in the curve where the effective frontal areas change. The jitter in the curve for very low compressive strengths is an artifact of the numerical integration routine.

In summary, for a compressive strength greater than 2..4 kPa the lander could operate nominally, while for a compressive strength less than 2 kPa, leading to Philae's baseplate touching the ground (but then effectively stopping further penetration), the 360° rotation capability of the landing gear would be compromised. Still, all experiments could be performed. Only for compressive strengths < 100 Pa (equivalent to tensile strengths of less than 5 ..10 Pa) the mission would be jeopardized.

6.4 Anchoring Force

A more difficult issue is to estimate the anchoring force with deployed harpoons. The harpoons are fired downwards with an initial velocity of ≈ 70 m s^{-1} and can penetrate up to 2.5 m into the soil. Upon retraction (by rewinding the tether), the harpoon's flukes unfold. Anchor tether tension can be set between 5 and 25 N in 8 steps.

The projected area of one harpoon with deployed flukes is $A_h = 2560$ mm^2. The circumferential length of this area is $s_h = 590$ mm. To estimated the minimum anchoring force $F_{\min}$, we can set $F_{\min} = A_h \sigma_c$ (5–25 N per harpoon for a compressive strength of 2–10 kPa at a 2.5 m maximum depth, d). The maximum anchoring force (shear failure along the penetration channel) $F_{\max}$ can be estimated as $F_{\max} = s_h d \sigma_s$ (about 1 kN at $d = 2.5$ m) The truth should lie somewhere in between these values; this needs to be investigated with a (finite element?) proper soil mechanical model.

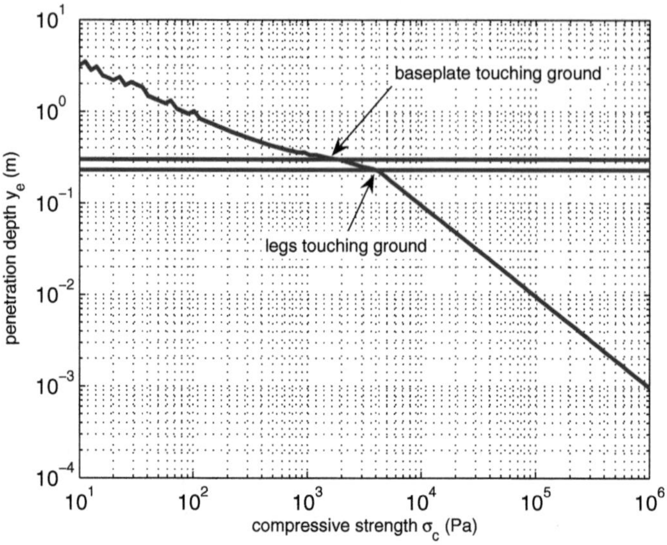

Fig. 7. Modeled penetration depths of Philae as a function of compressive strength.

7 Comet Strengths: Conclusion

We conclude that, unfortunately, neither DI nor other comet observations seem to provide yet firm data on the strength of cometary material. Theoretical considerations and laboratory measurements for weakly bound aggregates and the few observational constraints we have for comets and cometary meteoroids lead us to estimate the quasi-static tensile (or shear) strength of cometary material in the dm- to m-range as of the order of 1 kPa, while the compressive strength is estimated as of the order of 10 kPa. With these data, the penetration of Philae upon landing on comet Churyumov-Gerasimenko in 2014 should lead to a nominal scenario; it is indeed better to land on a soft (energy absorbing) surface than on a hard slab of ice.

References

1. A'Hearn, M.F. et al. (2005), Deep Impact: Excavating Comet Tempel 1, Science **310**, 258
2. Peplow, M., Comet reveals crumbly guts. Deep Impact results suggest Rosetta lander is in for a rough time, news@nature.com, published online, 8 September 2005, doi: 10.1038/news050905-15, http://www.nature.com/news/2005/050905/full/050905-15.html
3. Ai, H.-A., and Ahrens, T.J. (2004), Dynamic Tensile Strength of Terrestrial Rocks and Application to Impact Cratering, Meteorit. Planet. Sci., **39**, 233–246.

4. Stewart, S. T., and T. J. Ahrens (1999), Correction to the dynamic tensile 1606 strength of ice and ice-silicate mixtures (Lange and Ahrens 1983), Proc. 1607 Lunar Planet. Sci. Conf., 30th, abstract 2037
5. Xu, Y. (2005), Explanation of scaling phenomena based on fractal fragmentation, Mech. Res. Comm. **32**, 209
6. Petrovic, J.J. (2003), Mechanical properties of ice and snow, Review, J. Mat. Sci. **38** 1
7. Asphaug, E. and Benz, W. (1996), Size, Density, and Structure of Comet Shoemaker-Levy 9 inferred from the Physics of Tidal Breakup, Icarus **121**, 225
8. Keller, H.-U., et al. (2006), Observations of Comet 9P/Tempel 1 around the Deep Impact event by the OSIRIS cameras onboard Rosetta, Icarus (accepted)
9. Lichtenegger, H. I. M.; Komle, N. I. (1991), Heating and evaporation of icy particles in the vicinity of comets, Icarus **90**, 319–325
10. E.H. Beer, M. Podolak ., D. Prialnik (2006), The contribution of icy grains to the activity of comets I. Grain lifetime and distribution, Icarus **180**, 473
11. Jorda, L., Lamy, P., Faury, G. , Keller, H. U. , Hviid, S., Küppers, M., Koschny, D., Lecacheux, J., Gutierrez, P., Lara, L. M. (2006), Properties of the DEEP IMPACT Dust Cloud Icarus (accepted)
12. Holsapple, K.A., Housen, K.R. (2006), Gravity or Strength? An Interpretation of the Deep Impact Experiment. 37th Annual Lunar and Planetary Science Conference XXXVII, March 13–17, 2006, League City, Texas, abstract no.1068
13. Housen, K.R., Schmidt, R.M., Holsapple, K.A. (1983), Craterejecta scaling laws - Fundamental forms based on dimensional analysis. J. Geophys. Res. **88**, 2485
14. Housen, K.R. and Holsapple, K.A. (2003), Impact cratering on porous asteroids, Icarus **163**, 102
15. Belton, Michael J.S. et al., The Internal Structure of Jupiter Family Cometary Nuclei from Deep Impact Observations: The Talps or Layered Pile Model, Icarus: submitted June 1 2006; Revised: July 27, 2006
16. Kührt, E., Keller, H. U. (1994), The formation of cometary surface crusts, Icarus, **109**, 121–132
17. Jessberger, H. L.; Kotthaus, M. , Compressive strength of synthetic comet nucleus samples, in Hunt J., Guyeme T. D., eds, ESA SP-302, *Proceedings of the International Workshop on Physics and Mechanics of Cometary Materials*, ESA Publications Division, Noordwijk, p. 141
18. GH Heiken, DT Vaniman, MF Bevan (eds.) The Lunar Source Book, (Cambridge University Press 1991)
19. Kemurdzhian, A. L.; Gromov, V. V.; Shvarev, V. V. In: *Soviet progress in space studies: The second decade of space flight, 1967–1977.* (Moscow, Izdatel'stvo Nauka, 1978) p. 352–380. In Russian.
20. Richter, L.O., Inferences of Strength of Soil Deposits Along MER Rover Traverses, American Geophysical Union, Fall Meeting 2005, abstract #P21A-0134
21. Toth, I., Lisse, C.M. (2006), On the rotational breakup of cometary nuclei and centaurs, Icarus **181**, 162
22. Greenberg, J.M., Mizutani, H., Yamamoto, T., 1995. A new derivation of the tensile strength of cometary nuclei: Application to Comet Shoemaker-Levy 9. Astron. Astrophys. **295**, L35–L38 (1995)
23. Sirono, S.-i. and J. M. Greenberg (2000), Do Cometesimal Collisions Lead to Bound Rubble Piles or to Aggregates Held Together by Gravity? Icarus **145**, 230–238

24. Klinger J., Espinasse S., Schmidt B., 1989, in Hunt J., Guyeme T. D., eds, ESA SP-302, *Proceedings of the International Workshop on Physics and Mechanics of Cometary Materials.* ESA Publications Division, Noordwijk, p. 197
25. Brownlee, D.E. et al., Surface of young Jupiter family comet 81P/Wild 2: View from the Stardust spacecraft, Science **304**, 1764 (2004)
26. Blum, J., R. Schräpler, et al. (2006), The Physics of Protoplanetesimal Dust Agglomerates I. Mechanical Properties and Relations to Primitive Bodies in the Solar System, Astrophysical Journal, **652**, 1768–1781
27. Wetherill, G. W.; Revelle, D. O., Relationships between comets, large meteors, and meteorites, In: *Comets* (University of Arizona , Tucson, AZ, 1982) pp. 297–319.
28. Trigo-Rodríguez and Llorca, Jordi, The strength of cometary meteoroids: clues to the structure and evolution of comets. Mon. Not. R. Astron. Soc (2006, in press) and erratum (2006, in press)
29. Mellor, M. (1975): A Review of Basic Snow Mechanics, Snow Mechanics Symposium; Proceeding of the Grindelwald Symposium, Grindelwald, Bernese Oberland (Switzerland) April 1974: International Association of Hydrological Sciences Publication No. **114**, 251–291,
30. Bar-Nun, A., Pat-El, I., Laufer, D. (2006), Comparison between the findings of Deep Impact and our experimental results on large samples of gas-laden amorphous ice, Icarus, **191**, 562–566
31. Davidsson, B. J. R. (2001), Tidal Splitting and Rotational Breakup of Solid Biaxial Ellipsoids. Icarus **149**, 375–383
32. Chokshi, A., A. Tielens,G. G. M. et al. (1993), Dust coagulation, Astrophysical Journal Part 1 **407**(2), 806–819
33. Terzaghi, K.: *Theoretische Bodenmechanik*, (Springer, Berlin/Göttingen/New York, 1954)
34. Craig, R.F.: *Craig's Soil mechanics*, 7th ed., (Spon Press, London, New York 2003)
35. Holsapple, K.A. and Michel, P. (2006), Tidal disruptions: Acontinuum theory for solid bodies, Icarus **183**, 331–348
36. Wong, J.Y., *Terramechanics and off-road vehicles*, (Elsevier, Amsterdam 1989)
37. Kömle, N.I., Kargl, G., Ball, A.J., Lorenz, R.D. (eds.), *Penetrometry in the Solar System.* Proceedings of the International Workshop of Penetrometry in the Solar System, Graz, Austria, 1999. Verlag der Österreichischen Akademie der Wissenschaften, (Wien 2001)
38. Johnson, J. B., A physically based penetration equation for compressible materials. In: Kömle, N.I., Kargl, G., Ball, A.J., Lorenz, R.D. (eds.), Penetrometry in the Solar System. Proceedings of the International Workshop of Penetrometry in the Solar System, Graz, Austria, 1999. (Verlag der Österreichischen Akademie der Wissenschaften, Wien 2001), pp. 73–85
39. Biele, J. (2002), The Experiments onboard the ROSETTA Lander, Earth, Moon, and Planets, **90**, 445–458
40. Ulamec, S. et al. (2006), Rosetta Lander – Philae: Implications of an alternative mission, Acta Astronautica **58**, 435–441
41. Davidsson, B.J.R., Gutiérrez, P.J. (2005), Nucleus properties of Comet 67P/Churyumov-Gerasimenko estimated from non-gravitational force modeling, Icarus **176**, 453–477

Rosetta/ROSINA and Chemistry in a Cometary Coma

S. N. Delanoye[1] and J. De Keyser[1]

Belgian Institute for Space Aeronomy, Ringlaan 3, B-1180 Brussels, Belgium
sofie.delanoye@aeronomie.be, johan.dekeyser@aeronomie.be

1 Introduction

As comets are believed to consist of primordial material, including volatile components, they can reveal a lot of information about the origin of our Solar System. Therefore, the ESA Rosetta mission wants to gain an insight in these fascinating objects by studying comet 67P/Churyumov-Gerasimenko.

The Rosetta Orbiter Spectrometer for Ion and Neutral Analysis, ROSINA, part of Rosetta, will serve to determine the elemental, isotopic and molecular composition of the comet's atmosphere and ionosphere, the temperature and the bulk velocity of the gas and the reactions of the neutrals and the ions in the cometary atmosphere and ionosphere. To answer all these questions, ROSINA needs very specific characteristics: a very wide mass range, a very high mass resolution, a very wide dynamic range and a high sensitivity. No single instrument can handle all these aspects. So, ROSINA contains three sensors: a Cometary Pressure Sensor COPS, a Reflectron Time-Of-Flight mass spectrometer RTOF, and a Double Focusing Mass Spectrometer DFMS [1].

To interpret the future data and to obtain information on the variability of the cometary composition as a function of the distance to the Sun and to the nucleus, models of the chemistry in a cometary coma will be necessary.

2 Database and Chemical Modeling

Modeling the chemistry of a comet is extremely complex since a large number of species and reactions are involved. Moreover, such a model needs to be very flexible as it is often necessary to change the reactions and reaction rates. Therefore, the goal of our project is to develop a database of species, reactions and reaction rates.

2.1 Database

The software consists of two databases: a particle database and a reaction database. The particle database contains the list of the involved species, as can be seen in Fig. 1. Each species is represented by a unique graphic representation and for each species the mass and the charge is defined. Not only molecules and atoms are included, also other species important in chemical reactions such as electrons and photons are added to the particle database.

This list of species is repeated in the reaction database, shown in Fig. 2. All species that are present in the reactions included in the reaction database are shown in the list. The coma of a comet is a complex chemical environment. Not only different species are present, but also electrons, photons, solar wind ions, etc. can take part in chemical reactions. As a result, different types of reactions can take place, such as photoreactions, ion-neutral reactions, electron impact reactions, etc. All the included reaction types are shown in a separate list in the reaction database. Each reaction type in this list is characterized by a generic reaction equation, such as: electron impact dissociation: $\mathsf{X} + \mathsf{e}^- \rightarrow \mathsf{U} + \mathsf{V} + \mathsf{e}^-$.

The reaction type list is used to construct real reactions, shown in another column of the reaction database. The generic reaction equation is changed into a real reaction equation, by filling in the species. And for each reaction, information about the reaction rate is added. Also, additional information such as a reference to the literature can be added. From this list of real reactions, the reactions that should be used during modeling can be chosen and this selection also forms a separate part in the reaction database, including space for a more distinct view of a selected reaction.

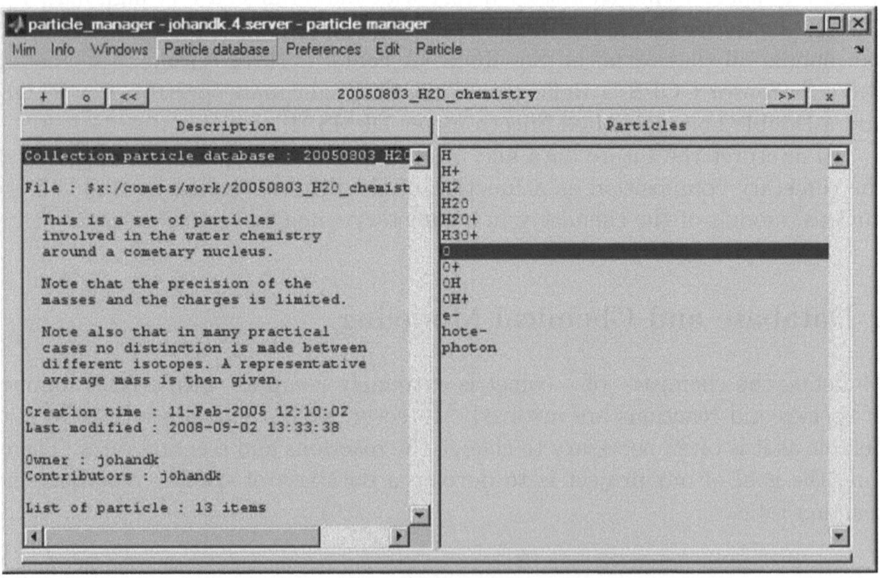

Fig. 1. Interface of the particle database.

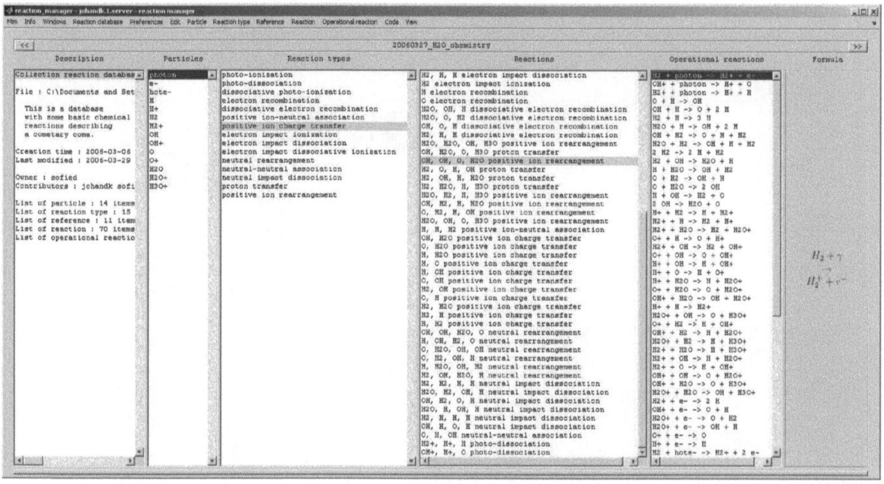

Fig. 2. Interface of the reaction database.

To enlarge the reaction database, existing databases from the literature, such as UMIST [2] and the ion-molecule reaction list of V. G. Anicich [3], have been added. As a result, the same reaction can appear different times, with reaction rates taken from different sources or valid in different temperature ranges. The database collects this information to get one fit of the reaction rate as a function of temperature. We use a smooth piecewise linear spline fit. Special attention is paid to the uncertainty estimates on the rate coefficients.

2.2 Modeling

Using the reaction database, the source-loss balance for each species in the selected reactions can be automatically generated. And these source-loss balances can be produced in C, Fortran and Matlab, so that they can be coupled to any other program to simulate the environment of a comet.

To test the particle and reaction database, we have also generated a simple model of a cometary coma. We assume a spherical nucleus with a constant gas production rate for a constant temperature and at a certain distance from the Sun. The expansion of the gas is also spherical. A constant density profile for the photons and a linear profile for the electrons is used. We limit the model to the inner coma, where no magnetic field is present. The model gives density profiles for each species as a function of the distance to the nucleus. Also, snapshot files are produced. One type of snapshot files shows what happens with all species at a certain point in the coma: which species are produced, which ones are destroyed, etc. The other type of snapshot files determines the important reactions for a certain species at a certain point in the coma. We have used this simplified model for a comet, made of only water. The snapshot file showing the different species at the outer border of the inner coma can be seen in Fig. 3.

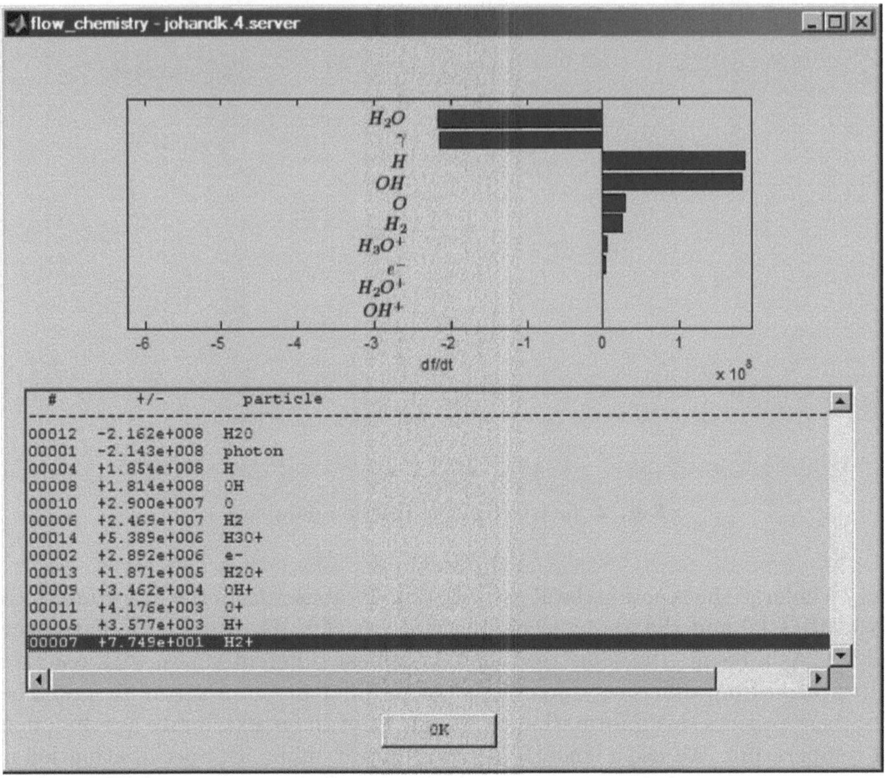

Fig. 3. Snapshot at the outer border of the inner coma, 3000 km in this case. The behavior of all the species can be seen.

3 Conclusion

The use of the reaction database can facilitate the modeling of the chemistry of a comet. It is a useful and flexible instrument, containing a large amount of information about the chemical behavior of species available in the cometary coma. Thanks to the simple generation of the source-loss balances, the database can be easily coupled to any simulation software. It can therefore be used to interpret the data obtained from cometary missions, such as Deep Impact and Rosetta. Moreover, the database can also be used in other applications. Any kind of reactions can be added, so the database can be adapted for the simulation of chemical reactions in environments, different from comets.

References

1. K. Altwegg: ROSINA's scientific perspective at comet Churyumov-Gerasimenko. In: *The New ROSETTA Targets*, ed by L. Colangeli et al. (Kluwer Academic Publishers, The Netherlands 2004) pp 257–269

2. Y. H. Le Teuff, T. J. Millar, A. J. Markwick: Astron. Astrophys. Suppl. Ser. **146**, 157 (2000)
3. V. G. Anicich: An Index of the Literature for Bimolecular Gas Phase Cation-Molecule Reaction Kinetics. JPL Publication 03-19, Pasadena, USA (November 2003)

High-Speed *R* -Band CCD Photometry of Comet 9P/Tempel 1

T. R. Mitchell, W. F. Welsh, and P. B. Etzel

San Diego State University, 5500 Campanile Dr. San Diego, Ca. 92182-1221, `tmitchel@sciences.sdsu.edu`

Abstract We present high speed *R*-band CCD photometry of comet 9P/Tempel 1 and the Deep Impact Event on 2005 July 4 UT. These data were obtained on the 1-meter telescope at Mount Laguna Observatory. On the night of the impact event we collected approximately two hours and fifty minutes of data. An exposure time of four seconds yielded a time resolution of 5.5 seconds and thus we are able to temporally resolve the rapid rise of the lightcurve due to the impact event. These data will be used in conjunction with data received from the flyby spacecraft, as well as other Earth-based observations, to give a better understanding of the evolution of the ejecta released by 9P/Tempel 1 as a result of the impact as well as information about the opacity of the dust cloud during and just after the impact event.

1 Observations

Observations were made on the 1-meter telescope at Mt. Laguna Observatory, located in Southern California, using a Kron *R*-band filter. The CCD, a 2048^2 Fairchild, was set up in a 2×2 binning mode with a subarray of 200×200 pixels, which enabled high-speed readout yielding a temporal resolution of 5.5 seconds. With this 2×2 binning the scale is 0.8" per pixel. The conditions at Mt. Laguna on July 4th were photometric with an average seeing of 2.8 arcseconds FWHM.

The observations began a little less than an hour before the impact occurred, and lasted for just under two hours post-impact. Observations were also made for an hour and a half on July 2 UT and for three hours on July 3 and 5 UT.

Due to the diffuse nature of the comet's coma and our short exposure times which produced significant noise in our field stars, we employed non-standard techniques in the calibration of our data. No sky annuli were used in subtracting out the sky brightness from the comet. Instead, the mode of the bottom quarter of each image (which did not contain the tail of the comet) was subtracted from the entire image. This eliminated any difficulties arising from the diffuse nature of the comet and its coma.

In correcting for extinction due to changes in airmass, we needed the extinction coefficient for the R bandpass at Mt. Laguna Observatory. We were able to empirically derive the extinction coefficient, $k = 0.078$ magnitudes/airmass, by plotting the airmass vs. magnitude for one of our comparison stars, that traversed the greatest change in airmass. The slope of this line was taken to be our extinction coefficient.

2 Discussion

The light curve shows a dramatic increase in brightness in the minutes following the impact, followed by a slow continual increase in brightness for the duration of our three hour observing window.

Looking through a 8.0" (4160 km) radius aperture, the data shows the first noticeable increase in brightness just before 5:52:21 UT. The increase in brightness is pseudo-linear for approximately 23 minutes, with a rate of change of 4.2% of the pre-impact flux per minute. The brightness continues to increase for the next 20 minutes at a rate of 1.1% of the pre-impact flux per minute. Although the increase in brightness continues for the duration of the observing window, it begins to level off 42 minutes after the impact with an overall increase in brightness of almost 2.5 times the pre-impact level as seen in Fig. 1.

Larger apertures show similar pseudo-linear increases in brightness. There is a less dramatic overall increase in brightness in the larger apertures, just over 1.5 times, but the brightness continues to increase throughout the duration of the observations. The smaller fractional increase in the larger apertures is most likely due to the inclusion of more of the coma in large apertures and lack of contrast between the dust already present in the coma and that excavated by the impact. In Fig. 2 the light curve of the impact there is no evidence of the

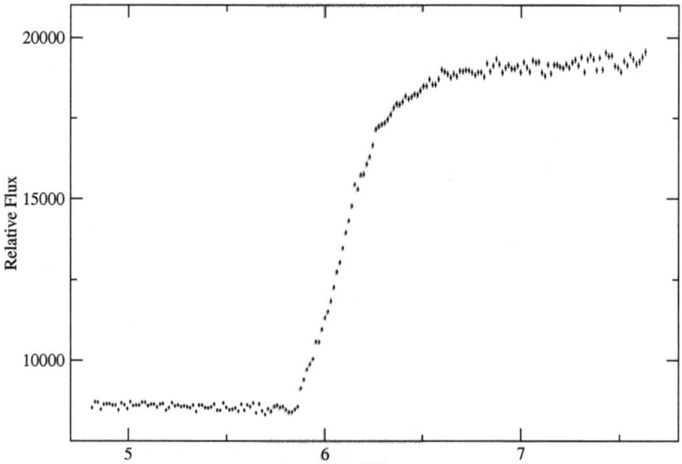

Fig. 1. 9P/Tempel 1; R-band light curve during the impact with 55 second temporal resolution, measured with a 10″ aperture.

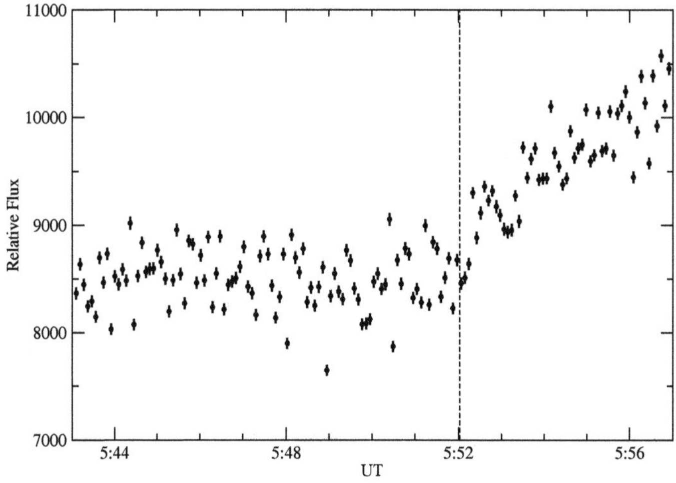

Fig. 2. 9P/Tempel 1; time of impact shown with dashed line.

initial flash at the time of impact. The coma was either too bright or the flash too dim to be detected in our observations.

Initial calculations of $Af\rho$ [1], a proxy for relative dust production, indicate that it increased by a factor of 2.5 times from its pre-impact value. Indicating the amount of dust produced as a result of the impact is more than twice that at quiescence.

The last 20 minutes of our data must be treated with caution due to airmass values in excess of six. Our final images were taken through an airmass greater than eight and the FWHM of our comparison stars increased substantially, to an average of 4", at the end of the evening.

Acknowledgements We would like to give special thanks to Robert W. Leach for his efforts in making the high-speed readout possible.

This research was enabled by the High Performance Wireless Research and Education Network (HPWREN) managed by Hans-Werner Braun and Frank Vernon of the University of California, San Diego (UCSD) with support from National Science Foundation Grant Numbers 0087344 and 0426879.

Reference

1. M. F. A'Hearn, D. G. Schleicher, P. D. Feldman et al. (1984): Astron. J. **89**, 579

How Tempel 1 Fits into the Ensemble of Comets: A Spectrosopic Perspective

U. Fink

Lunar and Planetary Laboratory, University of Arizona, Tucson Az, 85721

1 "The Whole Text"

During the period 1985–2004 we have obtained spectra of a sizable number of comets using our CCD spectrograph at the 61″ Catalina Observatory Telescope near Tucson. I am in the midst of re-reducing all of these observations on a consistent basis. This work is not yet complete and thus I stress that the present contribution is a preliminary summary of the results obtained so far. A number of comet spectra require additional review and a fair amount of reduction and analysis remains to be done. Data for about 92 comets are being reduced, most of them with multiple observing dates. The spectral range observed is ≈ 5200 – –10200 Å, which encompasses emissions by C_2, NH_2, O I (1D), and CN. About 50 comets show the whole suite of emissions. Data for 39 of these comets, observed from 1985 to 1993 April have been published earlier [4]. A number of additional comets were added in the dissertation of Michael Hicks [5]. The three comets 2P/Enke, 73P/SW3, and 81P/Wild2 were observed at two apparitions. Essentially no differences in their production rate ratios were found. A number of bright comets or important spacecraft targets were observed over an extended interval: 1P/ Halley, C/1996 B2 Hyakutake, 46P/ Wirtanen, 81P/ Wild2, Hale-Bopp, and 43P/ Wolf/Harrington.

Four taxonomic groups of comets were found:
(1) Normal (Halley type) 70%
(2) Low C_2, normal NH_2 (Tempel 1 type) 22%
(3) Low C_2 and low NH_2 (G-Z type) 6%
(4) Low C_2, low CN, normal NH_2 (Yanaka type) 2%

We point out that the group of low C_2 comets was also found by A'Hearn et al. [1]. However, they could not distinguish between normal NH_2 abundance comets and those with low NH_2 since they do not observe the NH_2 molecule. There is in addition the odd comet Yanaka (1988r) which displayed normal NH_2 emissions but showed no detectable C_2 or CN emissions. Our new reductions incorporate a number of differences from our earlier published results. First the g factors for NH_2 were corrected to account for the presence of even and odd bands [2]. This essentially doubled the production rates of NH_2. Secondly the Haser model corrections were changed to better account for the fraction of emissions lost by using the edges of our slit for sky subtraction. For comets with low C_2 abundances a substantial contamination by NH_2 occurs in the C_2 band complex that we use ($v = -1$), for which a correction needs to be applied. Finally we reduced all of our data using both the scale lengths that we determined from comet 1P/Halley [3], and those used by A'Hearn in their 1995 summary paper on production rates of comets from photometric observations [1]. This allows a straightforward comparison between the two data sets. A summary of the reduction parameters used is shown in Table 1.

1. $$Q(s^{-1}) = \frac{L(\text{photons}^{-1}) \text{ x } v\ (\text{km s}^{-1})}{g \text{ x } \Lambda_d\ (\text{km})} = \frac{L \text{ x } 1(\text{km s}^{-1})}{g_0 \text{ x } \Lambda_{d0}}$$

2. $Q_{H2O}\ (s^{-1}) = 16.0\ L_{OI}$ (photons s^{-1})

3. **Scale Lengths used:**

Species	Fink et al. (1991,1996) Parent km	Daughter km	A'Hearn et. l (1995) Parent km	Daughter km	Typ. Prod Rate Ratios
C_2	58000	58000	22000	66000	2.57
NH_2	4900	62000	--	--	--
H_2O (OI,OH)	80000	--	24000	160000	3.18
CN	28000	320000	13000	210000	2.05

4. **g- Factors used at 1AU (g_0); $g=g_0/r^2$**

Species	Band	10^3 Photons s^{-1}
C_2	Δv=-1	57
NH_2	0,10,0	2.65
	0,8,0	2.47

Table 1. Summary of reduction parameters used.

We will now show how Tempel 1 fits into our large spectroscopically observed groups of comets. We will show that Temple 1 clearly falls into the category of low C_2 and normal NH_2 comets. We observed Tempel 1 on four dates during its 1994 apparition. A summary of our observations is given in Table 2.

The production rates in the above table are in units of 10^{25} molecules s^{-1}. The ratios are expressed in parts per thousand with respect to water. Standard deviations are shown to get a feeling for the internal scatter of the data. The production rate ratios for those dates were weighted according to their SNR and quality of each of the observations and an average production rate ratio with respect to water was determined. (This is shown in Table 3 below and is plotted in Fig. 2.) Examples of our spectra are given in Fig. 1 above.

All spectra are ratios of the comet divided by solar analog stars matched in airmass. The spectra are plotted using the original CCD columns. They go roughly from 5200 – –10200 Å. The large noise excursions at both ends of the spectra are caused by the signal of the comet and comparison star going to zero.

9P/Tempel1

Run/Date	r AU	delta AU	Aperture arc-sec	Indv Reductions	Production Rates E25 mol/s C2	NH2	H2O	CN	Afr cm
94BMar17	1.83	0.89	31.5x5.0	Red 1 40 m	0.91	3.61	1510	1.82	281.0
			Rat. Wrt H2O		**0.60**	**2.39**	**1000.00**	**1.21**	**186.09**
94CApr12	1.70	0.73	31.5x5.0	116 min	0.99	5.56	1710	2.56	297.0
			31.5x5.0	116 min FF	0.94	5.00	1510	3.43	
			61.5x5.0	116 min	0.81	4.77	1950	2.93	292.0
			61.5x5.0	116 min FF	0.31	4.91	1610	3.14	
			Average (4)		**0.76**	**5.06**	**1695.00**	**3.02**	**294.50**
			Std Dev.		0.31	0.35	188.59	0.37	3.54
			Fract. Dev.		0.41	0.07	0.11	0.12	0.01
			Rat. Wrt H2O		**0.45**	**2.99**	**1000.00**	**1.78**	**173.75**
94DJun13	1.51	0.77	31.5x5.0	50 min	2.10	4.48	1670	2.32	235.0
			31.5x5.0	50 min FF	1.33	4.50	1720	2.05	
			61.5x5.0	50 min	1.35	4.95	2810	1.97	243.0
			61.5x5.0	50 min FF	1.14	4.20	1940	2.17	
			Average (4)		**1.48**	**4.53**	**2035.00**	**2.13**	**239.00**
			Std Dev.		0.42	0.31	529.81	0.15	5.66
			Fract. Dev.		0.29	0.07	0.26	0.07	0.02
			Rat. Wrt H2O		**0.73**	**2.23**	**1000.00**	**1.05**	**117.44**
94DJun13	1.51	0.77	31.5x5.0	70 min Early Red.	1.92	5.05	1710.00	2.64	240.00
			Rat. Wrt H2O		**1.12**	**2.95**	**1000.00**	**1.54**	**140.35**
94DJun14	1.51	0.77	31.5x5.0	20 min Early Red.	1.71	3.17	691.00	1.67	221.00
			Rat. Wrt H2O		**2.47**	**4.59**	**1000.00**	**2.42**	**319.83**

Table 2. Summary of our Tempel 1 observations.

The S-shaped feature in the middle is due to imperfect wavelength registration of the telluric O_2 A-band. The smooth line under the middle spectrum shows how we fit a continuum to extract the emission intensities. Emissions can be seen from C_2, NH_2, atomic oxygen (O 1D), and CN. It is evident from the spectra that C_2 emission is very week giving occular proof that this is a C_2 poor comet.

In Fig. 2, we plot the log of the ratios of C_2/H_2O versus log of the ratio NH_2/H_2O. The comets with normal C_2/H_2O ratios, what one might call the Halley type, are shown by solid diamonds. Below this is the group of C_2 deficient comets indicated by open squares. This group is in turn divided into two groups with the group of three comets at the left also being deficient in NH_2. Comet Tempel 1 is drawn as a solid square and falls in the middle of the group that is C_2 deficient but has "normal" NH_2 abundance. The solid lines are least squares fits to the data and are drawn on the figure mostly as a guide to the different groups. Not much significance should be attached to their slopes since there is a

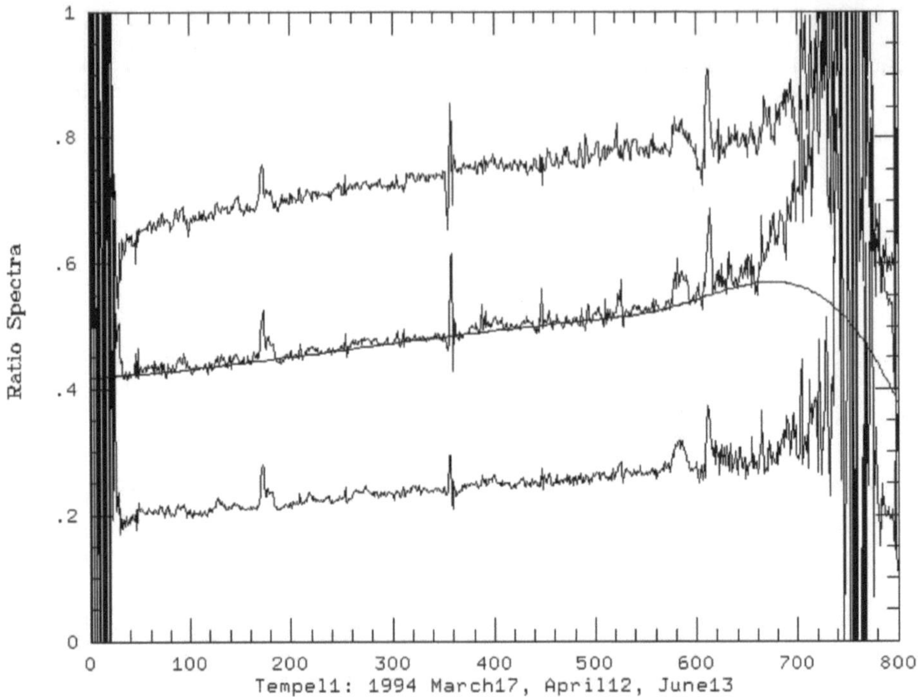

Fig. 1. Examples of our spectra of Tempel 1.

fair amount of scatter among the data points. It is also not entirely clear that the low C_2 comets form a distinct separate group rather having than a continuum between them and the comets with 'normal' C_2.

The following table gives a preliminary summary of the data used for the plot in Fig. 2. Column1 gives the name of the comet. Columns 2,4,6,8 give the abundance ratios with respect to water, while columns 3,5,7 and 9 give the log10 of these numbers. The log10 ratios are listed to allow ready comparison with the photometric production rates of A'Hearn et al. [1] who use a log10 scale to report their production rate ratios with respect to OH. (To convert OH to an H_2O production rate a factor of 0.85 for OH/H_2O can be used). A "p" in column 10 indicates that the comet is periodic. As can be seen from the sporadic distribution of the periodic comets there appears to be no correlation of dynamical groups with taxonomic type. As mentioned earlier A'Hearn et al. (1995) used different scale lengths from those employed by us. All of our data were reduced using both our scale lengths and those of A'Hearn et al. [1]; (their scale lengths are listed in Table 1). In Table 3 production rates using those scale lengths are listed in columns 11–19 and are labeled MAH. In comparing the production rates calculated using our and MAH scale lengths, I was surprised that there was little significant departure of these ratios across the whole gamut of our comets observed, although they encompass a wide range of essentially random

Comet	Ratios wrt H2O (per thousand)											Using MAH scale lenghts						MAH
	C2	log	NH2	log	CN	log	Afr	log		C2/CN	log	C2	log	CN	log	C2/CN	log	1995
Halley type (typical C2, CN , and NH2)																		
2P/Encke Av.	3.81	-2.42	1.08	-2.98	2.20	-2.66	10.86	-26.97	p	1.73	0.24	4.73	-2.33	3.41	-2.47	1.39	0.14	
P/Bradfield	3.74	-2.43	1.85	-2.73	1.13	-2.95	50.14	-26.30	p	3.33	0.52	4.64	-2.33	1.74	-2.76	2.66	0.42	
Mueller (1991h1)	3.68	-2.43	1.00	-3.00	1.18	-2.93	13.11	-26.88		3.12	0.49	4.56	-2.34	1.83	-2.74	2.50	0.40	
Hale-Bopp	3.58	-2.45	2.95	-2.53	1.69	-2.77	692.11	-25.16		2.12	0.33	4.44	-2.35	2.61	-2.58	1.70	0.23	
141P/Machholz 2	3.37	-2.47	1.50	-2.82	1.11	-2.95	19.10	-26.72		3.04	0.48	4.18	-2.38	1.72	-2.76	2.43	0.39	
1P/Halley Av pre/post	3.20	-2.50	2.69	-2.57	1.68	-2.78	157.38	-25.81	p	1.91	0.28	3.97	-2.40	2.60	-2.58	1.53	0.18	
C/Hartley-Good	3.13	-2.50	2.37	-2.63	1.47	-2.83	46.70	-26.33		2.13	0.33	3.88	-2.41	2.28	-2.64	1.70	0.23	
C/1995 Y1 Hyakutake	2.99	-2.52	1.43	-2.84	1.38	-2.86	18.10	-26.74		2.16	0.33	3.70	-2.43	2.14	-2.67	1.73	0.24	
Hyakutake 1996 B2	2.60	-2.59	2.89	-2.54	1.68	-2.78	229.33	-25.64		1.55	0.19	3.22	-2.49	2.60	-2.59	1.24	0.09	
103P/Hartley2	2.59	-2.59	3.44	-2.46	1.96	-2.71	68.49	-26.16	p	1.32	0.12	3.22	-2.49	3.03	-2.52	1.06	0.03	
C/Helin-Roman-Alu	2.58	-2.59	2.69	-2.57	1.99	-2.70	45.23	-26.34		1.30	0.11	3.20	-2.49	3.08	-2.51	1.04	0.02	
C/Yanaka (1989a)	2.56	-2.59	1.10	-2.96	3.42	-2.47	50.56	-26.30		0.75	-0.13	3.17	-2.50	5.30	-2.28	0.60	-0.22	
109P/Swift-Tuttle	2.47	-2.61	1.01	-2.99	1.24	-2.91	120.52	-25.92	p	1.99	0.30	3.06	-2.51	1.93	-2.72	1.59	0.20	
46P/Wirtanen	2.33	-2.63	2.44	-2.61	1.44	-2.84	59.07	-26.23	p	1.62	0.21	2.89	-2.54	2.23	-2.65	1.29	0.11	
C/Neat C/2001 Q4	2.33	-2.63	2.22	-2.65	1.40	-2.85	131.33	-25.88		1.66	0.22	2.88	-2.54	2.18	-2.66	1.33	0.12	
C/Levy (1990c)	2.32	-2.63	1.91	-2.72	2.43	-2.61	204.09	-25.69		0.96	-0.02	2.88	-2.54	3.76	-2.42	0.77	-0.12	
62P/Tsushinshan 1	2.30	-2.64	2.13	-2.67	2.02	-2.69	52.43	-26.28	p	1.14	0.06	2.86	-2.54	3.13	-2.50	0.91	-0.04	
C/Austin	2.27	-2.64	1.78	-2.75	1.75	-2.76	20.55	-26.69		1.30	0.11	2.82	-2.55	2.71	-2.57	1.04	0.02	
P/McNaught-Russel	2.18	-2.66	1.78	-2.75	1.43	-2.85	22.07	-26.66	p	1.53	0.18	2.70	-2.57	2.21	-2.65	1.22	0.09	
23P/Brorsen-Metcalf	2.10	-2.68	1.02	-2.99	1.05	-2.98	7.34	-27.13	p	2.00	0.30	2.61	-2.58	1.63	-2.79	1.60	0.20	
122P/deVico	2.04	-2.69	1.28	-2.89	0.82	-3.08	15.75	-26.80	p	2.48	0.39	2.53	-2.60	1.28	-2.89	1.98	0.30	
Zan-Brew	2.00	-2.70	1.27	-2.90	1.19	-2.92	24.14	-26.62		1.68	0.23	2.48	-2.61	1.84	-2.73	1.34	0.13	
7P/Pons-Winnecke	1.97	-2.71	2.24	-2.65	1.02	-2.99	52.93	-26.28	p	1.92	0.28	2.44	-2.61	1.59	-2.80	1.54	0.19	
10P/Tempel2	1.96	-2.71	1.92	-2.72	1.15	-2.94	34.40	-26.46	p	1.70	0.23	2.43	-2.61	1.78	-2.75	1.36	0.13	
58P/Jackson-Neujmin	1.90	-2.72	4.36	-2.36	2.66	-2.57	60.82	-26.22	p	0.71	-0.15	2.35	-2.63	4.13	-2.38	0.57	-0.24	
Mueller (1993p)	1.88	-2.73	1.30	-2.89	0.81	-3.09	219.20	-25.66		2.33	0.37	2.33	-2.63	1.25	-2.90	1.86	0.27	
55P/Tempel-Tuttle	1.88	-2.73	2.43	-2.62	1.30	-2.89	25.45	-26.59	p	1.45	0.16	2.33	-2.63	2.01	-2.70	1.16	0.06	
C/Stonehouse 1998H1	1.70	-2.77	2.53	-2.60	2.32	-2.64	137.41	-25.86		0.74	-0.13	2.11	-2.68	3.59	-2.44	0.59	-0.23	
93P/Lovas 1	1.57	-2.80	1.30	-2.88	1.91	-2.72	144.12	-25.84	p	0.82	-0.09	1.94	-2.71	2.96	-2.53	0.66	-0.18	
24P/Schaumasse	1.53	-2.82	1.85	-2.73	1.26	-2.90	31.74	-26.50	p	1.21	0.08	1.90	-2.72	1.95	-2.71	0.97	-0.01	
1996 B1 Szczepanski	1.49	-2.83	2.52	-2.60	1.71	-2.77	44.28	-26.35		0.87	-0.06	1.85	-2.73	2.65	-2.58	0.70	-0.16	
Shoe-Levy (1991a)	1.42	-2.85	1.80	-2.74	0.95	-3.02	89.30	-26.05		1.50	0.17	1.76	-2.76	1.47	-2.83	1.20	0.08	
80P/Peters-Hartley	1.37	-2.86	1.93	-2.71	1.34	-2.87	31.93	-26.50	p	1.03	0.01	1.70	-2.77	2.07	-2.68	0.82	-0.09	
6P/d'Arrest	1.27	-2.90	2.37	-2.63	1.10	-2.96	45.88	-26.34	p	1.15	0.06	1.57	-2.80	1.71	-2.77	0.92	-0.03	
22P/Kopff	1.24	-2.91	2.41	-2.62	1.88	-2.73	95.45	-26.02	p	0.66	-0.18	1.53	-2.81	2.92	-2.54	0.53	-0.28	
Number	**35.00**			(~70%)														
Average	**2.32**	-2.65	**2.02**	-2.72	**1.57**	-2.83	**87.75**	-26.28		**1.63**	0.17	2.88	-2.54	2.44	-2.61	1.18	0.07	0.06
Median	**2.27**	-2.64	**1.92**	-2.72	**1.43**	-2.85	**50.14**	-26.30		**1.55**	0.19	2.82	-2.55	2.21	-2.65	1.27	0.10	
Std.Dev	0.73	0.14	0.75	0.16	0.56	0.14	121.57	0.43		0.68	0.19	0.91	-3.04	0.87	-3.06	0.55	0.19	0.10
Max	3.81	-2.42	4.36	-2.36	3.42	-2.47	692.11	-25.16		3.33	0.52	4.73	-2.33	5.30	-2.28	0.89	0.42	0.29
Min	1.24	-2.91	1.00	-3.00	0.81	-3.09	7.34	-27.13		0.66	-0.18	1.53	-2.81	1.25	-2.90	1.22	-0.28	-0.09
Tempel1 type (low C2 <1.0; normal NH2)																		**<-0.18**
45P/Honda-Mrk-Pajd	0.94	-3.03	1.80	-2.74	1.64	-2.78	34.54	-26.46	p	0.57	-0.24	1.17	-2.93	2.55	-2.59	0.46	-0.34	
Naka-Nishi-Machlz	0.90	-3.05	2.69	-2.57	2.15	-2.67	80.45	-26.09		0.42	-0.38	1.11	-2.95	3.34	-2.48	0.33	-0.48	
81P/Wild2	0.84	-3.08	3.03	-2.52	2.83	-2.55	300.00	-25.52	p	0.30	-0.53	1.04	-2.98	4.39	-2.36	0.24	-0.62	
C/1997 T1 Utsunomia	0.79	-3.11	1.88	-2.72	1.24	-2.91	152.43	-25.82		0.63	-0.20	0.97	-3.01	1.92	-2.72	0.51	-0.29	
Machholz (1994r)	0.72	-3.14	1.81	-2.74	2.69	-2.57	275.48	-25.56		0.27	-0.57	0.89	-3.05	4.18	-2.38	0.21	-0.67	
Takamizawa (1994i)	0.69	-3.16	2.16	-2.67	2.38	-2.62	205.36	-25.69		0.29	-0.54	0.86	-3.07	3.69	-2.43	0.23	-0.63	
9P/Tempel1	0.65	-3.19	2.57	-2.59	1.36	-2.87	149.68	-25.82	p	0.48	-0.32	0.80	-3.09	2.11	-2.68	0.38	-0.42	
104P/Kowal2	0.64	-3.20	2.17	-2.66	1.91	-2.72	72.19	-26.14	p	0.33	-0.48	0.79	-3.10	2.96	-2.53	0.27	-0.57	
67P/Chury-Gera	0.60	-3.22	1.86	-2.73	1.52	-2.82	280.20	-25.55	p	0.40	-0.40	0.75	-3.13	2.36	-2.63	0.32	-0.50	
71P/Clark	0.52	-3.28	2.05	-2.69	1.14	-2.94	140.88	-25.85	p	0.46	-0.34	0.65	-3.19	1.76	-2.75	0.37	-0.43	
19P/Borelly ??	0.23	-3.63	2.21	-2.66	1.63	-2.79	194.04	-25.71	p	0.14	-0.84	0.29	-3.54	2.53	-2.60	0.11	-0.94	
Number	**11.00**			(~22%)														
Average	**0.68**	-3.19	**2.20**	-2.66	**1.86**	-2.75	**171.39**	-25.84		**0.39**	-0.44	0.85	-3.07	2.89	-2.54	0.29	-0.53	-0.61
Median	**0.69**	-3.16	**2.16**	-2.67	**1.64**	-2.78	**152.43**	-25.82		**0.40**	-0.40	0.86	-3.07	2.55	-2.59	0.34	-0.47	
Std.Dev	0.20	0.17	0.40	0.08	0.58	0.13	89.15	0.29		0.14	0.18	0.24	-3.61	0.90	-3.05	0.11	0.18	0.35
Max	0.94	-3.03	3.03	-2.52	2.83	-2.55	300.00	-25.52		0.63	-0.20	1.17	-2.93	4.39	-2.36	0.27	-0.29	-0.21
Min	0.23	-3.63	1.80	-2.74	1.14	-2.94	34.54	-26.46		0.14	-0.84	0.29	-3.54	1.76	-2.75	0.16	-0.94	-1.22
Giacobini-Zinner Type (low C2 and low NH2)				(~6%)														
21P/Giacobini-Zinner	0.73	-3.13	0.38	-3.42	1.66	-2.78	92.80	-26.03	p			0.91	-3.04	2.58	-2.59	0.35	-0.45	
73P/Schwass-Wach 3 Av	0.75	-3.15	0.61	-3.23	2.53	-2.60	185.58	-25.75	p			0.92	-3.03	3.92	-2.41	0.24	-0.63	
43P/Wolf/Harrington	0.98	-3.01	0.52	-3.29	3.14	-2.50	439.87	-25.36	p			1.22	-2.92	4.87	-2.31	0.25	-0.60	
C/Yanaka type (lowC2, low CN, normalNH2)				(~2%?)														
C/Yanaka (1988r)	0.17	-3.78	1.48	-2.83	0.17	-3.77	29.77	-26.53				0.21	-3.69	0.26	-3.58	0.79	-0.10	

Table 3. Production rates and taxonomic groups of 50 comets observed spectroscopically.

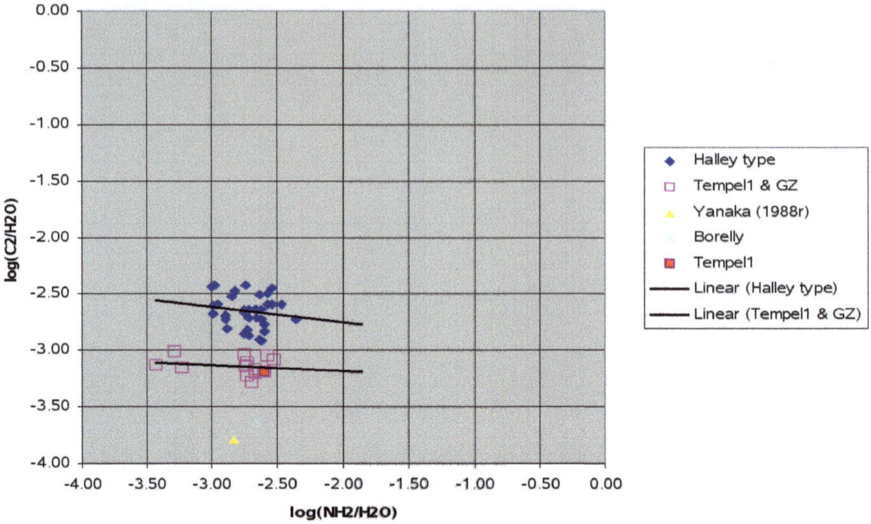

Fig. 2. Log of the ratios of C_2/H_2O versus log of the ratio NH_2/H_2O.

heliocentric and geocentric distances. The average ratios, listed in Table 1, are 2.57, 3.18, and 2.05 respectively for production rates of C_2, H_2O and CN, determined using our scale lengths versus those of A'Hearn et al. (1995). We give in column 19 the average log (C_2/CN) ratio of 0.06 ± 0.10 reported by A'Hearn et al. (1995) for the Halley type comets. We get an average of 0.07 ± 0.19. For the carbon-depleted comets A'Hearn et al. (1995) gives an upper limit of < -0.18 with an average of -0.61 ± 0.35 while we get an average of -0.53 ± 0.18. We consider this excellent agreement between the two data sets once they are both analyzed using the same scale lengths. A'Hearn et al. (1995) do not observe NH_2 and so no comparison with this group of comets nor the unusual object Yanaka (1988r) can be made.

References

1. A'Hearn Michael .F., Millis R., Schleicher D., Osip D., Birch P., 1995. The ensemble properties of comets: results from narrowband photometry of 85 comets, 1976–1992. Icarus 118, 223–270.
2. Arpigny Claude, 1994 Physical chemistry of comets: Models, uncertainties, data needs. AIP Conf. Proc. 312, 205–238.
3. Fink Uwe, Combi M., and DiSanti M. 1991. Comet P/Halley spatial distribution and scale lengths for C_2,CN,NH_2 and H_2O Astrophys. J. 383, 356–371.
4. Fink Uwe and Hicks M. 1996. A survey of 39 comets using CCD Spectroscopy Astrophys. J. 459, 729–743.
5. Hicks (1997)

Dust Evolution of Comet 9P/Tempel 1

J. Pittichová[1,3], Y. Fernández[2] and K. J. Meech[1]

[1] Institute for Astronomy, University of Hawaii jana@ifa.hawaii.edu
[2] University of Central Florida, Department of Physics yan@physics.ucf.edu
[3] Astronomical Institute of Slovak Academy of Science

Abstract We present wide-field g, r, and i-band images of comet 9P/Tempel 1, target of the Deep Impact mission, taken with the MegaCam CCD (FOV 1°×1°) on the 3.6-m Canada-France-Hawaii Telescope on Mauna Kea. Our observations on UT July 3–6, 2005, covered the night prior to impact to ensure that we have data for the pre-impact dust environment. The two nights following impact allowed us to monitor the inner coma for changes in grain size properties as a result of the impact. We created a deep composite image (from three hours of observing time) for each night. Our morphological studies can establish the pattern of dust emission from the nucleus, which is determined by the surface distribution of discrete sources of dust on the rotating nucleus and their temporal evolution. The scientific questions we address are: How does post-impact dust differ from pre-impact dust? Is there any compositional difference between surface and subsurface grains? Does surface mantling change this? From the relative photometry we observed that some of the impact ejecta leaves our smallest aperture within the first hour after the impact. The light curves through different apertures implies that the ejecta was traveling roughly 200 ms^{-1} projected on the sky. Fifty hours after the impact, the core of the comet was within a few tenths of a magnitude back to its pre-impact brightness. Also by our last night the morphology of the coma had begun to return to its pre-impact shape.

1 Introduction

Deep Impact was the first planetary mission to carry out direct experimentation on a cometary body. Comets represent the most pristine accessible remnants of Solar system formation, but over time the exterior mantle of these objects has been highly altered by orbit-induced solar insolation fluctuations. Ground, space, and in-situ observations have not been able to easily determine the physical and chemical characteristics of the primordial material.

The Deep Impact mission delivered a 364-kg impactor to comet 9P/Tempel 1 at 10.3 kms^{-1} at UT 5:45 (spacecraft time) July 4, 2005 [1, 2]. The impact removed the evolved material and excavated into the pristine interior. This event released dust and volatiles that were observable for only a few days after impact (Fig. 1).

We decided to obtain visible-wavelength, high spatial-resolution imaging of the comet before, during, and after the impact. Our goals were: (1) determine compositional and physical differences between pre-impact and (presumably more primordial) post-impact dust grains, and (2) monitor the near-nucleus coma immediately after impact to watch the spatial evolution of the ejecta.

Fig. 1. 9P/Tempel 1 selected individual images before and after the impact.

2 Observations

Data were obtained with the 3.6-m Canada-France-Hawaii Telescope on Mauna Kea, using the wide-field MegaCam CCD camera with g, r, and i-band Sloan Digital Sky Survey filters. Most of the data are in r-band. On the four nights, 73, 124, 58, and 74 r-band images were obtained. Exposure times were 30 seconds except on July 4 when they were only 10 seconds. The CCD frames were reduced with standard routines in IRAF (Image Reduction and Analysis Facility). All the images were reduced using flat field images taken during the evening and morning twilight sky and cleaned of bad pixels and cosmic rays. We measured the magnitudes of as many as possible (usually 20–30) field stars of equal or greater brightness to the comet on each frame in order to do relative photometry. After correcting the measured magnitudes for extinction, we used the deviations of the field star magnitudes in each frame from their nightly average values to correct for frame-to-frame extinction in the comet's measured signal.

The photometry reduction and the comet flux from each CCD frame was extracted using circular aperture photometry routine "PHOT" in IRAF software. This routine automatically finds the centroid of the image within the user-specified photometry aperture, with the sky background determined in an annulus lying immediately outside the photometry aperture (for stellar images), or selected from an average of many sky positions outside the extent of any coma. A sky annulus with an inner radius of 8.0–10.0$''$ and width of 4.0–5.0$''$, was used to measure the average of the sky background. The average of many sky background positions around the object allowed us to reject any bad pixels or field stars found in the individual sky annulus. We used different aperture radii ranging from 1″ to 4″ for comet photometry as you can see on Fig. 4, and from 4″ to 5″ for the sky annulus and field stars, which encompassed 95% of the entire signal. We were trying to choose the smallest possible aperture to minimize the contamination from coma, while including the most light from the nucleus with respect to the mean FWHM.

3 Analysis and Results

The representative sample of images from impact night (UT July 4, 2005) is shown in Fig. 2. There are images at nine different times; the first is before impact and the next eight are after impact. At each time stamp, there is a pair of images, with the original image at top and an enhanced image at bottom. The enhancement involved simply dividing out a ρ^{-1} model coma. Exposure times were 10 seconds, through SDSS r-band filter. North is up and East is left. Each image is 37.5″ across. The pixel scale at MegaCam is 0.187″, which is 122 km at the comet on this date. A little white bar between the top and bottom images corresponds to 2000 km at the comet. The spread of the ejecta toward the southwest was evident in less than an hour after impact.

We followed the evolution of the ejecta over several days, as shown in Fig. 3. Images are shown from UT July 3, 4, 5, and 6, 2005, all through the SDSS r-band filter. Each image is 1.25′ across. As before, from each image a ρ^{-1} model coma has been removed. Notice how the character of coma has changed on the impact night (second image on top). The ejecta is expanding outward, and is still noticeable on the two post-impact nights. However by the last night the morphology of the coma is beginning to return to its pre-impact shape. This is similar to what has been reported by several observers [2].

In addition to morphological analysis, we measured relative photometry of the comet within various apertures during our observations. In future work we will provide absolutely-calibrated magnitudes. For now a reasonable airmass correction of 0.06 mag airmass^{-1} has been assumed. In the top panel of Fig. 4 we show the light curve from impact night (UT July 4). Various aperture radii were used for the photometry, ranging from 1.0″ to 4.0″ in steps of 0.5″. Note how some of the impact ejecta leaves the smallest aperture within the first hour. We estimated that the ejecta was traveling roughly 200 ms^{-1} projected on the sky. In the bottom panel of Fig. 4 is our light curve from all four nights. Note that there was some cirrus problems on July 6, so at this point we assumed that the brightest point in the batch is the valid measurement of the comet's magnitude. Fifty hours after the impact, the core of the comet was within a few tenths of a magnitude back to its pre-impact brightness (dashed lines). The change in magnitude between pre-impact and impact+50 hours is listed for each aperture next to its attendant dashed line, and is from −0.01 mag for 1.0″, −0.14 mag for 2.0″, and −0.17 mag for 3.0″ and 4.0″, respectively. Note that this does not take

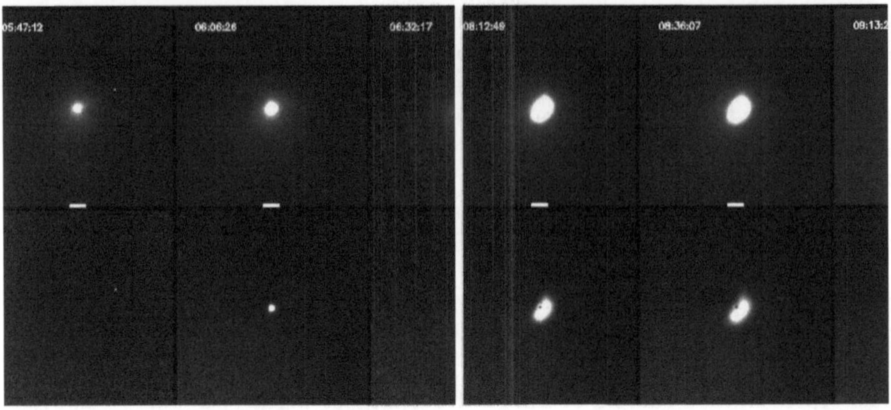

Fig. 2. Sequence of individual (*top*) and enhanced (*bottom*) images before – first image and after the impact – next 8 images.

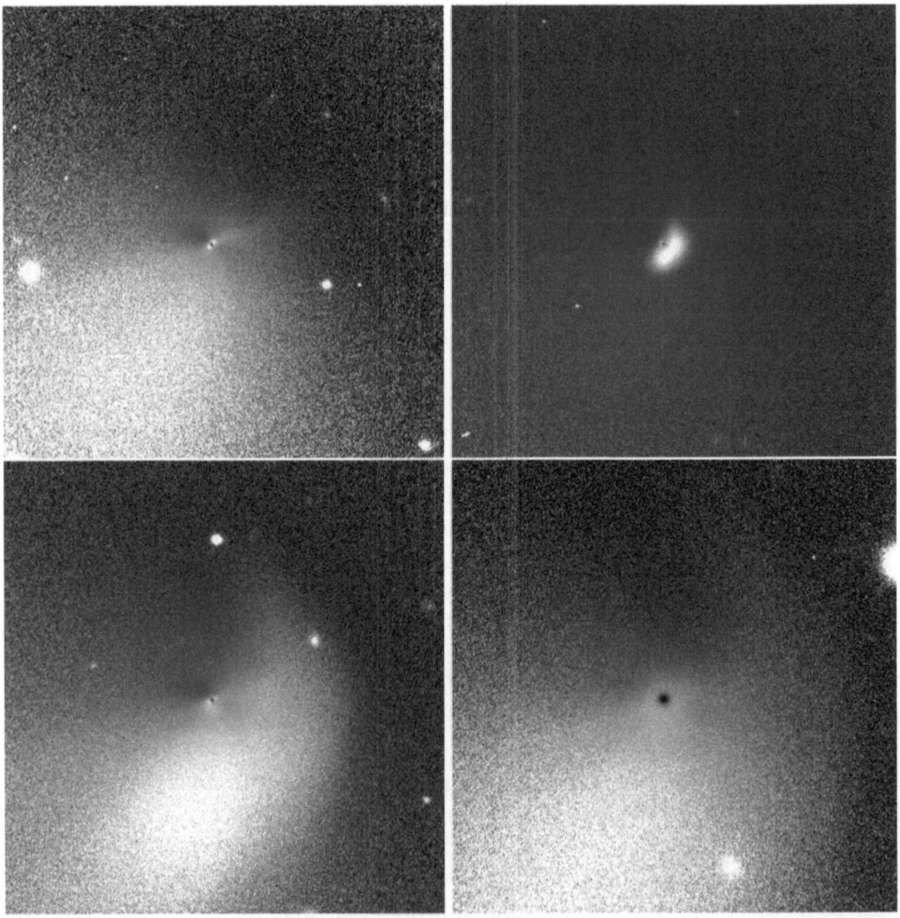

Fig. 3. Images of comet 9P/Tempel 1 on July 3, 4, 5, and 6, 2005 where ρ^{-1} model coma has been removed.

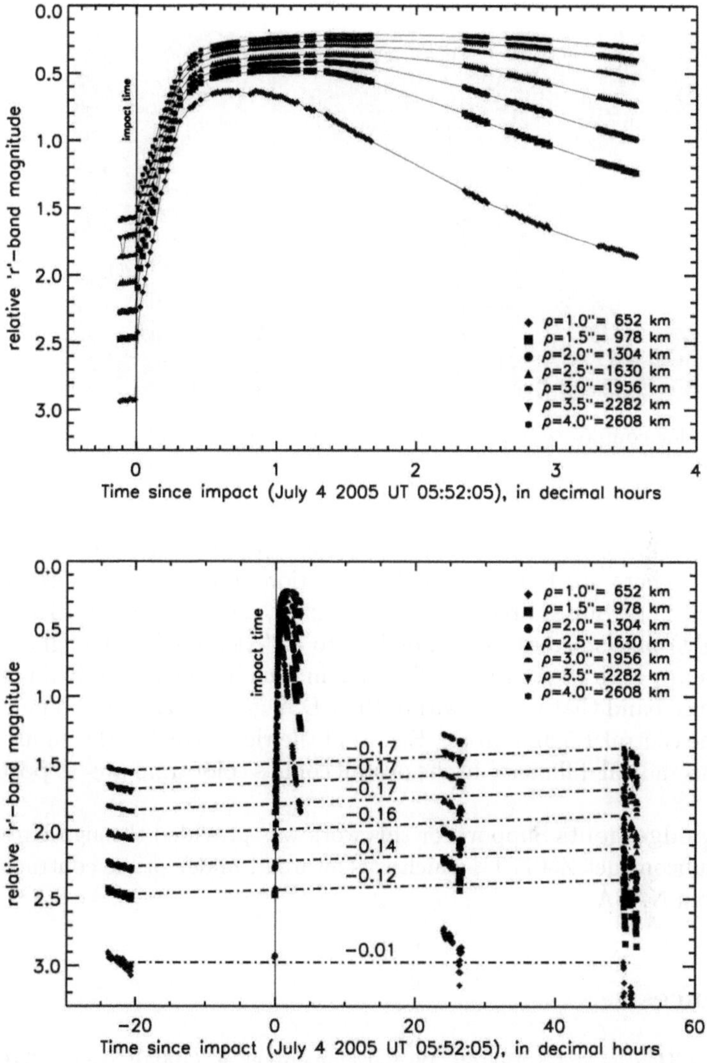

Fig. 4. Relative r-band photometry of the comet during the impact night (*top*) and on July 3, 4, 5, and 6, 2005 (*bottom*) for different aperture radii The bottom line is for $1.0''$ aperture radii with $0.5''$ step toward the top line, which is for $4.0''$ aperture radii.

into account the changing geometry of the comet between UT Jul 3 and UT Jul 6. The heliocentric distance and phase angle changed trivially over these days but the geocentric distance increased from 0.889 AU to 0.905 AU, indicating in the absence of any other effects that the comet would fade by a few hundredths of a magnitude.

Color composite images from UT July 3 (night before the impact) are shown in Fig. 5 on the left and UT July 5 (night after the impact) on the right. We

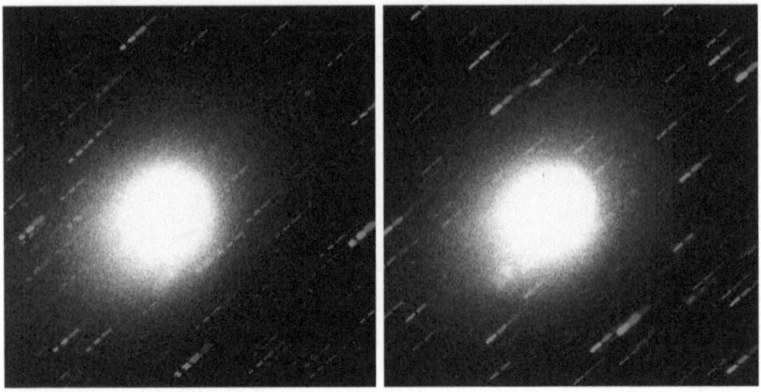

Fig. 5. Color composite images from UT July 3 (*before*) and UT July 5 (*after*) the impact.

combined SDSS g, r, and i-band images, mapped to blue, green, and red, respectively. There are 150 seconds of integration on each filter. North is up and East is left. The Sun's directions is PA = 290°. The images are 1550 (left) and 1600 (right) pixels across, corresponding to 4.7′ (left) and 5.0′ (right). There is some blue fringe in the sun-ward direction in each image, probably due to the C_2 fluorescence band that resides within the CDSS g-bandpass filter. The ejecta lies within the central 1.5 arcmin (cf. Fig. 3) of the right panel in this figure; there is no obvious radical difference in the overall coma's color from pre- to post-impact.

Acknowledgements Support for this work was provided through University of Hawaii subcontract Z667702, which was awarded under prime contract NASW-00004 from NASA.

References

1. M.F. A'Hearn, M.J.S. Belton, W.A. Delamere, et al. (2005): Science **310**, 258.
2. K.J. Meech et al. (2005): Science **310**, 265.

One Month of Near-IR Imaging Photometry of Comet 9P/Tempel 1

Y. Mori[1], T. Sekiguchi[2], S. Sugita[1], N. Matsunaga[1], H. Fukushi[1], N. Kaneyasu[3], T. Kawadu[4], R. Kandori[2], Y. Nakajima[2], and M. Tamura[2]

[1] The University of Tokyo moriyuki@impact.k.u-tokyo.ac.jp
[2] National Astronomical Observatory of Japan t.sekiguchi@nao.ac.jp
[3] Department of Physics, Meisei University
[4] Department of Astrophysics, Kyoto University

Abstract We performed a month-long near-IR photometric and imaging observations of comet 9P/Tempel 1 in South Africa IRSF/SIRIUS. Our monitoring of comet brightness shows that a sustained comet activity and outbursts were not triggered by the impact. No color change was observed within 3σ 12 hours after the impact. This suggests the color change found by other IR observations by the impact may be a short-period phenomenum. Our observations of ejecta distribution detected large change in ejecta trajectory due to the solar radiation pressure.

1 Introduction

Comets are thought to be primitive bodies in the solar system because they spend most of their lives far away form the Sun and they are less differentiated. It is very important to know the nature of comets in order to understand the formation of the solar system. However, the nature of a cometary nucleus is not known well. Spacecraft observations of the surfaces of cometary nuclei have revealed that there is little volatile on the surface and that active areas from which gas and dust are released is only a fraction of cometary surface [1]. Primordial and volatile material must exist underneath the surface dust mantle layer, but there was no datum about material in a cometary nucleus. NASA's Deep Impact (DI) mission was conducted in order to explore the interior of a cometary nucleus [2]. The DI spacecraft successfully hit the target comet 9P/Tempel 1 with 370 kg cooper-based impactor on UT 4 July [3]. Before and after the impact, comet 9P/Tempel 1 was observed intensively with telescopes on spacecrafts and the ground [4].

We carried out a month-long near-IR photometric and imaging observations in South Africa. Such a long-term observation is important because most of other observations covered only a short period of time (one to two days after the impact). Thus, this will allows us to detect possible long-term phenomena induced by the DI collision. The objective of our observations was to see how the impact changed the comet activity. We monitored the comet brightness which reflected dust production rate and occurrence of outbursts. Duration of dust release which might be induced by the impact was expected and it would suggest that excavation reached the region where volatile material might exist. Occurrence of periodic outbursts was also expected when the comet rotated and the impact site was illuminated by the Sun if volatile material was exposed on the surface. We also observed color and ejecta distribution in order to see whether dust size or composition was changed by the impact.

2 Observations and Data Reduction

Near-IR photometry of 9P/Tempel was carried out for 16 nights, covering from 7.5 days before the impact to 24.5 days after (i. e., 27 June to 28 July). Observations were performed in near-infrared wavelengths J (λ=1.25 μm), H (1.65 μm), K_s (2.15 μm) simultaneously with the IRSF (Infrared Survey Facility) 1.4m telescope of Nagoya University and its near infrared camera SIRIUS instrument at Sutherland South African Astronomical Observatory (SAAO: 32.38°S, 20.81°E). SIRIUS, the Simultaneous three-color InfraRed Imager for Unbiased Survey, has three sets of 1024 × 1024 pixel HgCdTe array detectors. The image scale used for our observations is $0''.45$ pixel^{-1} which gives a field of view of $7'7 \times 7'7$.

Two exposure series were obtained each observing day. Individual dataset is composed of 15 ditherd frames, exposure time of each frame is 10 or 30 seconds depending on the sky conditions. Sky frames were obtained after acquisition of every dataset with the same exposure time and number of dithers as object frames. Dark frames and twilight flatfield frames were obtained at the beginning and end of observation nights. Photometric calibrations were conducted using the standard star 9150, 9155 in the faint near-infrared standard star catalog [5]. Average seeing size was $1''.25$.

Data reduction and measurements were done using NOAO's IRAF (Imaging Reduction and Analysis Facility) software package. After subtraction of the dark frame, flatfielding was adopted for all images. Sky images were created by median of 15 ditherd frames and subtracted from corrected object frames. After these operations, one set of object frames were combined by taking a median.

3 Results and Discussion

3.1 Brightness Variation

Brightness variation of the comet was monitored before and after the impact. Each measurement of the comet flux was obtained with aperture photometry. In order to find the comet flux, photon counts within each nucleocentric aperture of 6,000, 9,000 and 12,000 km from the nucleus were subtracted by the sky value measured on the circular ring with the radius of 100,000 km from the nucleus. After the calibration using the flux of the standard star, photometric results were calibrated in corrected magnitude, which was free from the effect of the difference of geocentric distance. Obtained corrected magnitude is shown in Fig. 1. Error in magnitude was estimated with $MERR = \sqrt{\sigma_1{}^2 + \sigma_2{}^2 + \sigma_3{}^2 + \sigma_4{}^2}$, where σ_1, σ_2 was photometric error of the comet and the standard star, σ_3 was the uncertainty in magnitude of the standard star listed in the catalog, and σ_4 was the standard deviation of the brightness of the standard star in nine frames taken each day.

One data point was rejected from each of 3, 4 and 16 July (−0.5, 0.5 and 12.5 days from the impact) because the weather of these days was bad. In order to reject one data point, relative photometry using stars in the same frame as the comet as standard stars was conducted. Data of magnitude of the stars were taken from 2MASS (The Two Micron All Sky Survey) data. In Fig. 1, the comet was brightened by 0.5–0.6 magnitude in each of the three wavelengths after the impact. However the high brightness did not last for a long time, it darkened to the pre-impact brightness about three days later. This result shows impact-induced gas ejection did not last for a long time; no active area was created on the comet nucleus by the DI collision.

Four events of brightness increase were observed on 27 June, 2, 11 and 25 July. These may be due to natural outbursts. The brightness increase on 2 July (−1.5 days from the impact) was certainly a natural outburst. This outburst was also observed by the DI spacecraft [3]. The outburst occurred at 8:34 (UT), 9 hours before our observation. This result shows that the effect of the outburst remained at least for 9 hours with aperture of 6,000 km. The small brightness increases on 11 and 25 July (7.5 and 21.5 days from the impact) might be also the effect of outbursts. Schleicher et al. [6] report photometric results in visible wavelengths till ~UT 22:00 10 July, but no outburst was observed [6]. Thus, outburst might have occurred during the two observations (i. e., between 22:00 10 July and 19:00 11 July). Observations with the DI spacecraft show that outbusts have periodicity; there are two active areas on the nucleus of the comet 9P/Tempel 1 [3]. One area is related to the phase of the rotational light curve which is called 'shoulder', from which most outbursts appear to occur. The outburst on 2 July was identified with occurrence on "shoulder". The other active area corresponds to the one peak in the rotational light curve, called 'peak B'. Using simply extended rotational light curve to 25 July, 'peak B' is about 2 hours before our observation on 11 July, and about 8 hours before our observation on 25 July. Both outbursts might occur in 'peak B'. The outburst on 11 July did not correspond to the impact point because the last time the impact point face the solar direction before our observation on 11 July was before the observation by Schleicher et al. [6]. The high brightness on 27 June (−7.5 days from the impact) may be due to the bad weather because the scatter in 9 measurements of the flux of the standard star was rather unstable, or lower elevation of the comet; airmass was about 1.6. In conclusion, outbursts which were induced by the impact did not observed during our observing period.

These results suggest that the excavation may not have reached the region where volatile material exists. However, many volatiles were detected by some observations after the impact. Detections of increase of H_2O [3, 7] and some super volatile like CO and C_2H_2 [8] indicate that inner material which exist under the dust mantle layer was ejected by the impact. The radial distribution of dust in the DI-induced plume observed by mid-IR also strongly suggest the presence of a strong gas-driven acceleration of dust particles [9]. In order to solve the contradiction among these results, we may need to adopt something new idea, for example, exposed volatiles were buried owing to collapse of the crater.

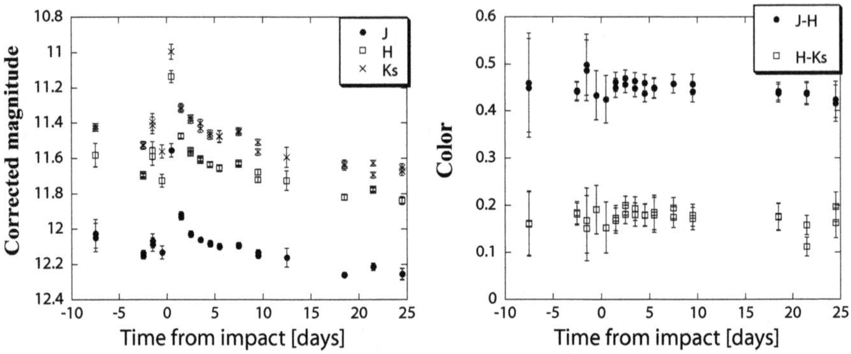

Fig. 1. *Left*: The variation of the comet brightness. Aperture radius is 9,000 km. Error range is 1σ. *Right*: The variation of the comet color. Aperture radius is 6,000 km. Error range is 1σ.

3.2 Color Variation

Using J, H, and K_s band images which were obtained simultaneously, color values of each datum were obtained. Figure 1 is the result with aperture of 6,000 km. The color change would occur if dust size or composition were changed by the impact. However, obvious color change larger than error range was not observed. J–H (i. e., the difference of J and H band magnitude) was 0.459 ± 0.009 before the impact and 0.446 ± 0.003 after on average. H–K_s was 0.171 ± 0.006 before the impact and 0.167 ± 0.009 after. From this result, ΔJ–H (i. e., the difference of J–H before and after the impact) was less than 0.012 and ΔH–K_s was less than 0.015.

However, near-IR observation with IRTF detected obvious color change between before and after the impact [10]. There are two big differences between our observations and observations with IRTF: time and area of observations. Our observations were conducted 12 hours after the impact and measured the color change within nucleocentric aperture with radius of 6,000 km. On the other hand, observations with IRTF were conducted immediately after the impact and measured close to the nucleus. It is possible that the color change disappeared by the time of our observations. If the color change was caused by water ice grains ejected by the impact, this fast disappearance of color change could be explained readily. There is another possibility that the area where we measured color was too large to detect color change, that is, normal component of dust coma except for ejecta was dominant. We measured the color in the area where ejecta component was dominant and found no color change within 3σ. From this result, 2nd possibility is excluded.

3.3 Ejecta Distribution

Ejecta distribution was obtained as a form of angular profile around the nucleus (Fig. 2). We did photon counting within each small circle which was placed on the same nucleocentric distance per 15°. These photon counts were converted to corrected magnitude in the same way. In order to see only ejecta component, obtained magnitude was subtracted by magnitude on 1 July which was a representative of pre-impact values.

On 4 July, ejecta spread across 180° on 6,000–9,000 km from the nucleus (Fig. 2). Although the center of ejecta distribution was zero on 4 July, the center moved to the Sun direction (i. e., 225° PA + 65°) on 5 July (Fig. 2). This result is consistent with the effect of solar radiation pressure. Ejecta released toward the Sun was slowed down by the solar radiation pressure. It stayed near the nucleus longer than ejecta toward the other direction. Thus the concentration of ejecta was observed in the Sun direction at the same distance. Such data on the trajectory of ejecta particles will allow us estimate the β value (the ratio of solar radiation pressure to momentum), which is directly related to the mean size of the ejecta dust. The mean size of the ejecta is an important parameter to determine the total mass of ejecta. The fact will help us estimate the crater size created by the DI collision. Because typical excavation depth is estimated to be about 1/10 of the diameter of gravity-controlled impact craters, we can further infer the depth of the cometary surface fresh material exposed by the impact. Such information on crater excavation depth if combined with the observation that no new active area was created on the cometary surface by this impact will place an important constraint on the vertical volatile distribution underneath the comet.

4 Summary and Conclusion

The comet was brightened after the impact, but the high brightness did not last very long. Outbursts which were induced by the impact were not detected by our monitoring observations. These results show that no active area was created on the comet nucleus by the DI collision. However, this does not simply mean that excavation did not reach the region where volatile material might exist. Some mechanism that can explain both no duration of dust release and release of volatile right after the impact are needed. Our observation detected no color change within 3σ 12 hours after the impact. This result suggests that

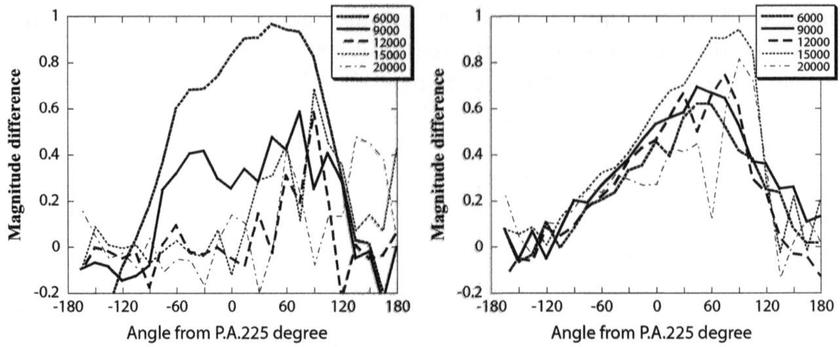

Fig. 2. Angular profiles around the nuclei of ejecta distribution on 6,000, 9,000, 12,000, 15,000 and 20,000 km from the nucleus on 4 (*left*) and 5 (*right*) July. The origin of the horizontal axis is 225°PA, the direction of the impact seen from the earth [9, 11]. The direction of the Sun is 65°(290°PA).

the color change which was detected by other observations right after the impact disappeared by the time of our observations. The color change was possible to be caused by water ice grains which were ejected by the impact. Our observations of ejecta distribution detected ejecta which spread across 180° 12 hours after the impact and the effect of solar radiation pressure. The data on the trajectory of ejecta particles will allow us to estimate the β value, which is directly related to the mean size of the ejecta dust.

References

1. H. U. Keller et al.: *Comet II* (Univ. of Arizona Press 2004) pp 211–222
2. M. F. A'Hearn et al. (2005): *Space Science Reviews* **117**, 1
3. M. F. A'Hearn et al. (2005): *Science* **310**, 258
4. K. J. Meech et al. (2005): *Science* **310**, 265
5. S. E. Persson et al. (1998): *Astron. J.* **116**, 2475
6. D. G. Schleicher, K. L. Barnes, N. F. Baugh (2006): *Astron. J.* **131**, 1130
7. M. Küppers et al. (2005): *Nature* **437**, 987
8. M. J. Mumma et al. (2005): *Science* **310**, 270
9. S. Sugita et al. (2005): *Science* **310**, 274
10. Y. R. Fernandez et al. (2006): *Deep Impact as a World Observatory Event*
11. T. L. Fernham et al. (2005): American *Astron. Soc., DPS meeting* **37** 711

Author Index

MIX
Papier aus verantwortungsvollen Quellen
Paper from responsible sources
FSC® C105338
FSC
www.fsc.org

If you have any concerns about our products,
you can contact us on
ProductSafety@springernature.com

In case Publisher is established outside the EU,
the EU authorized representative is:
Springer Nature Customer Service Center GmbH
Europaplatz 3, 69115 Heidelberg, Germany

Printed by Libri Plureos GmbH
in Hamburg, Germany

